D1218458

The New England Mill Village, 1790–1860

Documents in American Industrial History
Michael Brewster Folsom, general editor

Volume 1
The Philosophy of Manufactures: Early Debates over Industrialization in the United States, edited by Michael Brewster Folsom and Steven D. Lubar, 1982

Volume 2
The New England Mill Village, 1790–1860, edited by Gary Kulik, Roger Parks, Theodore Z. Penn, 1982

This book was set in VIP Optima and Baskerville by DEKR Corporation and printed and bound in the United States of America.

Library of Congress Cataloging in Publication Data

Main entry under title:

The New England mill village.

(Documents in American industrial history; v. 2)
Bibliography: p.
1. Textile industry—New England—History—Sources. 2. Textile factories—New England—History—Sources. 3. Textile workers—New England—History—Sources. 4. Cities and towns—New England—History—Sources. 5. Villages—New England—History—Sources. 6. New England—Social conditions—Sources. I. Kulik, Gary. II. Parks, Roger N. III. Penn, Theodore. IV. Series.
HD9857.A11N48 307.7′6 81-23665
ISBN 0-262-11084-9 AACR2

The New England Mill Village, 1790–1860

Edited by Gary Kulik, Roger Parks, Theodore Z. Penn

The MIT Press
Cambridge, Massachusetts London, England

Merrimack Valley Textile Museum
North Andover, Massachusetts

Contents

General Editor's Preface

To most students, teachers, and the general public the industrial origins of the United States remain largely unknown. Aside from a few textbook cliches about "fathers" and "birthplaces" of the "Industrial Revolution," we learn little about this central aspect of our past. Scholarly literature does exist, but primary materials are not easily available. Teaching industrial history too often resembles teaching Shakespeare with only secondary studies for texts, learning about the works from what the critics quote and scholars argue. Conventional historiography, when it does touch on the industrial past, often disappoints. Social, cultural, economic, political historians—even labor and business historians—often do not understand the technology itself. Historians of technology, with significant exceptions, traditionally have confined their study to machines abstracted from the human origins and consequences of technical innovation.

Just why this should be so—why the most highly industrialized and technology-dependent society should ignore the very source of its world power—is a cultural and intellectual problem well worth study in its own right. What concerns us more immediately is one practical consequence of this problem and one practical step toward its solution. Much of the documentary record in social, cultural, and political history is available to the nonspecialist in published collections. This is not true for industrial history. The abundant primary materials remain, for the most part, in archives. Our aim is to stimulate inquiry into these materials by publishing a series of collections which draw upon archival resources, make available the most important and representative documents, and demonstrate the ways in which greater familiarity with the industrial record can illuminate our material past.

The second volume of the series Documents in American Industrial History covers the first characteristic form of American industrial community, the small textile mill villages that grew along the lesser rivers of New England between 1790 and 1860. Two subsequent volumes will deal with the rise of the first industrial *cities* of New England—such as Lowell, Lawrence, and New Bedford—which differed significantly from the earlier manufacturing villages and eclipsed them. Volume 3 covers the New England textile city during the "Yankee Era" up to the Civil War, when most of the "operatives" were native-born farm women. Volume 4 considers the period of maturity and decline, from the Civil War to the Great Depression.

Anticipated steps beyond these volumes are documentary collections on other major industries of the region: paper, glass, and especially the shoe industry, which was second only to textiles. There are plans also for volumes on the coal industry, the iron and steel industry, and the industrialization of the South.

Note: When portions of text have been deleted from the documents, an asterisk appears in the margin. Note numbers in the margin refer to editorial comments, which appear at the end of the documents.

Acknowledgments

The editors wish to thank the following organizations, collections, and libraries for allowing us to photocopy documents for publication in this volume: Baker Library, Harvard Business School; Connecticut State Archives and Library; Jacob Edwards Library, Southbridge, Massachusetts; Haverhill Public Library, Haverhill, Massachusetts; Massachusetts State Archives; Research Library, Old Sturbridge Village; World of Work Project, Museum Education Department, Old Sturbridge Village; Library of the Rhode Island Historical Society, Providence, Rhode Island; Slater Mill Historic Site, Pawtucket, Rhode Island; The National Museum of American History, Smithsonian Institution; and the Library of Congress. In addition the editors wish to thank the following individuals for their valuable assistance in preparing documents for publication in this volume: Etta Faulkner, librarian, Old Sturbridge Village; Charles Pelletier, archeology volunteer, Research Department, Old Sturbridge Village; Caroline Sloat, social historian, Research Department, Old Sturbridge Village.

The general editor wishes to acknowledge the special contributions to this volume made by the two institutions which have done the most to preserve and enhance the history of the New England mill village, Old Sturbridge Village at Sturbridge, Massachusetts, and Slater Mill Historic Site in Pawtucket, Rhode Island. Much of the scholarship that forms the core of this documentary collection was done by the staff of Sturbridge Village's Research Department, principal among them Roger Parks, then director of the department, and Theodore Z. Penn. The research program at Sturbridge Village especially concerned the industrial history of central Massachusetts and northern Connecticut. Materials in this collection on labor history and the Rhode Island origins of the American textile industry are largely the result of research conducted by Gary Kulik during his tenure as curator at Slater Mill Historic Site.

Introduction

"The manufacturing operations of the United States," wrote Zachariah Allen in 1829, "are carried on in little hamlets, which often appear to spring up in the bosom of some forest, around the water fall which serves to turn the mill wheel."[1] Allen thus expressed a widespread optimistic conviction that industry might come to this country in a benign form, free from the fetid slums and blighted landscapes of the already notorious factory cities of the English midlands.[2] This early aspiration for a pastoral compromise with manufacturing has never quite been laid to rest during the intervening hundred and fifty years. Allen's vision of widely scattered rural mill hamlets has been revived in the mill villages designed by the firms of Frederick Law Olmstead and McKim, Mead and White,[3] the "greenbelt" movement of New Deal planners, and more recently in the ideas of the advocates of "small is beautiful." A distant echo of the call to blend the factory and the country can even be heard in the rhetoric of real estate speculators who build industrial "parks."[4]

For most of our history as an industrial nation, however, such a vision of pastoral manufactures has been at odds with the dominant forces of the American economy and culture. When Zachariah Allen wrote of "little hamlets" in 1829, the city of Lowell, Massachusetts, had already been founded to exploit the great power of the Merrimack River and was on its way to becoming the first of the great murky cities that would make America a world power. Yet Allen knew what he was talking about. He himself was a manufacturer, the proprietor of a hamlet such as he described, in the vicinity of Providence, Rhode Island. In 1829, most American manufacturing was still conducted in small company-owned villages along the smaller streams and rivers of the Northeast. By 1840, when Lowell had become a city of 20,000, most of the 700 cotton mills in New England still more or less fit Allen's vision. As the agricultural historian Percy W. Bidwell has written, at that time "it would have been difficult to find 50 out of the 479 townships in Southern New England which did not have at least one manufacturing village clustered around a cotton mill, an iron furnace, a chair factory or a carriage shop."[5] Many agrarian towns had several such villages. The town of Thompson in northeastern Connecticut, for instance, had seven mill villages within its 59-square-mile area, six of them producing cotton goods and the other woolens.[6]

Nor were such villages confined to New England. Anthony F. C. Wallace's recent study of textile manufacturing communities in southeastern Pennsylvania has drawn attention to the growth of small local industries in that region.[7] An article in *Niles' Weekly Register* in 1823 records the existence of what its author considered model factory villages along the Brandywine River in Delaware: ". . . a continual succession of mills and elegant houses and comfortable cottages, with pretty extensive fields and gardens, wrested from the late rock covered wild. . . ."[8] By the 1840s, mill villages had appeared in the deeper South.[9] However, New England remained the major source and example of small-scale industrial development throughout the period. To be more precise, it was *southeastern* New England, where such development was often called the "Rhode Island System" because it emanated originally from factories and machines after the pattern of Samuel Slater's works in the Providence area in the 1790s.

In large measure this volume is devoted to clarifying the definition of this distinctive form of American industrial organization, so that its particular contribution to the history of manufactures may be better understood. Later volumes in this series will document subsequent developments, starting with the textile and shoe cities that grew in the middle decades of the nineteenth century. The documents we have gathered and edited in this volume reveal how New England mill villages were built and what they looked like and who the owners and managers were, how they perceived themselves, and how they lived. These documents define the workers, the conditions under which they labored, lived, and died, their views of their work, and their early efforts at labor organization. The period covered here is roughly the half-century from 1790 to 1860, when mill villages were most important in American manufacturing.

Most of the documents in this collection concern the cotton textile industry. The cotton industry was the first to become fully mechanized. It employed more workers (50,000 in New England by 1840) than any other branch of manufacturing except the nonmechanized shoe industry. It also had the greatest immediate impact, both technologically and socially, on the rural society in which it was established. It employed a diversity of workers, including many children and young women, and was, in fact, the first large-scale employer of women outside the home.

In the cotton industry, American industrial technology had its real beginning. In 1790, the young English mechanic Samuel Slater built the first successful waterpowered machinery for spinning cotton in the United States, for the firm of Almy & Brown of Providence. From then until the embargo of 1807, the industry spread slowly to other parts of Rhode Island, southern Massachusetts, and eastern Connecticut.[10] From 1807 through the War of 1812, when the United States was cut off from much of its profitable foreign trade and from imported British textiles, the number of cotton-spinning mills in New England grew rapidly. Secretary of the Treasury Albert Gallatin, in a survey of manufactures, found twenty-one cotton mills operating or under construction in New England in 1809.[11] By 1815, there were nearly 170 such factories in the general vicinity of Providence alone.[12]

The technology of these early village mills was a blend of native ingenuity and foreign expertise. American colonists had been developing mill sites in the New World since the 1630s, and new cotton mills often joined gristmills or sawmills in utilizing an existing dam. American millwrights had no problem erecting a barnlike wooden or stone building suitable for textile manufacture, and they were adept at installing a waterwheel to power the machinery. But the machinery itself was wholly British in design at first and slow in crossing the Atlantic. The Arkwright spinning frame had been in operation in England for twenty years before Samuel Slater built his in Rhode Island. Likewise Samuel Crompton's spinning mule, the next major advance in textile technology, had been developed in England for two decades before two other immigrants, John Slater (Samuel's brother) and Samuel Ogden, independently introduced it here at the turn of the nineteenth century.[13]

Although at this time the British were well into experiments with various kinds of power looms, American textile mills depended on spinning technology for another fifteen years. The great expansion of the industry after the embargo of 1807 was based on the relative efficiency of even primitive waterpowered spinning machinery over hand spinning, and in the absence of competition from British goods American millowners lacked incentive to innovate further. The end of hostilities was a disaster for the American textile mill village. The introduction of yet further new British technology then offered a reprieve, if not permanent salvation.

Aided by the widespread adoption of power looms after 1814, New England cotton manufacturing was gradually and successfully reorganized. Before then textile mills only spun yarn by powered machinery. Much of this yarn was sold outright, but some was "put out" on consignment by mill owners to be woven at home by local weavers. On occasion hand weavers were set up to work in the spinning mill itself.[14] Such arrangements, however, began to change after 1814, when a group of wealthy Boston investors headed by Francis Cabot Lowell introduced power looms in their first factory at Waltham, Massachusetts. This machinery, based on designs that Lowell had been able to study in Britain just before the war, was designed to produce large quantities of coarse, inexpensive grades of cloth that could compete successfully with imports. By 1820, a number of small New England manufacturers, hurt by renewed foreign competition, had followed suit and begun to put power looms into their mills. The loom that came into general use, however, was not Lowell's but an improved design that had been developed in a small Rhode Island mill village machine shop.[15]

With the help of this new technology the cotton industry began to expand once more. Hundreds of new mills were built in New England during the 1820s and 1830s. During those decades the Boston merchants who had experimented with the new power loom in Waltham went on to build Lowell, Massachusetts, as the first industrial city in the United States. Several other such cities based on the Lowell model had their beginnings then as well; however, most of the new mills built between 1820 and 1840 were still in factory villages where small-scale manufacturing continued to thrive.[16]

Compared to Lowell, the typical mill village was quite small. In 1840, Lowell had grown to a population of more than 20,000, a third of whom worked in the city's 26 cotton mills. These mills had a capacity of 166,000 spindles.[17] In contrast, a typical cotton manufacturing village employed some 100 workers and had at its center one or possibly two cotton mills with a total capacity of about 1,000 spindles. Near the cotton mill was often a sawmill or gristmill that had served the needs of local farm families for generations and had been taken over by the manufacturing company when it acquired the waterpower site. To shelter the families whose children supplied much of its labor force, the company made use of any housing that had been standing on the land it had bought and also built new tenement houses. One-, two-, and four-family tenements were common. A nearby tavern sometimes doubled as a boardinghouse for single employees, who were mostly male supervisors and skilled workers during the early years of the industry. (The arrangements provided for single men by the Ware Manufacturing Company are spelled out in one of the documents included in this collection.) In early mill villages there was seldom anything resembling the rows of brick boardinghouses that were characteristic of the factory cities of northern New England.

Some companies staked out garden plots that mill village families could rent to raise part of their own food. Pasturage to support a family's cow could also be rented in some villages. In the larger villages, company-owned farms might supply firewood, meat, and other commodities. In general, the distance between farm and factory was not great, and there is some evidence that both workers and managers moved back and forth between the two means of livelihood.[18]

In contrast to the corporations of the great factory cities, which paid cash wages and made no provision for the supply of consumer goods to their employees, nearly all mill villages had a company store, where company-grown produce and other goods might be purchased on credit against future wages. The evidence is that such village company stores were opened out of necessity rather than to gouge a captive clientele. The population of the new village was often too far from established stores for convenience and too small to attract an independent shopkeeper. The records of the Slater and Tiffany Mill, from which the Davis family accounts included in this collection are drawn, indicate that the proprietors had no clear sense of how much of their profits came from manufacturing and how much, if any, came from their store. Mill owners, however, were not without some self-interested motives in establishing company stores. Zachariah Allen put the matter bluntly when he wrote that company stores provided workers "with the necessaries of life in order to retain them for service."[19]

Some village mill owners provided schools for children whom they employed. The first was Samuel Slater's Sunday School, established in Pawtucket in the 1790s as essentially a secular institution for teaching reading, writing, and arithmetic and modeled on similar English schools. Other mill owners convinced the local town to establish a district school nearby. Like rural schools that operated only during those times when children were not needed to help on the farm, mill village schools were in session only a few months a year.

To encourage social order and regular behavior, village mill proprietors often gave land to any religious denomination willing to organize and build a church. The Baptist Fiske family of Fiskdale, Massachusetts, went a step further and provided its workers with a church of the owners' choice.[20] The Unitarian agent of the Ware (Massachusetts) Manufacturing Company, prevailed upon his board of directors to sponsor a Unitarian meetinghouse.[21]

In economic organization the mill village and the factory city were very different. The capital that built Lowell and the other large industrial centers of New England was raised primarily within the mercantile community of Boston and was invested in manufacturing corporations in which a dozen or many scores of individuals had a stake as owners who seldom participated in management. Providence merchants, on the other hand, played a prominent role in financing southern New England mill villages, though some of the initial capital came from the communities in which the mills were built. Characteristically a small village mill was organized as a partnership, not as a corporation, with at least one partner active in management. Frequently this was a local artisan with the technical skills necessary to maintain and repair textile machinery.

A village mill could be built or bought for as little as $10,000. The Lowell corporations were capitalized at $400,000 and up. Because of the relatively small capital required, occasionally a village mill employee was able to rise to the ranks of the mill owners. William Fisher, whose memoirs are included in this collection, was one such entrepreneur of humble origin. No worker in Lowell is known ever to have risen to corporate management. Village mills were often as marginal as they were small in economic scale, and William Fisher was among a great many proprietors who failed. Although the Lowell corporations were required to shut down operations and miss dividends in "hard times," none went bankrupt during the entire nineteenth century.

The work force in cotton mill villages was ethnically homogeneous in the early nineteenth century. It consisted mainly of New Englanders of British extraction and some British immigrants. Beginning with Samuel Slater, British immigrants for many years provided a significant number of skilled workers required in textile manufacturing.

The overwhelming majority of production workers were native born, and this would remain so until the late 1840s and early 1850s. An analysis of the birthplaces of textile mill workers in several Connecticut towns in the 1850 census shows that the majority even at mid-century had been born either in the town in which they worked or in an adjoining one.[22] By then, however, Irish and French Canadian workers were increasing in number. The years from 1845 to 1860 witnessed the transformation of the New England textile mill work force, as immigrant workers came to replace the native born in virtually every production task.[23] The final document in this collection suggests that this was not just a "natural" process, but a demographic shift encouraged by mill owners for their own purposes.

The majority of the first workers were children. Most were part of a system based on the employment of whole families. Some were children of skilled workers and supervisors but also, especially in the early 1800s, children of widows and unskilled agricultural laborers, who were landless or nearly so, from nearby communities. One mill owner, Smith Wilkinson of Pomfret, Connecticut, wrote, "[I]n collecting our help, we are obliged to employ poor families, and generally those having the greatest number of children, those who have lived in retired situations on small and poor farms, or in hired houses, where their only means of living has been the labour of the father and the earnings of the mother, while the children spent their time mostly at play." Wilkinson claimed that he provided year-round work for families who would otherwise have had only seasonal income in agriculture, and he believed that factory labor imposed a "restraining influence" on people who "are often very ignorant, and too often vicious."[24] Claims such as these were self-serving, but rural America did contain a surplus force of poor, seasonal workers to whom work in textile mills may have seemed a reasonable alternative to vagrancy, want, or hardscrabble farm toil. The fathers of mill village families were not always hired in the mills; some continued to work as artisans or agricultural laborers in the neighborhood. Children might go into the mills as young as seven or eight, but in 1850, at least, when information about ages and occupations begins to be more readily available from census sources, few apparently started work before they were twelve or thirteen.

Both on farms and in mills, labor was long and began early in life, but industrial work was very different from agricultural work. The demands of power machinery fundamentally altered the nature and pace of work. The adjustment that children and other workers had to make to machine production was only one and perhaps not the most important change they confronted. They also had to come to terms with new forms of authority. Some had certainly known dependence before, but now their dependence would be tied to a wage payment that would fluctuate with changing market forces. This shift from an economy based upon independent proprietorship of the land to an economy increasingly organized around wage labor was one of the most important and deeply felt historical changes in the structure of American social experience, and sharp social conflicts arose from this change in structure.

The efforts of mill owners to reduce wages were the primary causes of early textile mill strikes. A newspaper account of one of the earliest of these strikes, which occurred in Pawtucket, Rhode Island, in 1824, is included in this volume. The Pawtucket-Providence area saw the beginning of the first regional efforts at trade union organization. The New England Association of Farmers, Mechanics, and Other Workingmen, organized in the early 1830s, spoke to the diffuse sense of injustice spawned by the factory system. Fear of an encroaching aristocracy of wealth that would threaten American liberties was widespread. The New England Association, largely dominated by artisans, never reached a substantial segment of textile mill work force, but it did help to articulate a generation of resentment and anger—anger over the unregulated damming of streams, the growing inequality in wealth and power, the proliferation of factory rules and other methods designed to inculcate new forms of work discipline, and increasingly the issue of child labor.

Opponents of the new factory system went so far as to argue on ethical, political, and economic grounds that the factory system should be discouraged altogether in the United States. (The first volume in this documentary series is devoted to these early debates over industrialization.) In these debates, the social, cultural, and technological character of the mill village allowed it to play an interesting role, for it provided a model of industrial organization that some argued was compatible with the interests of Americans who believed the nation needed manufacturing but could have it without the social ills that had accompanied industrialization in Great Britain and Europe. In the selection that opens this volume, Zachariah Allen stated the problem vigorously and the solution complacently. "God forbid that there ever may arise a counterpart of Manchester [England] in the New World." In contrast to the fetid British industrial center, the American mill village seemed to Allen to provide a pastoral interlude: "In most of the manufactories in the United States, sprinkled along the glens and meadows of solitary watercourses, the sons and daughters of respectable farmers who live in the neighborhood of the works find for a time a profitable employment."

Most Americans, even advocates of industrialization, expected the United States to remain an essentially agrarian society. They believed that the population would continue to be primarily rural, that manufacturing would be a minor economic function, and that the country would avoid many of the problems Americans had witnessed or heard about in industrial Europe. Even proponents of large-scale manufacturing, when they built Lowell and the other cities modeled after it, felt constrained to devise a new system that would substitute the temporary and carefully supervised work force of New England farm "girls" for that of a permanent proletariat.

The hope that manufacturing could be confined to small units of production in rural villages proved illusory, as did the premise on which the "mill girl" experiment was based at Lowell. Yet it was in modest and obscure places with names like Slatersville and Fisherville, Globe Village and Phoenixville, Swamp Factory and New Manchester that the United States began its move toward industrial self-reliance before New World counterparts of the old Manchester made their appearance.

Unlike the picturesque center villages of rural New England, few mill villages have survived intact. Harrisville, New Hampshire, is a notable exception. Limited by insufficient waterpower and capital, and perhaps devastated when the factory was burned or swept downstream by a freshet or bypassed by railroads, many manufacturing villages disappeared long ago. In a number of rural New England towns today, little more than a ruined dam and parts of mill foundations remain as evidence that mill villages were ever part of the landscape. Other villages survived the rise of the great textile cities only to experience a decline when the textile industry shifted decisively to the South in this century. Village mills and tenements that remain are usually incorporated in towns that look unrecognizably different from their early nineteenth-century form—often overrun by suburbs or swallowed by larger working-class industrial neighborhoods.

To recapture what the now vanished mill village once looked like and what happened there is one aim of this volume, but the documents provide contradictory testimony.[25] Excerpts from Thomas Man's acerbic *Picture of a Factory Village* are echoed by a former Sturbridge, Massachusetts, resident who wrote home in 1837 to inquire about a local mill village in which he had once worked: "Have you been down at all Westville since your return? . . .[I]f the hard times have a tendency to make it appear more frightful than it used to I pity whoever is obliged to inhabit it."[26] A Sturbridge town historian twenty years later wrote of the same village in a different manner:

> A flourishing state of things is here again witnessed. . . . It is obvious, here are advantages for a very considerable enlargement of business. The village, although the ground is very uneven, is pleasant, being surrounded with beautiful scenery. In the summer season, when nature is robed in her most beautiful attire, few places present more inviting rural attractions. The river, gliding through the midst of the village, is the crowning beauty of the scene. The sheet of water, rolling over the dam, furnishes the villagers with uninterrupted music, as well as a constant display of sparkling gems. Connecting these natural advantages with the pleasantness of the place, it may be safely calculated there will be an increasing thrift and prosperity.[27]

The complex and problematical contributions of the New England mill village to the American landscape were, thus, a reflection of our uneasy accommodation to the factory system. The variety of often contradictory historical responses is what we hope makes this collection an opportunity for reflection and insight into the national past as well as a rich gathering of information about a now unfamiliar kind of classic American community. (RP/MBF)

Notes

1. Zachariah Allen, *The Science of Mechanics. . .* (Providence, 1829), p. 352; see excerpt reprinted in this collection.

2. Volume 1 in this series, *The Philosophy of Manufactures,* reprints the key documents in the debate over industrial progress and the national welfare.

3. Leland M. Roth, "Three Industrial Towns by McKim, Mead & White," *Journal of the Society of Architectural Historians* (December 1979), pp. 317–347.

4. Although Leo Marx does not discuss the phenomenon of the mill village, his analysis of the "pastoral" "middle landscape" in American thought is pertinent here. *The Machine in the Garden* (New York: Oxford University Press, 1964).

5. Percy W. Bidwell, "The Agricultural Revolution in New England," *American Historical Review,* volume 26 (July 1929), p. 686.

6. Caroline Ware's chapter, "Starting the Industry" (in *The Early New England Cotton Manufacture*), which is reprinted in this collection, gives an overview of the textile mill village and its general characteristics. William R. Bagnall, *The Textile Industries of the U.S.* (Cambridge, Mass.: 1893) compiles detailed accounts of a great many village firms founded during the years covered by this collection; his account of the Pomfret Manufacturing Company is reprinted here. Jonathan Prude's is the best recent general study: "The Coming of Industrial Order: A Study of Town and Factory Life in Rural Massachusetts, 1813–1860," Ph.D. dissertation, Harvard University, 1976.

7. Anthony F. C. Wallace, *Rockdale: The Growth of an American Village in the Early Industrial Revolution* (New York: Knopf, 1978).

8. *Niles' Weekly Register,* Nov. 29, 1823, p. 196.

9. See Broadus Mitchell, *The Rise of Cotton Mills in the South* (Baltimore, 1921).

10. For good recent studies of early industrial development in the Rhode Island region, see James E. Conrad, "Evolution of Industrial Capitalism in

Rhode Island, 1790–1830; Almy, The Browns and the Slaters," Ph.D. dissertation, University of Connecticut, 1973; Barbara Tucker, "Samuel Slater and Sons: The Emergence of an American Factory System, 1790–1860," Ph.D. dissertation, University of California, Davis, 1974; and Gary Kulik, "The Beginnings of the Industrial Revolution in America, Pawtucket, Rhode Island, 1672–1829," Ph.D. dissertation, Brown University, 1980.

11. See excerpts from Gallatin's *Census of Manufactures* reprinted in this collection.

12. See *Transactions of the Rhode Island Society for the Encouragement of Domestic Industry in the Year 1861* (Providence, 1862), pp. 73–77, reprinted in this collection.

13. The most comprehensive work on American waterpower technology is Louis C. Hunter, *A History of Industrial Power in the United States, 1780–1930* (Charlottesville: University Press of Virginia, 1979, for the Eleutherian Mills-Hagley Foundation). Slater Mill Historic Site has recently (1980) installed a replica of an early nineteenth-century breast wheel of a kind that powered most small village mills. The best study of the transfer of British textile technology to the United States is David Jeremy's *Transatlantic Industrial Revolution: The Diffusion of Textile Technology, 1790–1830* (Cambridge: MIT Press, 1981). On the early British mechanics in Rhode Island, see Brendan F. Gilbane, "Pawtucket Village Mechanics—Iron, Ingenuity, and the Cotton Revolution," *R.I.H.S.* 34 (February 1975), pp. 3–11.

14. See Gallatin documents in this collection for an indication of the way a manufacturer might combine hand weaving in his mill with "putting out" of yarn to be woven in the home.

15. Zachariah Allen stresses the contribution of Rhode Island innovations in early power loom design in his manuscript *History of New England Textile Manufacturers,* excerpts from which are printed for the first time in this collection.

16. For documentation of the survival of the mill village, see Richard Candee, "New Towns in the Early Nineteenth Century: The Textile Industry and Community Development in New England," (unpublished research study, Research Department, Old Sturbridge Village, Sturbridge, Massachusetts, 1976), p. 2, 29ff; see also Gary Kulik and Julia C. Bonham, *Rhode Island, an Inventory of Historic Engineering and Industrial Sites* (Historic American Engineering Record, Washington, D.C., 1979).

17. U.S. census, 1840, manuscript schedules for Middlesex County, Massachusetts.

18. William Fisher, a village mill owner in Connecticut whose memoirs are printed in this collection, recalls without comment that a former partner of his turned from manufacturing to farming.

19. Zachariah Allen, unpublished manuscript memoir, Allen Papers, Rhode Island Historical Society, p. 279. This view of the company store modifies that of Norman Ware, *The Industrial Worker, 1840–1860* (Boston, 1924), p. 75.

20. The Baptist meetinghouse, erected in the center village of Sturbridge, Massachusetts, in 1832, was moved to nearby Fiskdale several years later. It now is part of the Old Sturbridge Village museum.

21. See the *Records* of S. V. S. Wilder and documents of the Ware Manufacturing Company printed in this collection.

22. U.S. census, 1850, Killingly, Thompson, and Pomfret, Connecticut. Census takers in 1850 were required only to list the state or country of birth; those who took the census for these three towns listed town of birth as well. Research Department files, Old Sturbridge Village.

23. See Rowland Berthoff, *British Immigrants to Industrial America* (New York: Russell & Russell, 1968); Jonathan Prude, "The Coming of the Industrial Order"; and Paul Buhle, "The Knights of Labor in Rhode Island," *Radical History Review,* volume 17 (Spring 1978), pp. 38–73.

24. Wilkinson's letter to George White is reprinted in this collection.

25. For a heavily illustrated account of the mill village as it appears in graphic documents, see Richard Candee, "The Early New England Textile Village in Art," *Antiques* (December 1970), pp. 910–916.

26. Jonathan G. Plimpton to Manning Leonard, January 6, 1837, Leonard Family Papers, Old Sturbridge Village Research Library.

27. George Davis, *A Historical Sketch of Sturbridge and Southbridge* (West Brookfield, Mass., 1856), p. 200.

The New England Mill Village, 1790–1860

I

Overview

Unidentified abandoned New England textile mill, late nineteenth century. Mill Village Collection, Research Department, Old Sturbridge Village.

Zachariah Allen

The Practical Tourist (1832)

Zachariah Allen (1795–1886) was a textile manufacturer in Rhode Island and a vigorous publicist for American industry. Excerpts from his writing on a number of subjects appear in this volume. In the mid-1820s he visited England in search of information about British manufacturing techniques. His disgust at the squalor of the industrial cities he visited there was mixed with admiration for the ingenuity of engineers and proprietors of the factories he toured. Like many of his American contemporaries, he fixed his attention on the "moral condition" of factory life, and he worked out a neat formula for squaring his ethical concerns with his manufacturing interests. His quasi-pastoral vision of the New England mill village reflects the hope of many early promoters that the industrial revolution might come to America without disrupting the traditional patterns of agrarian culture. (MBF)

Zachariah Allen, *The Practical Tourist* (Providence, 1832), vol. 1, pp. 153–155.

The manufacturing operations of the United States are carried on in little villages or hamlets, which often appear to spring up as if by magic in the bosom of some forest, around the water-fall which serves to turn the mill wheel. These manufacturing villages are scattered over a vast extent of country from Indiana to the Atlantic, and from Maine to Georgia. A stranger, travelling in the United States, commonly forms but an imperfect estimate of the extent of manufacturing operations carried on in the country. Where steam engines are in use instead of water power, the laboring classes are collected together, to form that crowded state of population, which is always favorable, in commercial as well as in manufacturing cities, to the bold practices of vice and immorality, by screening offenders from marked ignominy. In the narrow circle of a small community or country village, the finger of public scorn and disapprobation is pointed at the vicious and forms a repulsive circle around them. Intercourse with their former companions becomes, in a degree, cut off, and they find not, in a village, a sufficient number of new ones of vicious character, to countenance their indulgence in a course of depravity. But in a place like Manchester, in addition to the vast manufacturing population, there is a great influx of strangers, of boatmen from the numerous canals which centre here, and also of the various workmen from the machine shops, founderies, and other subordinate manufactories, forming a population as before stated, of nearly 150,000 persons.

In most of the manufactories in the United States, sprinkled along the glens and meadows of solitary water-courses, the sons and daughters of respectable farmers, who live in the neighborhood of the works, find for a time a profitable employment. The character of each individual of these rural manufacturing villages, is commonly well scanned, and becomes known to the proprietor, personally; who finds it for his interest to discharge the dissolute and vicious.

The proprietor of a manufactory in Manchester has many hundred persons daily entering his gates to labor, of most of whom he does not even know the names. He rarely troubles himself with investigations of their conduct whilst they are without the walls of his premises, provided they are reported to be regular at their labor whilst within them. The virtuous and vicious females are thus brought into communion without inquiry and without reproach. The contamination spreads, and the passing traveller is induced to pause at the sight, like Southey, in his letters of Espriella, to denounce the sources of present wealth, however overflowing and abundant, whilst the enriching stream is contaminating, and undermining the best interests of man. Whilst he sees plenty scattered over a smiling land, and every prospect pleases, he may sigh on finding that "only man is vile." God forbid, however fondly the patriot may cherish the hope of increasing the resources of his country by opening and enlarging the channels of national industry, that there ever may arise a counterpart of Manchester in the New World.

It may be intended as a blessing that an all-wise Providence has denied to the barren hills of New-England the mines of coal, which would allow the inhabitants to congregate in manufacturing cities, by enabling them to have recourse to artificial power, instead of the natural water power so profusely furnished by the innumerable streams, that in their course to the ocean descend over beds furrowed in the rocks of an iron bound country.— Whilst a cold climate and an ungrateful soil render the inhabitants from necessity industrious, thus distributed in small communties around the waterfalls, their industry is not likely to be the means of rendering them licentious; and of impairing the purity of those moral principles, without which neither nations nor individuals can become truly great and happy.

Samuel Slater

Census of Manufactures, Returns from Rhode Island (1833)

In early 1832 the U.S. Congress mandated Secretary of the Treasury Louis McLane (1786–1857) to conduct a census of manufacturing establishments in the United States. McLane delegated the task to prominent businessmen in each state, who in turn delegated the task to industrial proprietors known to them in each town or district. Samuel Slater, whose illustrious career is considered in other documents in this collection, was appointed to conduct the census for the state of Rhode Island. When he submitted his returns, he took the opportunity to reflect upon the present condition of manufacturers in the state and generally. McLane printed Slater's essay along with the data Slater had gathered. Other excerpts from McLane's report appear elsewhere in this volume. (MBF)

Samuel Slater, "Returns from the State of Rhode Island," in [Louis McLane], *Documents Relative to the Manufacturers in the United States, Collected and Transmitted to the House of Representatives,* volume 1 (Washington, 1833), pp. 927–931. Reprinted by Augustus M. Kelley, 1969.

MANUFACTURES.

RETURNS FROM THE STATE OF RHODE ISLAND.

Document 8.—No. 1.—*General Summary for Rhode Island.*

The mill manufactures of this State are located chiefly in the three counties of Providence, Kent, and Washington, on the waters of the Blackstone, Moshasuck, Monasquatucket, Pawtuxet, and Pawkatuck rivers, and their tributary streams. The four first named rivers fall into Narragansett bay, and the last into Long Island sound. The most considerable of them would, in the western country, be designated as a creek. Its course, from its head waters to the place of embouchure, does not, perhaps, exceed sixty miles, and the volume of water which it discharges would be inconsiderable in comparison with that discharged by what is called a creek; yet these streams, more steady in their volumes than those of the western country, and descending, in their short courses, an elevation of from two to four hundred and fifty feet to the tide waters of the bay and sound, furnish, with their tributaries, innumerable cascades, and a power of propelling machinery almost incalculable in amount. The operative capacity thus offered has been improved by human art, in some instances, to the extent of constructing artificial works for the retention of superabundant waters during the winter and spring freshets, to be again turned into the river in the summer and autumn. Many streams which, with their natural flow, would drive the works upon them only ten or eleven months of the year, have been thus made to operate them throughout the year. The power thus offered by nature, thus improved by art, and actually applied to the propulsion of machinery of all sorts within the territorial limits of the State, may be computed as equal to the labor of *more than twelve thousand horses*, working eight hours in each day. The amount of water power still unoccupied is very great. With such a redundancy of this natural agent, the more expensive one of steam has been, so far, very little used, and only under peculiar circumstances of location and business. Four cotton mills are now driven by steam, and two more, to be operated by that agent, are in progress. Most of the bleacheries and calenders, and some machine shops, foundries, and other works for iron, make use of the same agent. But the extent of this power used in Rhode Island does not exceed, exclusive of steamboats, the power of eight hundred horses.

The works operated by these powers include almost every branch of manufacture hitherto introduced into this country. Cotton, woollen, and iron, with the kindred and collateral operations which they require, are, however, the most important. Only one glass house has been established, which gives flattering indications of ultimate success.

The most remote date of these mill works cannot be stated with certainty. Very early after the settlement of the State, water mills for corn were established on the Moshasuck, within the limits of the present city of Providence, and at the falls on the Blackstone, now within the limits of North Providence. The iron work of these mills was imported from England At subsequent dates the manufacture of scythes, plough shares, and other implements of husbandry, was begun upon the Blackstone, and of anchors and other heavy ship iron, at Pawtucket and Warwick. The great anchor shop, at the latter place, was established by the family of the revolutionary General Nathaniel Greene. Some time during the war of 1755, a steam engine was imported by some merchants of Providence, and set up at the Hope Furnace, in Scituate. This furnace was the first in this country which produced cannon and hollow ware. The cotton manufacture was begun in 1790; but, up to the commercial restrictions consequent on the unjust measures of France and Great Britain, against the trade of this country, that business was confined to few hands. In the years 1806–'7, and to the end of the second war with Great Britain, the increase of this branch of business was rapid. The check which it received on the repeal of the war duties, the subsequent revival of the tariff of 1816, its actual progress to the present time, are well known. In Rhode Island this and almost every other branch of manufacture has been carried on by joint stock companies, few of which are incorporated. In the cotton trade are one hundred and nineteen establishments, or about one hundred and forty mills for spinning and weaving, five bleacheries and calenders, and two printing establishments, with a fixed capital (about one-half in lands and fixtures, and one-half in machinery of all sorts) of five millions six hundred thousand dollars, and a circulating capital of near two millions; 239,000 spindles, 5,856 power looms and other machinery, implements, and tools in proportion. The amount of wages in the cotton business, for 1831, was computed at about twelve hundred thousand dollars, the quantity of raw cotton consumed 10,415,000 pounds, at an aggregate cost of 1,050,000 dollars; 3,253 barrels of starch; 66,000 gallons of sperm oil; 44,500 pounds leather; 286 tons iron and steel; 4,110 tons fossil coal; 3,240 cords firewood, are among the articles consumed in the manufacture, exclusive of the fuel and provisions of the work people.

The manufactures of woollens, iron, glass, and other substances, cannot be estimated with the precision and certainty indicated by the queries. Their aggregate amount in capital, wages, and numbers employed, may be, when compared with the cotton manufacture, as about five to nineteen; thus making the aggregate capital in all mill manufactures in Rhode Island more than 8,000,000, the persons employed by, and supported at them about 24,000, and the amount of wages paid about 1,850,000 dollars. With still less confidence can we attempt any general estimate of profits. The profits of branches of business which are well known and even skilfully followed, must, in the nature of things, depend upon the comparative skill and industry of different individuals engaged in those branches. A greater contrast takes place when the competition is between persons some of whom are well skilled, others of whom have no skill in the business pursued. Some of the persons who embarked in the cotton business in 1790 were skilled in the business, and they realized, in the first fruits of their enterprize, those remunerating profits which commonly reward the successful projectors of a new trade, required by the wants of the community. Of the investments in manufactures consequent upon our commercial restrictions, a few only were successful, while the greater number, managed without the skill, prudence, and economy which such undertakings require, brought ruin upon all concerned in them. Mr. Whipple's answers to the questions proposed on this point are very full, and may be referred to with much confidence. Investments made subsequent to the war with Great Britain have been managed with more skill and economy than were those made before the war; but it questionable whether the average profits on those investments, including re investments and actual issues in the form of cash dividends, come up to the average of profits in commerce, shipping, or agriculture.

On the subject of profits, the theoretical economists of this country have been widely misled from the actual state of things here, by the scholastic distinctions of European economists. Because the latter, taking the actual relations of older and richer communities, have distinguished between capitalists, or those who furnish money and employers, or those who hire money, the former have applied the same classification to this country, which, being a younger and a poorer community, has no such distinct and independent relations or classes. As a general rule, all our able bodied citizens gain the whole, or a part of their own and the living of their families, by labor. The richest of them labor for themselves, and employ others to labor for them. They employ the surplus of profits over the consumption of themselves and families as a capital for some business, which themselves follow; or they lend that surplus at an interest which, with the profits of their personal labor, make up their income. If a man, therefore, accumulates a surplus of profits or income over his expenses, he will employ that new capital in his business, if by so employing it he can realize a new profit beyond the interest which he can obtain by lending it to another. If he now employs a thousand dollars, and, with his personal labor in the business, makes one hundred dollars a year more than interest and the expenses of his family, and has another thousand which may be as well employed, he will put that into the business also, because it will make his gross income twice as much as it was before, his family expenses remaining the same, and perhaps require no more of his own labor than his business now requires. The income, therefore, of nearly all the working classes in this country, which comprehend its whole population, is compounded of interest on money; profits on capital employed, and wages of labor. When, therefore, we inquire as to the rate of profits on capital, strictly so called—that is to say, the amount of product which remains after payment of interest, on one hand, and of the wages of labor on the other—we shall find it much less than superficial observers and theoretical economists have supposed. There must always be some inducement beyond the legal or common rate of interest, to tempt a man who has money in stock, or on loan, to transfer that money to a new employment. He will do so perhaps if morally certain of a remuneration of one or two per cent. beyond interest. He will do so, most surely, if, in the triple capacity of money lender, master, and laborer, he can, by employing his money himself, realize a gross profit, which, rateably divided between three individuals, holding the same relation to each other, could not remunerate either of them, but which, coming to him entire, would be twice or thrice the common rate of interest on his money.

It is in this triple capacity of money lender, employer, and laborer, that our most successful manufacturers have succeeded. They furnished money for the business and superintended its operation in person. They employed their families in the labors of the business, and, to the extent of this saving of the wages of superintendence and labor, realized the gross profits of manufacture. Instances of ultimate failure among manufacturers of this description are very rare. Yet would their gross profits fall very far short of a fair remuneration to each, if those profits should be divided among three distinct classes of persons, such as our theorists have supposed. Less successful than the above description of manufacturers, but more so than another description to be mentioned hereafter, have been those who, having no money of their own, but having industry, skill, and prudence, have manufactured on borrowed capital; and, by personal superintendence and hard labor, have saved a considerable proportion of the gross profit of manufacture, after payment of the interest on their loans. Many of these have succeeded in replacing the money loaned, and saved their mills; but this saving consists entirely in the saving by them of wages. Allow them the common rate of wages which has been paid to others for similar services, with a very small profit over interest on the money which they hired, and their mills would not pay the account. Least successful of all have been those who, themselves engaged in other pursuits, have invested the nett profits of their own business in manufacturing, and left the latter to the superintendence of others. A large proportion of these investments has been wholly lost through the waste, profusion, unskilfulness, and want of fidelity of agents, superintendents, and other servants, hired at large salaries, and having no interest in the ultimate prosperity of the concern. Many of these adventurers have not only lost the whole amount of their investments, but have been called upon for the debts contracted by their agents, in some instances to the extent of their whole property. Without, therefore, making any definite conclusion as to the precise amount of gross profits on manufacturing business, when conducted by the two first mentioned classes of adventurers, we may safely affirm, that the aggregate of those profits, if rateably divided according to the legal rate of interest, the common rate of profit on capital employed, and the usual rewards of labor in other callings, among three distinct and independent classes of individuals holding the relation of money lenders, employers, and laborers, would afford to neither of those classes that fair and common remuneration which other branches of business afford, and that nearly all the prosperous manufacturing concerns in this State have depended for their prosperity on an union, in the same persons, of the three capacities which have been mentioned.

*

We are asked, if the capital now invested in fixtures and machinery for the cotton manufacture can be employed in any other business? If the cotton manufacture should be proscribed and destroyed, the machinery and implements, constituting about one-half in cost and value of that capital, must be totally lost. The mills, mill sites, hydraulic works, and dwellings, now used in the business, might be made available in other branches of manufacture, if any such branches should be more favorably dealt with by the Government. But would there be any others more fortunate in this respect than that of cotton? It is fair to presume, that a manufacture which affords a domestic market for one-fifth of the cotton crop of the country, is not an object of *peculiar* hostility with those who are unfriendly to the protective policy; and that, if that should be destroyed, all the other manufactures of the country must share its fate. These works, therefore, could not be made available for any other purpose, and would be suffered to go to ruin. The 24,000 people now employed at the manufacturing villages in Rhode Island, and plentifully supported in ease and contentment, must, in that event, betake themselves to the cultivation of a sterile and ungrateful soil, at home, or seek, abroad, perhaps under more wise and paternal, though less freely constituted governments, that protection to their meritorious poverty and diligence which, under this Government, they will have been forbidden to expect! Deprived of her manufactures, the State will have nothing left with which to maintain her usual exchanges with other States. The comparative sterility of her soil will not permit her to compete with them in agricultural products for exportation. There is now no *foreign carrying trade* to employ her shipping; her coasting trade will have been annihilated with her manufactures; and the poverty of the rural population of this and the contiguous States, now supplied with imported goods from her markets, would, in such a state of things, reduce her direct foreign trade for consumption to what it was before the adoption of the federal constitution.

*

If the people of this country desire to be a civilized and powerful people, they must cultivate and promote those arts of life which form the elements of civilization and power. An exclusively pastoral or agricultural nation can never be formed into a polished or a powerful community. It is by diversity of employments, alone, that the various wants of a civilized population can be supplied, or wealth accumulated, or national institutions established and maintained. A young and growing community, provided with an ample supply of food and of the raw materials of manufacture, must diversify its products by the successive and constant introduction of new employments, or the redundancy of its products, in employments already introduced and established, will force upon its agriculture an undue proportion of its teeming population; thereby making redundant its production of commodities, which, being of a perishable nature, can least endure redundancy, thereby inflicting upon the community a loss of the labor of a part of its hands; an evil similar to that experienced in old and declining societies by the exemption from labor of a large proportion of their numbers.

Drawing of textile mills, Willimantic, Connecticut, by Edward Lewis Peckham. Sketch Book, 1846, Rhode Island Historical Society.

Caroline F. Ware

Launching the Industry (1931)

The earliest years of the American cotton textile industry, from Samuel Slater's arrival in 1789 to the boom period of the embargo and War of 1812, are described in this chapter from Caroline F. Ware's *The Early New England Cotton Manufacture*. Ware's study, published in 1931, is dated in some respects. Recent research has revealed that Almy, Brown & Slater initially produced both yarn and handwoven cloth rather than simply yarn as Ware believed. The firm closed its weaving and finished shops about 1796 but began weaving again—this time through the putting-out system—in response to the demand for cloth generated during the years of embargo and war, 1807 to 1815. Despite these qualifications, Ware's work has never been superseded and remains the classic study of the subject. (RP/GK)

Caroline F. Ware, *The Early New England Cotton Manufacture* (Boston and New York: Houghton Mifflin, 1931; reprinted New York: Johnson, 1969), chapter 2 (pp. 19–38); reprinted here by permission. See also James Conrad, "The Evolution of Industrial Capitalism in Rhode Island," Ph.D. dissertation, University of Connecticut, 1973; and Gary Kulik, "The Beginnings of the Industrial Revolution in America, Pawtucket, Rhode Island, 1672–1829," Ph.D. dissertation, Brown University, 1980.

LAUNCHING THE INDUSTRY

THE cotton industry in its initial stages swiftly justified Hamilton's faith in manufactures as a force to bind the union together. It vindicated, too, his insistence that ways and means could be found if the will to develop mechanical production were present. In its earliest form, it contained all the elements of the industrial revolution; during its first decade, it developed a characteristic organization and showed itself clearly as part of a national rather than a purely local economy; by 1807 it had made a distinct place for itself in the community and had taken firm root. In the years between Slater's arrival in America in 1789 and the passage of the Embargo Act, December, 1807, the industry adopted English machinery, imitated England in the type of labor employed, took advantage of the cheapened product of the cotton gin, and gained for itself a wide and expanding market. At each step the lead was taken by Almy and Brown of Providence, the first successful machine spinners in America.

This firm was an old Quaker mercantile house trading out of Providence and Newport to colonial ports, the West Indies, and European shores. Its members, Moses Brown and his son-in-law, William Almy, had mastered the technique of making money from trade and had inherited or accumulated great wealth, but they had not exhausted their spirit of experiment and enterprise. Brown had, moreover, a strong bent toward public service, which led him to endow the New England Friends' School and the Rhode Island College which now bears his name as Brown University. These were just the people to use some of their extra resources for an experiment which offered the triple attractions of difficulty, possible profit, and patriotic service.

At the time of Slater's arrival in America, they were trying out a variety of manufactures. In 'different cellars of dwelling houses' they were running a stocking frame, spinning wool, and using one of the first fly-shuttles in the country for weaving. 'Being desirous of perfecting, if possible, the business of the cotton manufactory so as to be useful to the country,' they had begun carding cotton, spinning it on hand jennies, and getting it woven upon linen warps into several types of coarse cloth. They had bought a number of crude and very imperfect American-made machines, a carding machine, a jenny and two spinning frames 'to work by hand after the manner of Arkwright's invention,' and were trying to improve them sufficiently to make them work.[1]

They were not alone in realizing the possible public advantage and private profit which the cotton manufacture might hold, or the only ones seeking to perfect its technique. A factory at Beverly, Massachusetts, set up in 1787 with horse-driven carding machines and spinning jennies, had been exempted from taxation and subsidized by the legislature, but four years later it was still operating at a loss. Of the investment of fourteen thousand dollars, about one third was reckoned 'sunk in waste of materials, extraordinary cost of first machines, and in maintaining learners and compensating teachers.' Even with the aid of the state land and lottery tickets amounting to about four thousand dollars, there remained a thousand-dollar deficit which was not being reduced as the net cost of producing cloth amounted to at least as much as the average price of the goods, three shillings and sixpence a yard. The partners explained that 'a want of skill in constructing the machinery and of dexterity in using it, added to our want of a general knowledge of the business we had undertaken, have proved the principal impediments to its success.' In 1791 they had not given up all hope of ultimate success, but they were ready to abandon the factory organization wherever possible in order to take advantage of 'the cheapness of household labor.' [2] A few years later the enterprise was abandoned as a failure.

Americans knew well that the English were spinning by

[1] Moses Brown to John Dexter, July 22, 1791, in Cole (ed.), *Hamilton Correspondence*, pp. 72–74; *Almy and Brown Papers*, Ledger A, 1789–93.

[2] George Cabot to Alexander Hamilton, September 6, 1791, in Cole (ed.), *Hamilton Correspondence*, p. 62.

water power but they were prevented from learning details of the process by the stringent English laws against the emigration of skilled workmen and against the export of models or drawings of machinery. In spite of public encouragement, all attempts to invent water frames of their own produced very inadequate results. The constructors of one such set of machinery, the Barr brothers, were granted a subsidy by the Massachusetts Legislature, their machines were dubbed the 'State's Models' and put on exhibition, but none of the frames copied from these models worked.[1] The frames which Almy and Brown bought were copies of these models but they were so far from satisfactory that Slater 'declined doing anything with them and proposed making a new one, using such parts of the old as would answer.'[2] This firm's good luck in discovering Slater, who knew all the details of the Arkwright frame, gave them the lead over all who were trying to work from purely American sources.

In the fall of 1789, they invited Slater to come to Providence, build a mill for them, install machinery of the English type, set the mill in operation, and manage it. Early in 1791 the new machinery, tended by nine children, turned out its first satisfactory yarn.[3] This was the real beginning of the cotton manufacture in America and this was the center from which it spread during the next twenty years. Although there were fewer spindles in Almy and Brown's first mill than in the Beverly factory, only seventy-two as compared with six hundred and thirty-six,[4] superiority of machinery made the Providence firm for several years the only successful machine spinners of cotton in the country. An early competitor set up a business in Paterson, New Jersey,[5] in 1794, but, according to Almy and Brown, this company always made a very inferior grade of yarn.[6] Three factories

[1] Bagnall, William R., *The Textile Industries of the United States*, Vol. I, 1639–1810 (all published), Cambridge, 1893, pp. 85, 86.

[2] Moses Brown to John Dexter, July 22, 1791, in Cole (ed.), *Hamilton Correspondence*, p. 73.

[3] *Ibid.* [4] Bagnall, *op. cit.*, pp. 96, 159. [5] *Ibid.*, p. 181.

[6] *Almy and Brown Papers*, Almy and Brown to E. Waring, November 9, 1808.

started independently in Connecticut, but at least one of them failed to perfect the water frame and probably both of the others relied chiefly on spinning jennies.[1]

It would have required a discriminating observer to recognize in Almy and Brown's first little shop the beginnings of the American factory system, but the characteristics of the industrial revolution were all there in elementary form. In place of the cellars of dwelling houses which had housed all their other experiments, they built a 'factory house' to avoid the 'ill conveniences' of the early locations[2] and to bring the workers together under supervision. Their child laborers were no longer furnished with clothes as apprentices but received pay as wage workers.[3] Many hand processes persisted, but the principal steps of carding and spinning were performed by machinery, and the machines worked successfully. Capital was furnished in the form of building supplies from the firm's retail store, pay from the same source in notes on the store, and some ready cash to purchase cotton and pay the overseers who would not accept store pay.[4] Almy and Brown themselves exercised slight oversight over the enterprise but left the active direction of production to their manager, Slater.

Only certain of the steps in the manufacture of cloth were performed in the factory. The baled cotton which the mill owner received direct from his agent in the south[5] first had to be opened and picked to loosen the fibers and to remove the specks of dirt. This was all done at home by children too small to work in the factory, who spread the cotton on a 'whipping machine,' a bed cord stretched across a square frame, and beat it with whip stocks until the dirt had fallen

[1] Norwich, probably 1790, and Bethlehem, before 1793, Bagnall, *op. cit.*, pp. 168, 197; East Greenwich, Cole (ed.), *Hamilton Correspondence*, p. 73.

[2] Moses Brown to John Dexter, July 22, 1791, in Cole (ed.), *Hamilton Correspondence*, p. 74.

[3] *Almy and Brown Papers*, Ledger A, 1789–93.

[4] *Ibid.* and *Slater Papers*, Almy and Brown, account with spinning mill, 1793–1803, cf. Appendix H.

[5] *Almy and Brown Papers*, Almy and Brown correspondence with Sam Maverick, Charleston, 1807–11, *passim.*

out and the material was light and fluffy.[1] Picking and beating was one of the processes which remained longest a hand operation even after it had been brought into the factory. It was still commonly done by hand in 1814 [2] and sometimes even after 1820.[3] In 1818 one mill operator advised his manager to have picking done by hand rather than to try any of the machines invented for the purpose as they did not do the work well and left specks of dirt which injured the cards.[4] Machine picking did not come into general use until the mid-twenties.[5]

The picked cotton was brought to the mill and spread by children upon a carding machine, two slowly moving cylinders full of wire pins which combed the fibers until they lay parallel. A second machine took the wad of carded cotton and formed it into a loose, soft roll, called 'roving,' which was then ready to be spun. The spinning machinery introduced by Slater and used in this and later mills was the simplest form of the Arkwright spinning frame driven by water power and tended by young children. The carding, roving, and spinning machines were all so simple to operate that the only adults needed in the mill were the overseers and repair mechanics. Almy and Brown, who started with nine children, were employing in 1801 over one hundred between the ages of four and ten.[6] They could not leave the children without at least one adult person present, so they put all their machinery into one room where they needed only a single overseer.[7]

The simple spinning machine operated by means of a

[1] Walton, Perry, *The Story of Textiles*, Boston, 1912, p. 208.

[2] *Boston Manufacturing Company Papers*, Accounts Current, 1814; *Slater Papers*, Slater Company Weave Book 1813–15, pp. 104, 113.

[3] Walton, *op. cit.*, p. 208.

[4] *Poignand and Plant Papers*, D. Greenough to Poignand and Plant, September 20, 1818. A picking machine had been invented in 1807.

[5] The Boston Manufacturing Company installed a picking machine in 1824. *Boston Manufacturing Company Papers*, Accounts Current B, Machinery, p. 77.

[6] 'Account of Journey of Josiah Quincy,' 1801, in *Massachusetts Historical Society Proceedings*, Series II, Vol. IV (1887–89), p. 124.

[7] *Almy and Brown Papers*, Almy and Brown to Silas Wood, October 27, 1814.

series of rollers revolving at different speeds. As the roving passed between them the rollers lengthened the thread while a 'flier' spindle simultaneously twisted it. The yarn which came from this machine was stout and fairly coarse, suitable for warps and for all types of heavy cloth.[1] There was significance for the American industry in the fact that it was this machine rather than the hand jenny which was first perfected here, for it made the small yarn mill rather than the merchant manufacturer the basis of industrial organization.

The hand jenny, which had been invented in England a few years before but was not patented until after Arkwright's frame, reproduced the process of hand spinning, while enabling a single worker to spin many threads at once. The spinner turned a wheel which operated anywhere from eight to one hundred and twenty spindles and gave twist to the yarn, while a carriage with a clamp for each thread took the place of the spinner's hand in pulling out the thread. The carriage then traveled back, performing the hand spinner's second process of winding the length of yarn on the bobbins. This machine could not easily be driven by power since it had to stop between the two operations. The yarn which it made was soft and excellent for filling but not suitable for warps.[2]

In England, hand jennies were extensively owned by former hand spinners who continued to work on cotton put out to them by merchant entrepreneurs. Sometimes a spinner installed several such machines in his attic, employed a number of hands to work them, and became, in a sense, a small manufacturer, though remaining dependent upon the merchants who furnished capital and put out materials.[3] After the invention of the mule, a hand-driven machine which combined the roller method of the Arkwright frame and the stretching process of the jenny for making stronger, finer thread, mule spinners worked in their

[1] Daniels, George W., *The Early English Cotton Industry*, Manchester, 1920, p. 80.

[2] *Ibid.*, pp. 79, 80.

[3] Unwin, George; Hulme, Arthur; and Taylor, George, *Samuel Oldknow and the Arkwrights*, Manchester, 1924, p. 70.

homes in the same way as jenny spinners.[1] The English industrial unit thus comprised the merchant capitalist and the many individuals or groups of workers dependent on him for work.[2]

American attempts to use hand jennies before the introduction of the water frame were unsuccessful and produced a condition in the cotton industry which contrasted both with England and with the early stages of the American woolen manufacture where jenny spinning was perfected before the water-driven frame.[3] The Philadelphia Society for the Encouragement of Domestic Manufacture set up jennies in its workshop in 1787,[4] the Beverly factory struggled to make such machines serve,[5] and a small factory in Norwich, Connecticut, tried them.[6] Almy and Brown also experimented with jennies but they, like all the others, found it impossible to spin yarn which could be used for warps and had to continue to use linen for that purpose. The Philadelphia Manufacturing Committee suggested the importation of cotton yarn from India to fill the need for warp thread.[7] The product of the water frame was, from the first, suitable for warps and also for filling in the coarse type of cloth which the Americans used. This cloth needed neither the soft thread which jenny spinning gave nor the fine mule-spun yarn. As a result, jenny spinning stood condemned by the early efforts to make it serve all purposes, mule spinning was never extensively introduced, and the Arkwright frame reigned supreme in the spinning field.

This frame was easily run by power. The process was continuous, the product fairly stout and capable of standing

[1] Daniels, *op. cit.*, pp. 117, 122.

[2] Unwin, Hulme, and Taylor, *op. cit.*, *passim.*

[3] Cole, *Wool*, Vol. I, pp. 110, 111, 224.

[4] Cf. description of jenny in *American Museum*, Vol. V, p. 225. 'One spinning machine, commonly called a jenny, with forty spindles, (which is a proper number,) will cost about thirteen pounds. One man or woman will work this machine, and will spin from four to six pounds of good yarn per day, of a suitable degree of fineness for good jeans, fustians, etc.'

[5] Cf. *supra*, p. 20.

[6] Batchelder, *op. cit.*, pp. 32–34.

[7] Moses Brown to John Dexter, July 22, 1791, in Cole (ed.), *Hamilton Correspondence*, p. 77.

some strain, and the parts of the machinery which had to be put in motion were easily controlled. The only skill needed for tending the machine was the ability to knot a broken thread, a facility easily acquired by children, in contrast to a certain amount of judgment necessary in jenny spinning and a high degree of skill in running a mule. The dominance of this frame, consequently, made for factory rather than household manufacture and developed the small, water-power spinning mill as the typical early American industrial unit. The American yarn mill required more capital investment than the English jenny workshop and much less than the merchant entrepreneur. It employed unskilled instead of skilled labor from the start. It never developed a class of artisans with a vested interest in the industry who would resist the further introduction of machinery. It remained a small unit characteristic of the American cotton industry until the power loom brought an entirely new form of organization after 1813.

The early manufacturers, well aware that their machinery was imperfect, were constantly on the lookout for mechanical improvements, and were eagerly bombarded with suggestions by inventive mechanics.[1] New devices, unfortunately, did not all work. A modest inventor, advertising in 1820, admitted that manufacturers had been induced to adopt all sorts of 'improvements,' many of which had proved failures, and were becoming cautious about trying others. He, therefore, assured readers that he would not think of urging his own contribution were he not persuaded of its unqualified merit.[2] The chance which every ingenious person had of hitting upon an invention of importance must have been a strong influence in making mechanics leave the employ of others to set up for themselves, since, so long as they were employees, the factory owner received the benefit of all that they devised. They were ordinarily bound by contracts which pledged them not to reveal or use for their own benefit the peculiarities of the company's machines

[1] Cf. for example, Felch, Walton, *The Manufacturer's Pocket-Piece, or the Cotton-Mill Moralized*, Medway, 1816, written to advertise a new spinning device.

[2] *Manufacturers' and Farmers' Journal*, October 19, 1820.

and which stated that if any of their inventions should be worth patenting, the company, not the individual, should be the holder of the patent.[1] As independent manufacturers they might be able to compete with others through the construction of more perfect machines.

In these circumstances, Almy and Brown became not only the pioneer manufacturers themselves but the trainers of future manufacturers. Mechanic after mechanic who had learned or practised in their plant left their employ and set up in business for himself. Slater built a mill of his own in Pawtucket in 1799 and ran this on the side while still superintending Almy and Brown's old factory in that town.[2] Former employees or associates started little mills in Pomfret, Connecticut,[3] Warwick, Rhode Island,[4] Coventry,[5] Cumberland,[6] and Scituate[7] in the same state, New Ipswich, New Hampshire,[8] and Greenwich[9] and Whitestown,[10] New York. These mechanics were backed by merchants, farmers, and professional men in and around Providence who put together small capitals and ventured into the business on Almy and Brown's pattern.[11] There were very few others in the early days, only one or two Englishmen — James Beaumont at Canton, Massachusetts in 1801,[12] John Warburton at Vernon, Connecticut, in 1802[13] and perhaps others, — with an occasional farmer like Benjamin Shephard of Wrentham, Massachusetts.[14] The old Providence firm was indeed the parent of the American cotton industry, for it was responsible, directly or indirectly, for the erection of most of the twenty-seven mills which the Secretary of the

[1] Cf. *infra*, p. 261

[2] Bagnall, *op. cit.*, p. 251.

[3] *Ibid.*, pp. 418, 420, Smith Wilkinson, 1806.

[4] *Ibid.*, p. 444, Natick Manufacturing Company, 1807.

[5] *Ibid.*, p. 404, Daniel Anthony, 1805.

[6] *Ibid.*, p. 277, Benjamin S. Walcott, 1802.

[7] *Ibid.*, p. 442, Hope Manufacturing Company, 1806.

[8] *Ibid.*, p. 368, Charles Robbins, 1804.

[9] *Ibid.*, p. 373, William Mowry, 1804.

[10] *Ibid.*, p. 502, Benjamin Walcott, 1808.

[11] *Ibid.*, passim.

[12] *Ibid.*, p. 269.

[13] *Ibid.*, p. 317.

[14] *Ibid.*, p. 172.

Treasury discovered in operation in Rhode Island, southern Massachusetts and eastern Connecticut in 1809.[1]

The yarn produced by these small mills was bleached or dyed and then distributed to be woven by housewives or master weavers or, later, by the mills themselves. A small proportion was put to a variety of other uses. Before the use of chemicals became common, bleaching and dyeing were sometimes done by women in their homes.[2] These processes were more often performed by skilled workmen who had their own shops and were taken over in other cases by the manufacturers who built 'dye houses' to supplement the work of their 'factory houses.'[3]

Where cloth was woven in homes for household use, the yarn was bought at retail from the local shopkeeper to whom the factory sold.[4] A number of master weavers, chiefly in the Philadelphia district[5] but also scattered through such varied places as Roxbury, Massachusetts,[6] Poughkeepsie, New York,[7] and Norfolk, Virginia,[8] purchased yarn wholesale directly from the factory. Some of these master weavers worked by themselves, others employed several fellow craftsmen. Merchant weavers whose business it was to get yarn woven in homes were not active before the War of 1812. There is only one reference in Almy and Brown's early correspondence to the delivery of yarn to others to be put out,[9] an arrangement which was

[1] Gallatin, Albert, 'Manufactures,' April 17, 1810, in *American State Papers, Finance*, Vol. II, p. 433.

[2] Walton, *op. cit.*, p. 208.

[3] Moses Brown to John Dexter, July 22, 1791 in Cole (ed.), *Hamilton Correspondence*, p. 77; *Almy and Brown Papers*, Almy and Brown to J. Brunson, January 3, 1804.

[4] Cf. *infra*, Chapter VII.

[5] *Almy and Brown Papers*, Almy and Brown correspondence with E. Waring, Philadelphia, passim.

[6] *Ibid.*, Robinson and Lemist to Almy and Brown, February 24, 1809.

[7] *Ibid.*, Almy and Brown correspondence with John Wintringham, 1804–1814, *passim.*

[8] *Ibid.*, Almy and Brown to E. Sturtevant, May 11, 1814.

[9] *Ibid.*, Almy and Brown to N. Rice, December 9, 1808. 'We are to allow 5 cents per yard for 7/8 shirting, 5½cents per yard wide, and 9 cents for 9/8 sheeting for weaving the same. The cloth to be made

quite common after the war when the class of merchant weavers assumed some importance in the organization of the industry.[1]

Spinning mills did little toward weaving or having cloth woven until after 1809. Almy and Brown tried putting some yarn out to be woven after they had been selling it outright for ten years,[2] but they did not find the experiment satisfactory. Factories, they felt, were at a disadvantage in trying to get weaving done on favorable terms.[3] The finishing of the woven cloth was done outside the mill, or in a separate establishment under the same management[4] until the power loom brought all processes together under one roof.

The labor supply of these yarn mills came chiefly from the farms of the vicinity whence children could easily be attracted by a wage of from thirty-three to sixty-seven cents a week.[5]

When the multiplication of mills within a narrow radius absorbed the supply of children immediately at hand, advertisements in such papers as the *Massachusetts Spy*, of Worcester, and the Providence *Manufacturers' and Farmers' Journal* offered attractive employment to large families, those with five or six children preferred.[6] The few skilled

to our satisfaction and delivered in Boston, thou to charge us back the yarn as per invoice and commission of 5% and to settle with or remit us the amount of account sale once in six months. Please to be particular and charge thy weavers to make good cloth.... If the cloth suits us and we do not get a sufficiency made at our factories, we shall want as much as thou can get made, but, as soon as thou can, send us a specimen of the cloth before thou puts out much of the yarn.'

[1] Cf. *infra*, p. 75.

[2] *Almy and Brown Papers*, Smith and Allison to Almy and Brown, June 10, 1801.

[3] *Ibid.*, Almy and Brown to John Wintringham, December 22, 1808.

[4] Moses Brown to John Dexter, July 22, 1791, in Cole (ed.), *Hamilton Correspondence*, p. 74.

[5] *Troy Company Papers*, Troy Company, Time Book, 1813–32; *Slater Papers*, Slater Company, Time and Wage Books, 1813–28; *Lippitt Papers*, Lippitt Company, Wage Book, 1825–27; advertisements of mills and mill sites in *Manufacturers' and Farmers' Journal*, 1820, *passim*, stating labor supply available in the vicinity.

[6] One factory advertised for large families in the *Massachusetts Spy*, May, 1818, March, 1820, August, 1821, October, 1822, June, 1824, February, 1828.

workers, chiefly mechanics and overseers, were often the fathers of these children. They came in answer to similar advertisements, which always specified that men with families of suitable quantity and age were most acceptable.[1] Wages of both children and adults varied with local labor conditions, tending in general to equal or slightly to exceed the current wage rate for other types of employment.[2] Payment, however, was never wholly in cash but at least half and sometimes entirely in goods at the company store, where the temptation to overcharge was often too strong for the owner.[3] Working hours, from dawn until dusk in the summer and nearly as long with the aid of candle or lamp light in winter, were of course universal in the naïve days when work was the normal occupation of even the very young.

Capital for these small mills was derived from sources almost as near at hand as the labor market. The bulk was furnished by Providence traders who were willing to experiment with a small part of their income, by lawyers and other professional men interested in developing their neighborhoods, by farmers who gave land and water rights in return for shares in the undertaking, and, later, from the earnings of the industry itself.[4] Most of the capital was tied up in land, buildings and machinery, leaving very little free for operating expenses.[5]

The greatest problem of the early manufacturers, once the technical difficulty had been overcome, was to build up a market. This problem was one which touched the whole nation more nearly than any other, for a wide-spread market meant a broad community of interest and reacted in turn upon every phase of national economic life.

To people unaccustomed to the use of cotton yarn, it seemed impossible that any one would buy the volume which the mills turned out. When James Beaumont started the first mill in Canton, Massachusetts, he took pride in

[1] Cf. *infra*, p. 212. [2] Cf. *infra*, pp. 236–45.

[3] Cf. *infra*, pp. 245, 246.

[4] Fuller, Oliver P., *The History of Warwick, Rhode Island*, Providence, 1875, pp. 189–270, *passim*; cf. *infra*, pp. 126–37.

[5] Cf. *infra*, p. 139.

showing off his machinery to the women who came to buy yarn. 'Now do tell! Lud-a-massy! Is that spinning?' they exclaimed. What amazed them most, however, was not the process but the quantity. 'What on earth are you going to do with all this yarn? You never will be able to sell it in this vast world.' [1]

At the time when the industry was started, market conditions were most unfavorable. A series of failures in England and Ireland had sent goods to America to be sold for what they would bring. The English agents added to their low prices very long credits, 'doubtless for the discouragement of the manufactory here,' as Moses Brown concluded. He added that 'this bait has been too eagerly taken by our merchants' and 'the quantities of British goods of these kinds on hand exceeding the market obstruct the sale of our manufacturies.' This policy was being carried so far that, 'when the actual sales of British goods fail, of the cotton manufacturies, they are sent and left here on commissions.' Moses Brown attributed these measures to a deliberate design on the part of the British to hamper the American industry, as they had tried to choke the beginnings of the industry in Ireland, and he predicted death to the American effort unless the government intervened with protection as the Irish government had done.[2]

Moses Brown's picture was a gloomy one but he went ahead with his machine spinning in spite of the presence of British goods. In truth, these goods had little or no effect on the industry which he proposed to develop, for it was cloth, not yarn, which the English were dumping on American shores. He knew of the complaints that there was no suitable yarn available and of the suggestion that the lack be remedied by importing from the East Indies.[3] He knew, too, that the Americans were unaccustomed to the use of cotton yarn and that his problem was primarily to create a demand for yarn for household use in place of, or in

[1] Beaumont, James, 'Reminiscences,' in Hurd, Hamilton, *History of Norfolk County, Massachusetts*, Philadelphia, 1884, p. 946.

[2] Moses Brown to John Dexter, July 22, 1791, in Cole (ed.), *Hamilton Correspondence*, p. 75.

[3] Cf. *supra*, p. 25.

addition to, wool and flax and only secondarily to persuade the public to prefer American to English cotton cloth.

Almy and Brown's correspondence shows the methods which these manufacturers used for introducing their product and the extent to which they were responsible for creating and building up the cotton yarn market.

They sold most of their first yarn in 1791 locally to Rhode Island weavers but sent some as far away as Norwich, Connecticut. For the next two years they supplied the same radius, extending it north to Charlestown, Massachusetts.[1] During these years their production was very small — a labor force of nine children could hardly turn out great quantities of yarn — and their customers were correspondingly limited. As soon as they got their new mill running in 1793, they began to trade with shopkeepers in commercial cities. Their books in 1798 show a large number of small accounts with these people but only six of their many customers disposed of two thousand dollars' worth of yarn during the whole of the ten-year period, from 1793 to 1803. The largest customer averaged less than six hundred dollars' worth a year.[2] The geographical range of their trade was, however, considerable.

By 1801 they furnished retailers in Portland, Newburyport, Marblehead, Salem, Boston, New Bedford, Nantucket, and the Rhode Island and Connecticut ports. They shipped a large proportion of their product to New York City, up the Hudson River to Albany and Hudson, and south to Philadelphia and Baltimore. Orders from these places came in faster than the manufacturers could fill them.[3] Cotton yarn was becoming an article of household use in the farming districts.[4] It was in demand for knitting stockings,[5] winding

[1] *Almy and Brown Papers*, Ledger A, 1789–93.

[2] *Slater Papers*, Almy and Brown account with spinning mill, 1793–1803.

[3] *Almy and Brown Papers*, letters received in 1801, *passim*.

[4] *Ibid.*, Daniel Waldo (Worcester), to Almy and Brown, April 27, 1801. 'The saving it will make in private families which have but few females, to purchase yarn instead of wool for domestic manufacture begins to be generally known in this neighborhood; the consequence is that the demand has increased very sensibly.'

[5] *Ibid.*, Samuel Tenney (Exeter), to Almy and Brown, September 14, 1801.

wire hat-frames, embroidering and making fringes,[1] as well as for cloth. New York City weavers, seeing the Providence yarns advertised in their papers, put in orders to supply the 'manufacture' which they were about to commence.[2] The demand was so great that a New York stocking manufacturer found that the factory could not supply him, tried New York stores, and finally had to send to Philadelphia to get Almy and Brown yarn at an advanced price.[3] Albany was the only place which made a discouraging report of market conditions in 1801. An agent there wrote: 'Cotton yarn seems dull on account of so much being here from other parts,' and, 'there is a great quantity in town, and the peddlers bring it along every day for sale from Connecticut.'[4]

In this wide market can be seen the early stages of a condition which caused Emerson to note a generation later that cotton thread held the union together.[5] Though the volume of output and sale was so small that any claim to its wide-spread influence appears ridiculous, nevertheless the cotton industry was, from the start, national in scope and as such played its part in the process of national consolidation. The predecessors of the cotton mills had all been strictly local. Local spinning bees, often using locally grown wool and flax, had spun woolen, linen and, rarely, cotton yarn for use in the immediate vicinity. Carding and fulling mills had also been essentially local affairs, scattered inland along small streams, serving limited rural communities and working on locally raised wool and locally woven cloth. Spinning mills, on the other hand, were local only in their supplies of labor and capital. Forced to import their raw material, they had to choose a location where they could conveniently receive cotton supplies by water. When they came to ship their finished product, the best method was again by boat. Their wares passed through much the same

[1] *Almy and Brown Papers*, H. Ten Brook (New York) to Almy and Brown, September 8, 1801.

[2] *Ibid.*, John Crichton to Almy and Brown, December 12, 1801.

[3] *Ibid.*, M. Trappell to Almy and Brown, May 20, 1801.

[4] *Ibid.*, J. Barney to Almy and Brown, December 21, 1801.

[5] Emerson, Ralph Waldo, *Journals*, edited by E. W. Emerson and W. E. Forbes, 10 vols., Boston, 1909–1914, Vol. VII, p. 201.

channels as imported goods, coming into the hands of dealers in 'foreign and domestic wares' in the commercial seaport towns.

The records of the Almy and Brown mill for 1801 give further evidence of the industry's development in its first decade. These letters show competition beginning with the peddlers who hawked Connecticut yarn in Albany [1] and the purchaser who sent a sample of Connecticut yarn to show Almy and Brown how he wanted his fine yarn bleached and spun.[2] They record success in the initial effort to introduce the new product and to build up a body of customers,[3] for they reveal the raising of prices [4] and the spinning of finer yarn.[5] There is evidence, too, of profit from the business sufficient to stimulate Slater to build his own mill in 1799 [6] and to make Almy and Brown experiment further by getting their yarn woven into coarse bed ticking for the Philadelphia market.[7]

Such was the situation at the beginning of the century. During the next years market conditions enabled the industry to grow steadily. Almy and Brown were able to increase their production continually,[8] and built new mills in 1806 and 1807.[9] It was during these years that many of the men who had been in their employ started mills of their own, believing that the progress of Almy and Brown's business promised a strong future for the industry.[10]

[1] Cf. *supra*, p. 33.

[2] *Almy and Brown Papers*, H. Ten Brook to Almy and Brown, May 23, 1801.

[3] *Ibid.*, John Benson and Company (Boston), to Almy and Brown, March 20, 1801. 'We are out of all those numbers and find that we lose custom by not having supplies more regularly.'

[4] *Ibid.*, H. Colt to Almy and Brown, May 29, 1801.

[5] *Ibid.*, William Brown to Almy and Brown, August 31, 1801.

[6] Bagnall, *op. cit.*, p. 251.

[7] *Almy and Brown Papers*, Smith and Allison to Almy and Brown, June, 10, 1801.

[8] *Ibid.*, Mill Balances, 1804–19; correspondence, *passim.*

[9] At Smithfield, Rhode Island, 1806, Warwick, Rhode Island, 1807. In addition to the original Pawtucket mill they had built a second mill at Warwick in 1794.

[10] Cf. *supra*, p. 27.

Increased production and market expansion kept such even pace with each other that first one and then the other is reflected in Almy and Brown's correspondence. At times the sale of yarn became so profitable that more and more traders carried the product. 'The venders of yarn are so multiplied,' wrote an old agent, annoyed by the presence of competitors, 'that I presume your sales are rapidly increasing, while ours are daily diminishing.' [1] While Almy and Brown's new mills were being built in 1806 and 1807, orders were outstanding which could not be filled until the new spindles were set in motion.[2]

At other times the manufacturers complained of being forced to retain old selling methods which they had hoped to discard. They had started by sending yarn on consignment, allowing a commission on all sold.[3] They had employed this method in order to make purchasers acquainted with their wares, hoping that, after this was accomplished, they would be able to abandon consignment in favor of direct sales [4] but, instead of being able to make the desired change, they had to keep on with the old system and in some instances were even forced to raise their commission rates.[5] They allowed very liberal credit terms wherever they hoped for a steady, large demand, showing especial generosity to master weavers [6] and in regions where they thought that weaving might become a local industry.[7]

In the effort to keep sales up to the increased production of their own and their neighbors' mills and to resist pressure from their agents to lower prices in face of competition, they strove to push their market farther north, farther

[1] *Almy and Brown Papers*, N. Gilman to Almy and Brown, January 17, 1804.

[2] *Ibid.*, Almy and Brown to Samuel Tenney, June 20, 1807.

[3] *Ibid.*, Letter Books, 1804, letters received, 1801; *Slater Papers*, Almy and Brown account with spinning mill, 1793–1803.

[4] *Ibid.*, Almy and Brown to E. Williams and Company, February 22, 1804; to J. Arnold and Company, June 26, 1804.

[5] Cf. *infra*, p. 161 ff.

[6] *Almy and Brown Papers*, Almy and Brown to John Wintringham, March 30, 1804.

[7] *Ibid.*, Almy and Brown to William Brown (Nantucket), March 31, 1804.

south, and into more remote rural areas. In 1803 they sent the junior partner, Obadiah Brown, to New York City, Baltimore, Philadelphia, Alexandria and towns in the neighborhood of each city, where he persuaded merchants to take small lots on trial.[1] They endeavored to enlarge the southern market, urging their cotton merchant at Charleston to try to spread the use of their yarn among the weavers of South Carolina, 'if there are any weavers in the part of the country where sent.' They hoped that by means of his exertion in introducing the yarn it would 'get into use and answer a valuable remittance for the raw material of which it is made.'[2] In the country districts they proceeded cautiously lest the extension of credit where payment was uncertain and a too great scattering of goods should bring loss.[3]

During these years other manufacturers in Massachusetts, Rhode Island, and Connecticut erected mills in response to the growing demand, until by 1807 Almy and Brown were distinctly disturbed at the rate at which the industry had grown, and feared that it might be 'overdone.' 'It is increasing in this part of the country,' they wrote, 'and so much so that, unless the sales are extended, ere long it will be difficult to sell the quantity made.'[4]

Before the addition of artificial stimulus in the form of embargo and war, the American industry had developed slowly and quite normally through the interaction of increased production and enlarged markets. After the preliminary stages of the first decade, it is impossible to say whether production or consumption dominated industrial development. The fact that a number of mills were started and abandoned and that some shopkeepers were reluctant to receive further consignments[5] suggests that the business was not universally profitable. Yet traders usually displayed an eagerness which shows that they found yarn a 'favorable article of trade,' however much they might

[1] *Almy and Brown Papers*, O. Brown's letters to the firm, 1803.

[2] *Ibid.*, Almy and Brown to S. Maverick, December 9, 1807.

[3] *Ibid.*, William Almy to Obadiah Brown, June 1, 1803.

[4] *Ibid.*, Almy and Brown to M. E. Hugen, July 27, 1807.

[5] *Ibid.*, Almy and Brown to J. Doberay, March 1, 1804.

complain that the commissions were not high enough to reward them for their pains. The manufacturers certainly did not dominate the market if they could not abandon the consignment system and dictate their own selling terms, yet they were in a sufficiently strong position to expand their production on a sound basis, especially after 1804.

The mills erected before 1807, moreover, proved their soundness by the test of survival. No available figures reveal accurately the fate of these early mills, but the report on manufactures made by the Secretary of the Treasury, McLane, in 1832, bears witness to the permanence of these early experiments. Incomplete though this report doubtless is, especially in its list of Massachusetts mills which falls short by fifty-eight of the number found in that state by the Friends of Domestic Industry in the previous year, it is the nearest we have to a full and accurate catalogue of the mills then operating.[1] From this report it appears that thirteen

[1] Cotton Mills Erected in New England [a]

	Me.	N. H.	Vt.	Mass.	R. I.	Conn.	Total
To 1805.......				1	1	2	4
1805.......				1	1		2
1806.......				1	1	1	3
1807.......					4		4
1808.......				1			1
1809.......	1			1		1	3
1810.......		1		3	2	2	8
1811.......				5	1	1	7
1812.......			1	9	4	3	17
1813.......	1			13	3	3	20
1814.......		3		12	4	6	25
1815.......		2		4	2	3	11
1816.......		2		1			3
1817.......		3			1		4
1818.......		1		1	1	1	4
1819.......			1		2		3
1820........		1		4		2	7
1821.......		1		2		1	4
1822......		1		3	1	2	7
1823.......	1	2		4	5	7	19
1824.......		4	1	2	4	5	16
1825.......		1		8	2	6	17

[a] Compiled from McLane's report by Clive Day, 'The Early Development of the American Cotton Manufacture,' *Quarterly Journal of Economics*, Vol. xxxix (1925), p. 452.

of the twenty southern New England mills in operation before the end of 1807 were still in existence in 1832. These had not all run without interruption, or reorganization, but they had never been abandoned. Sixty-five per cent seems a large proportion, considering the very modest character of these establishments, and confirms the view that before 1807 mill development had been sound and the industry's position secure.

COTTON MILLS ERECTED IN NEW ENGLAND — *Continued.*

	Me.	N. H.	Vt.	Mass.	R.I.	Conn.	Total
1826		1		7	6	2	16
1827		1	1	4	5	4	15
1828		7	1	7	3	13	31
1829		3		6	2	5	16
1830		2		3	7	4	16
1831		4	3	10	8	6	31
1832	2			4		7	13
Total dated	5	40	8	117	70	87	327
Total undated ..	1	2	2	75	49	14	143
Total enumerated in McLane Report	6	42	10	192	119	101	470
Total enumerated by Friends of Domestic Industry [b]	8	40	17	250	116	94	525

[b] Friends of Domestic Industry, Report of Committees, Convention at New York, October 26, 1831, Baltimore, 1832, p. 112.

II

Introduction of Cotton Textile Technology

The human experience of the industrial revolution cannot be fully grasped without some understanding of the mechanical innovations that were the initiating or organizing factors in early industrial life. In this section we gather documents that explain the new textile technology and indicate how the men who sponsored and put those innovations into operation themselves understood the principles and practice of textile manufacture. The focus is on the major figures of industrial Providence, Rhode Island—Samuel Slater, the Wilkinsons, and Zachariah Allen. This emphasis is appropriate because the Providence area dominated technological development in New England during most of the period covered by this collection. It was both the originating point of the American textile industry and a continuing center of machine building and experimentation.

After Richard Arkwright's machinery for spinning cotton yarn was introduced in the early 1790s, the new technology spread through northern Rhode Island and into nearby areas of Massachusetts and Connecticut. Within two decades Providence was surrounded by a widening ring of factory villages. The spread of this technology is, however, not easy to document, for its medium was the practiced hands and quick minds of skilled mechanics moving from factory to factory rather than the written word. The growth of the industry was aided by an expanding network of partnerships, often built on family connections, which included men personally conversant with the running of factories and manipulation of textile machinery.

In its origins the new textile technology was English. Samuel Slater was followed to this country by many other British mechanics, who played a continuing role in improving methods. Among them was

Samuel Ogden, whose writing is represented elsewhere in this collection. Ogden was one of the immigrant mechanics responsible for introducing Samuel Crompton's spinning mule to the United States between 1803 and 1805. Such men worked alongside native-born carpenters and blacksmiths to create a new generation of machine builders who greatly expanded and transformed the American cotton textile industry during the 1810s and 1820s.

The decline of the industry after the resumption of trade with England at the close of the War of 1812 turned the attention of progressive factory owners toward more aggressive and purposeful management of technology. The introduction of the power loom at Waltham in 1814 by the Boston Manufacturing Company marked both a great advance in technology and a shift of textile technology to Boston-based northern New England centers. But the vigor and contribution of Providence-based machine shops remained, most significantly in their quick contributions to advance in power loom design.

The distinctiveness of Rhode Island's contributions comes through clearly in the memoirs of Allen and the manufacturers who wrote for the *Transactions* of the Rhode Island Society for the Encouragement of Domestic Industry, reprinted in this collection, and who felt keenly that popular historical accounts and Boston corporate publicity slighted their work. Such overemphasis on the technology of the Waltham-Lowell system of textile manufacture, treated in detail in volume three of this series, has continued in the historical accounts of American industry, and one of the main functions of this volume is to right this imbalance and to help explain the full nature of textile innovation in the New England mill village. (TZP/MBF)

Cotton Spinning (1828)

It is extremely difficult to assess the role played by technical articles and books in transferring knowledge about the textile industry of Great Britain to the United States. Surely such publications acquainted a broad public in a general way with the processes involved in converting cotton into cloth, but it is doubtful that anyone unfamiliar with the intricacies of textile machine building and operation could learn enough from those terse descriptions and simplified drawings to actually construct and run a machine. Instead, technical knowledge moved from east to west across the Atlantic Ocean in the minds and hands of skilled mechanics.

The following selection was excerpted from *The American Journal of Improvements in the Useful Arts, and Mirror of the Patent Office,* printed in Washington, D.C., in 1828 and edited by I. L. Skinner. It was chosen because of its careful explanation of Sir Richard Arkwright's "great principle of cotton spinning, viz. the drawing by rollers, which extends the fibres in so perfect a manner." The spinning machine illustrations that accompany the text are of a late eighteenth-century version of an Arkwright waterframe. Machines such as this were little used after the first decade of the nineteenth century.

The American Journal of Improvements in the Useful Arts, and Mirror of the Patent Office in the United States, ed. I. L. Skinner, Volume I (Washington, D.C., 1828). The material on cotton spinning is excerpted from pp. 277–288. For a more complete description of manufacturing cloth from cotton, see the books of James Montgomery: *The Carding and Spinning Master's Assistant . . .* (Glasgow, 1832) and *A Practical Detail of the Cotton Manufacture of the United States of America . . .* (Glasgow, 1840), or the article on cotton manufacture in *The Cyclopaedia; or Universal Dictionary of Arts, Sciences, and Literature,* ed. Abraham Rees (London, 1819; 1st American edition, Philadelphia, 1825).

This article was twenty years out of date by the time it was printed in Skinner's volume. Apparently, the text and illustrations were copied from Edward James Wilson's, *The Artist's and Mechanic's Encyclopaedia . . .*, published previously in Britain (Newcastle-upon-Tyne, n.d.). It is possible that the article was commissioned for the encyclopedia or picked up from a still earlier work. In any event, this selection demonstrates the difficulty of assessing the part played by printed materials in transferring knowledge of cotton spinning from Britain to America. (TZP)

The mule is so called, because it is a compound of Hargraves' jenny and Arkwright's twist frame. These three machines, although they have been variously modified and greatly improved, are the elements of the spinning apparatus in all the great factories of Europe and America. A full conception of their importance, however, cannot be had, while we confine our views to the mere article of spinning. We must see their connexion with the preparatory process, the carding machine and power looms. The amount of all these improvements, in little more than half a century, is so great, that no one can contemplate them without astonishment. Kepler or Newton would be as much lost in them, as Pythagoras or Archimedes, although they lived two thousand years apart.

Although not necessary for those conversant with them, and not fully intelligible to those who have never seen the inside of a factory, yet for the satisfaction of some of our general readers, we will give sketches of the carding and roving machines, and also of a spinner, leaving the loom for a future number.

CARDING MACHINE.

See plate XXIII.

Fig. 1 is a section. Here B B is a large cylinder, turned rapidly round by an endless strap on the pulley A; the surface of the cylinder is covered with cards the sheets of leather for which are glued or nailed on in strips or sheets parallel with its axis, and disposed in such a direction that when it revolves in the direction of the arrow the teeth upon it go with their points forward, so that if a lock of cotton was held against them, it would be drawn inwards upon the teeth. The cylinder revolves under an arch *c c*, lined with the same kinds of cards which are shown in fig. 2; the teeth are disposed to meet those of the cylinder. D is a second cylinder of cards, the teeth meeting the first, its motion being taken by a large bevelled wheel *f*, on the end of its spindle from a small pinion on the end of an inclined axis *r*, which at the other end receives its motion by a pair of equal bevel wheels from the spindle of the great cylinder B B. Before the cylinder at *b*, are a pair of fluted feeding rollers, between which the cotton passes, and is delivered to the cylinder, the cotton is spread out upon a feeding cloth *e* which traverses constantly round two rollers *g h*, one of which is turned by means of a pinion from the feeding rollers; these receive their motion from a wheel on the end of the cylinder D by means of bevel wheels, and an inclined spindle not seen in the figure, but its direction is shown by the dotted line *a a*. The cotton is taken off the cylinder D in a continued fleece by the mechanism described in figs. 1 and 3, which is called the comb, or taker off. This is a rod or iron bar *i i*, situated parallel to the axis of the cylinder, and cut on the lower edge with fine teeth like a comb; it rises and falls parallel to itself by being united to two rods *k*, which are guided by two levers *l l;* the lower ends of the rods *k k* are, as shown in fig. 1, jointed to two small cranks *m* formed on a spindle which is turned by a pulley *n* with an endless strap from a pulley fixed on the main axis close behind the great pulley A. Now, by the motion of these cranks, the rod *i i* rises and falls, and at the same time moves a little to and from the surface of the cylinder D. By this motion it scrapes downwards between the teeth thereof, and in consequence removes the cotton from them the whole length of the cylinder at once, and the motion of the crank is so quick, that by the time this piece of cotton, so detached from the great cylinder D, has moved with the cylinder as much as its own breadth, the crank makes another stroke, and in consequence the second piece detached from the teeth adheres to the first, the third adheres to the second, and so on. The cotton is thus stripped, or skinned off the cylinder, in a continued and connected fleece. This fleece, as is shown in fig. 1, is received upon a plain cylinder E, which is turned slowly round, by means of pulleys and a band from the pulley dotted round the centre of the cylinder D. F is a small roller gently pressing upon the fleece, to make it lap evenly upon the surface of the cylinder E, as the same turns round, and takes it up when stripped off the surface of the cylinder D.

G is the main drum, or wheel, which turns the machine; it is fixed on a spindle extending the whole length of the mill, being suspended by brackets like *o* from the ceiling, and turning forty or fifty machines in a row.

I K, fig. 1, are two small cylinders, called urchins; they are covered with cards and revolve, so that their teeth act with the teeth of the great cylinder, through proper openings left between the top bars or rails, composing the arch C C; the urchins are turned slowly round in the direction of the arrows, by means of a band *x x* from a wheel on the spindle of the cylinder D.

To explain the action of this machine, we must give some idea of the nature of the operation of carding; the card may be compared to a brush, made with wires instead of hairs, stuck through a sheet of leather, the wires not being perpendicular to the plane, but all inclined one way, in a certain angle, see fig. 2, where R and S are these sheets of leather for a pair of cards, and T T, or V V, represent the teeth or card wires respectively belonging to each. Beneath is a view of one wire insulated, showing the two teeth, with their bend in the shank, called the knee bend, by which they are inclined to the leather, in the manner before mentioned. Now we may conceive, that cotton being stuck upon the teeth of one of these cards, another may be applied to it, and combed or scraped in such a direction, as to strike the cotton inwards upon the teeth, rather than tend to draw it out. Of the consequences of a repetition of the strokes of the empty card in this direction, upon the fall, one is a more equal distribution of the cotton, upon the surface of the teeth, and in doing this the fibres are combed and laid straight. Then, if one card be drawn in an opposite direction over the other, it will, in consequence of the inclination of its wires, take the whole of the cotton out of the card, whose inclination is the contrary way.

The cotton being spread out evenly upon the feeding cloth *e*, and advancing with the cloth, it is thrown in between the fluted feeding rollers *b*, which deliver it gradually and equally to the cylinder, by this it is carried round until it meets the urchin I, which is turning round very slowly, and its teeth meeting the teeth of the great cylinder, takes off part of the cotton therefrom, and carries it round till it meets K, which moves so as to take the cotton off from I, and return it again to the great cylinder. The object of thus transferring it is to obtain a more regular and equable distribution, than the feeding or spreading the cotton upon the cloth *e*, will make upon the cylinder. The great cylinder thus receiving the cotton, carries it round and cards it against the teeth lining the arch C C; in this process it becomes more equally distributed over the teeth in the cylinder, and gets carded; in so doing the cotton continues in this manner, hanging sometimes in the teeth of the cylinder, and sometimes in those of the arch, but slowly advancing till it comes to the cylinder D, whose teeth meeting those of the cylinder, and turning round very slowly take the cotton from it in a very regular and even film spread over its whole surface. This film it carries round to the comb *i*, by which it is detached, as before described, and lapped round the cylinder E, which continues to lap up the fleece upon it, until it has made fifteen or twenty turns, and of course as many revolutions of the fleece round its circumference. The attendant then breaks it off; by dividing it at one part and spreading it out straight, it will form a fleece called a lap, which is the length of the circumference of the cylinder, and consisting of fifteen or twenty thicknesses By this admirable contrivance great regularity is obtained in the thickness of the lap, because, if at any one part, the fleece produced by the machine is thinner or thicker, than it ought to be, in consequence of any irregularity in the spreading of the cotton wool upon the cloth, previous to carding, such irregularity will have no sensible effect upon the ultimate thickness of the lap, because it is composed of thirty or forty strata, and there is no probability that the inequalities of these several strata will fall beneath each other, but every chance that they will be equally dispersed through the whole, and thus correct each other. The lap, when taken off, is laid flat on a cloth, with which it is rolled up and conveyed to a second carding machine, and spread out upon its feeding cloth: in this machine it undergoes exactly the same operation as in the first, and the fleece is detached from the cylinder D in the same manner, but instead of going to the lapping cylinder E, as we have described, it is gathered up as shewn at X in fig. 3, into a tin funnel marked *p;* it then passes between a pair of rollers *q r*, which compress and flatten the fleece in its contracted state, into a pretty firm and connected sliver or band Y. and delivered into a tin can. The lowest of these rollers *r* is situated upon a spindle *s*, extending across the frame, and turned by a pulley *t* upon the end of it, which is connected by an endless band, with the pulley *z* upon the spindle of the cylinder D.

By these means, the fibres of cotton are disentangled from all knots, and the whole is reduced from the entangled and matted wool, to a regular and equable sliver or band, which is conveyed away in the tin can, to the drawing frame, which we shall next describe by the help of fig. 5, which is a section of the operative part of the machine taken through its middle; EEEE represent four of the perpetual slivers or endless cardings we have just described, entering into it; let A represent the section of a roller, whose pivot does not turn, in a pivot hole, but in the bottom of a long narrow notch B, cut in an iron standard; *a* is the section of another iron roller, whose pivot is retained in the same notches at each end, while the roller itself lies or rests on the roller A below it. The surfaces of these rollers are fluted lengthwise like a column, only the fluting are very small and sharp, like deep strokes of engraving, very close together; it is plain, that if the roller A be made to turn slowly round its axis, by machinery, in the direction as expressed by the dart, the roughness of the flutings will take hold of the similar roughness of the upper roller, and carry it round also in the direction of the dart, while its pivots are engaged in the notches B, which they cannot quit. If, therefore, we introduce the end F of the cotton sliver, or band EF, formed by the carding machine, it will be pulled in by this motion, and will be delivered out on the other side, considerably compressed by the weight of the upper roller *a*, which is of iron, and is also pressed down by a piece of brass, which rests on its pivots, or other proper places, and is loaded with a weight C. There is nothing to hinder this motion of the riband thus compressed between the rollers, and it will therefore be drawn through from the cans. The compressed part, after passing through, would hang down and be piled up on the floor as it is drawn through, but as it is not permitted to hang down in this manner, it is brought to another pair of sharp fluted iron rollers, K and L. Supposing this pair of rollers to be of the same diameter, and to turn round in the same time and in the same direction with the rollers A, *a*, it is plain that K and L would drag in the compressed riband and would deliver it on the other side, still more compressed. But the roller K is made, by the wheel work, to turn round more swiftly than A. The difference of velocity at the surface of the rollers, is, however, very small, not exceeding one part in twelve or fifteen. But the consequence of this difference is, that the skein of cotton will be lenghthened in the same proportion; for the upper roller pressing on the under ones with considerable force, their sharp flutings take good hold of the cotton batween them. Since K and L take up the cotton faster than A and *a* deliver it out, it must either be forcibly pulled through between the first roller, or it must be stretched a little by the fibres slipping among each other, or it must break.

When the extension is so very moderate as we have just now said, the only effect of it is merely to begin to draw the fibres (which at present are lying in every possible direction) into a more favorable position for the subsequent extensions

The fibres being thus drawn together, the cotton is introduced between a third pair of rollers O P, constructed in the same way, but so moved by the wheel-work, that the surface of O moves nearly, or full three times as fast as the surface of K, the roller P being also well loaded they will take a firm hold of the cotton, and the part between K and O is nearly or wholly trebled in its length, or the sliver is extended to almost four times the length in which it enters between A *a*. After the sliver has passed through the three pairs of rollers it is conducted through a tin funnel H, being drawn forward, by a pair of rollers R S, this contracts it into a regular sliver, and it is delivered at G into a can. The upper roller S, is merely pressed down upon the under one by its own weight, and therefore compresses it but a little, though sufficiently together with the contraction produced by the funnel H, to unite the four slivers EEEE, which enter together into one which passes out at G, between the rollers RS. These rollers do not draw or extend the cotton; their velocities being accurately adapted to take up the four slivers as fast as they come through the other pairs, and by drawing them all together through the funnel H, to unite the four into one, and the slight pressure of the rollers compresses them into a firm and connected sliver, which though compounded of four, is only the same size as any one of the four put in, because it is drawn out to four times the length. The effect of the machine has only been to straighten and lay the fibres parallel to each other; for the motion which the drawing produces among them always tends to extend each individual fibre to its full length, and it is necessary to unite several slivers together, as the drawing would reduce the sliver to such a small size, that it would not bear sufficient extension without separating and breaking across.

Figs. 4 and 6 explain the wheelwork, which communicates the motion from one roller to the others, fig. 4 being a view of the wheel-work at one end of the rollers, and fig. 5 the wheels at the opposite end. The motion is given to the whole machine by a strap and pulley D, fig. 5, on the end of the pivot of the roller O. On the opposite end of this roller, the small wheel *g*, fig. 6, is fixed and turns *h*, which is mounted only on a stud, and carries with it a pinion *i*, this turns a wheel *k*, on the end of the pivot of the roller K; now as the wheel *g* is larger than *h*, the latter will move much slower, and as *i* is much smaller than *k*, this will move slower than either; the proportions are so adapted, that K and *k* will only turn once for three times of O *g*, but the proportions vary in different mills. On the other end of the roller K, a pinion *f*, fig. 4, is fixed, this turns an intermediate or connecting wheel *e*, and thus gives motion to the wheel *d*, fixed on the end of the roller A, which has the roller *a* over it; now the wheel *f* being, in its diameter and number of teeth, to *d* as twelve to fifteen, of course the relative velocities of the rollers A and K will bear that proportion as before stated.

The rollers R S are turned by means of a strap from a pulley on the pivot of the front roller O.

The reader will by this perfectly comprehend Sir R. Arkwright's great principle of cotton spinning, viz. the drawing by rollers, which extends the fibres in so perfect a manner.

As the drawing frame takes in four slivers E E E E, and draws them into one at G, this is repeated four times over, by passing the sliver as many times through the machine; therefore, by this process the sliver is drawn out ($4 \times 4 = 16 \times 4 = 64 \times 4 = 256$) to 256 times the original length, as produced by the carding machine.

In this state the sliver presents a most beautiful appearance, being extremely regular in its size, and all the fibres being drawn so straight, that it bears a beautiful glossy or silky appearance. After this preparation of the sliver it must be reduced in size to a small thread, this is done at two operations, the first called roving, and the next spinning. The general effect of the spinning process is, to draw out this massive sliver, and to twist it as it is drawn out; but this is not to be done by the fingers pulling out as many fibres of the cotton at once, as are necessary for composing a thread of the intended fineness, and continuing this manipulation, regularly across the whole end of the riband, and thus as it were nibbling the whole of it away. The fingers must be directed, for this purpose by an attentive eye; but in performing this by machinery, the whole riband must be drawn out together and twisted as it is drawn; this requires great art and very delicate management, it cannot be done at once, that is, the cotton sliver cannot be first stretched or drawn out to the length that is produced; from the tenth of an inch of the sliver, and then twisted. There is not cohesion enough for this purpose, it would only break off a bit of the sliver, and could make no farther use of it, for the fibres of cotton are very little implicated among each other in the sliver, because the operation of carding and drawing has laid them almost parallel, in the sliver; and though compressed a little by its contraction in the card, from a fleece of twenty inches to a riband of two, and afterwards compressed between the rollers of the drawing frame, yet they were so slightly that a few fibres may be drawn out without bringing many others along with them. For these reasons, the whole thickness and breadth of two or three inches, is stretched to a very minute quantity, and then a very slight degree of twist is given it, viz. about two or three turns in the inch, so that it shall now compose an extremely soft and spungy cylinder, which cannot be called a thread or cord, because it has scarcely any firmness, and is merely rounder or slenderer than before, being stretched to about thrice the former length. This is called a roving, and the operation is performed in the roving frame which is shown in figs. 7. 8, and 9, the first being a front elevation, and the other a cross section, the reduction of the sliver is effected by rollers in the same manner as the drawing frame, but only two ends being put through together, instead of four, the size is of course reduced: but this reduction renders it so delicate, that it is necessary to give it a slight twist to render it sufficiently cohesive to bear handling.

The machine contains three heads or frames *A A* of rollers, each of which receives four ends or slivers from the can, BB fig. 8, which are those brought from the drawing frame, and enter between the back rollers *a*, and are drawn out from the cans between them, and the other rollers *b* to the proper degree of fineness, but which varies with the quality of the yarn which is to be spun. Each of the slivers after passing through the rollers is received into a tin can D, through a small funnel E, at the mouth of which the can is set up in a frame *d d e f*, called the skeleton. It is supported on a pivot at bottom, and is kept in rapid motion by a band, working on a pulley fixed at the bottom of the skeleton; the neck of the funnel E, is guided by a collar to keep the whole steadily upright, as it revolves. The rollers of this machine act in the same manner as those of the drawing frame, but have only two pairs of rollers instead of three; they are turned round by means of contrate wheels *g*, on the end of each, which are worked by pinions on the tops of as many vertical spindles *k*, which at their lower ends have pulleys turning the skeletons, by means of bands; the spindles *k* are turned round by a strap *l*, which passes round and is common to all, when the machine is ever so long, the strap receives its motion, by passing round the drum F, the spindle G of which is turned by the mill, the drum also receives other straps as at *m*, to turn other frames in different directions; I is the sliding coupling box, by which the drum can at any time be detached from its spindle, raising it by the lever K, and then its points do not touch. The arms of the drum, which being fitted on a round part of the spindle does not turn with it, but on letting down the box I, which is fitted on a square, it is put in motion, and also the other machinery. By a similar contrivance any one of the spindles *k* can be detached, and the two skeletons which it turns will then stand still while the cans D are removed.

The manner of action in this machine is easily gathered from the description, the slivers pass two together, through the rollers, and are reduced or drawn out therein to the proper degree of fineness; then falling into the funnels E, of the revolving cans, they are by the rapid motion thereof twisted round; because the centrifugal force disposes the cotton to lay round the inside of the can in a regular coil, forming as it were a lining of cotton, to the whole of the interior surface; and by this means the end of the roving becomes in a manner attached to the can, and is twisted round by its motion so as to form a coarse loose thread with a very slight twist, and a very soft and open substance. Such is the state of the roving as prepared by the roving frame. All the preceding processes are to be considered as the preparation; and the operation of spinning is not yet begun. These preparations are the most tedious, and require more attendance and hard labor than any subsequent part of the business. For the slivers from which the rovings are made, are so slight and bulky, that a few yards only can be piled up in the cans set to receive them from the carding and drawing. A person must therefore attend and watch each roller of the drawing and roving frames to join fresh slivers as they are expended. It is also the most important department in the manufacture; for as every inch will meet with precisely the same drawing and same twisting in the subsequent parts of the process, therefore, every inequality and fault of the sliver, indeed of the fleece as it quits the finishing card, will continue through the whole manufacture in a greater or less degree, being only diminished not corrected by the drawing, doubling, &c. It is evident that the roving produced by these operations must be exceedingly uniform; the uniformity really produced exceeds all expectation; for even although there be some small inequalities in the carded fleece, yet these are not matted clots, which the card could not equalize, and only consist of a little more thickness of cotton in some places than in others. This inequality will first be diminished by the lapping of the fleece in the breaking card; and when such a part of the sliver comes to the first roller of the drawing frame, it will be rather more stretched by the second. That this may be done with greater certainty the weights of the first rollers are made very small, so that the middle part of the sliver can be drawn through while the outer parts remain fast hold.

As a preparation for spinning, the rovings must be wound upon bobbins from the cans D of the roving frame, which are taken away from the skeletons as soon as they are filled, and carried to the winding machine, fig. 10, which, however, only shews the operative parts of the machine, the frame being omitted. The chief part of it is a cylinder A, which is turned round by a winch handle B; the bobbins *a a* on which the rovings *b b* are to be wound rest with their weight upon the surface of this cylinder, and are carried round by it with great rapidity, and wind up the rovings, which are guided by pins projecting from a rail *d d*, which has by the machine a slow traversing motion from one end of the bobbing to the other, and thus lays the cotton regularly on the whole length. The bobbins are merely put loosely on a wire *e*, and can quickly be changed for others when they are filled, they are then carried to the spinning frame (*see fig.* 11 *and* 12 *of plate* 24.) the former being a front view and the other a side section. In both of them, A represents the bobbins filled with rovings which is to be spun into thread; they are set up in a rack or frame over head, and are conducted down at *a a* through rollers *b c d*, which are the same as the drawing frame, and extend it in length 10, 12, or 16 times, accordingly as the yarn which is to be spun requires to be finer or coarser. This is delivered out to the spinning apparatus or spindles: these are straight steel arbors, on the lower end of which the pulleys, or hafts, as they are called, receive the bands *f* for turning them. These spindles are mounted in a frame common to them all, which consists of two rails B C, the lower one supporting the points or toes of the spindles, and the other having bearings for the cylindrical parts of each spindle, and a wire staple is fixed over each to keep them up to their bearings. Above this bearing the spindle is only a straight cylindrical wire, and on the upper end of it the fork or flyer *h* is fastened either by screwing it on, or it is stuck fast on by friction, which is sufficient to carry it about. The two arms or branches of the flyer.are sufficiently distant for them to revolve round clear about the bobbin *k*, which is fitted loosely upon the cylindrical spindle, and with liberty to slide freely up and down upon it. The weight of the bobbin is supported by resting on a piece of wood attached to a rail M, which has a slow rising and falling motion, equal in extent to the length of the bobbin between its shoulders, by which means the thread as it comes through the eye, is formed at the ends of either of the branches *h* of the flyer, and is wound by the motion thereof upon the bobbin. It becomes equally distributed throughout its length, giving it a cylindrical figure instead of keeping all the thread at one part like a barrel, as would happen if the bobbin did not rise and fall. The spindles are constantly kept in rapid motion by the machine, and twist the fibres round each other the instant their ends come out, before the rollers leave the other ends, or they would fall to pieces; being drawn out so fine, that the cohesion of the fibres is insufficient to bear any thing, and the twist given to the roving is entirely lost; for it was at first only one turn in one or one and a half inches in length, and this one and a half inch being by the draught of the rollers drawn out 10 or 12 times the length, the twist of one turn in this length is imperceptible, and adds no strength whatever to the roving, so that it is necessary the spindle should by the connexion of the thread passing down from the rollers to its flyer, give a twist to the fibres the instant they come through the rollers, which they do by the thread being conducted down from the rollers through the eye formed at the end of either of the branches of the flyer, which revolves with the greatest rapidity along with the spindle, and then gives the twist to the thread; the bobbin does not partake of the motion of the spindle but is retained by the friction of its lower end resting on the piece of wood *l*, and this is increased by a washer of leather put under it. This friction gives such a resistance to the motion of the bobbin, that the motion of the flyer running round it will lay the thread evenly upon it as fast as the rollers suffer it to come forwards.

The motion of the whole machine is communicated in the same manner as the roving frame by a vertical spindle D to a drum E which receives a strap *n* for one frame, and a similar one *o* for another. The former of these straps extends the whole length of the machine, turning all the vertical spindles *p* on both sides of the frame by means of pulleys on the lower ends of them.—Each of these vertical spindles puts in motion four spindles and the rollers belonging to them; the former by the bands *f*, which go round the wheel *r* upon the splindles *p*, and the rollers it turns by a pinion at the top of each, turning a contrate or face wheel *t* on the end of each roller.

It is to be observed that the frame, fig. 11, is in practice extended to contain 40 or 60 spindles on each side instead of four, and one of the verticle spindles *p*, is provided for every four spindles, but the strap *n* is common to them all. The wheel work for turning the rollers is shewn in fig. 13, and needs no explanation, being the same with those already described. The rise and fall of the rail *m*, and all the bobbins upon it, is thus produced; they are both suspended from the opposite ends of a horizontal lever L L M, which has a third arm M proceeding from it, which bears against the surface of a part N, which is a wheel of that figure turning slowly round, and thus moving the lower L L M and producing an alternate rise and fall of all the bobbins in the frame. The heart is turned round by a wheel R, fig. 11, on the end of its spindle worked by a pinion upon a spindle S, which also carries a wheel T, and this is turned round by means of a worm cut upon the main spindle of the frame.

The drum can at any time be detached from its spindle, and then the whole frame will stand still; for this purpose the spindle D passes through the drum E, a circular fitting, so that it slips freely round within it without giving motion to the drum, except when it is cast into gear; this is done by two locking bolts *w*, shewn by dotted lines passing through the drum, and both fixed into a collar or socket *x*, fitted to slide up and down the spindle. It has a groove formed round it, in which a fork at the end of a lever is received, so that the forked lever embraces the piece *w* in the groove, and when lifted up raises the two locking bolts with it, and unlocks the drum from the spindle D by withdrawing the locking bolts from their contact with an arm *f*, which is fixed fast on the spindle beneath the drum, and therefore turns with it; but the locking bolts being let down that their ends may project through the drum and intercept the cross arm *f* of the spindles, the drum and all the machinery is put in motion. In like manner each of the pulleys of the vertical spindles *p* which receive the great strap *n* are fitted to slip round on their spindles *p*, but can at any time be united thereto to give them motion by a locking box bayonet *z*, which is cast in or out of action at pleasure by a small lever in exactly the same manner as the locking of the principal drum; therefore, by this lever any four spindles can be detached from the machine at pleasure, and their motion stopped to change the bobbins when they are filled with thread, which is then finished, and requires only to be reeled off the bobbins for the weaver or other purpose.

The general machinery of the cotton mill, by which the various engines described are set in motion, is as follows: the moving power, whether a fall of water or a steam engine, is, by intervening wheels, adapted to its nature, made to turn round a vertical shaft, which passes through all the stories or floors of which the mill consists; in each of which it is furnished with a horizontal toothed wheel which gives motion to a vertical wheel, to which is attached a horizontal shaft going across one end of the floor, which gives motion to two or more other horizontal shafts, according to the breadth of the building, which run the whole length of the story. These again give motion to small vertical shafts which sustain the large drums that set the spinning frames in motion. The horizontal shafts have also drums on them, from whence bands proceed by which the carding engines and drawing machines are turned. What is said of the general arrangement of the millwork can only be understood in a general sense, for the number and position of the horizontal shafts set in motion by the vertical shaft must vary according to the nature of the buildings, and the disposition of the frames in each floor of them. Where it can be done, it is best to have the vertical shaft placed in the middle of the building, with the horizontal shafts proceeding from both sides of it at every floor, for then the horizontal shafts sustain less of that twisting motion which is very injurious to them, and to which they would be more liable if of the whole length of the building.

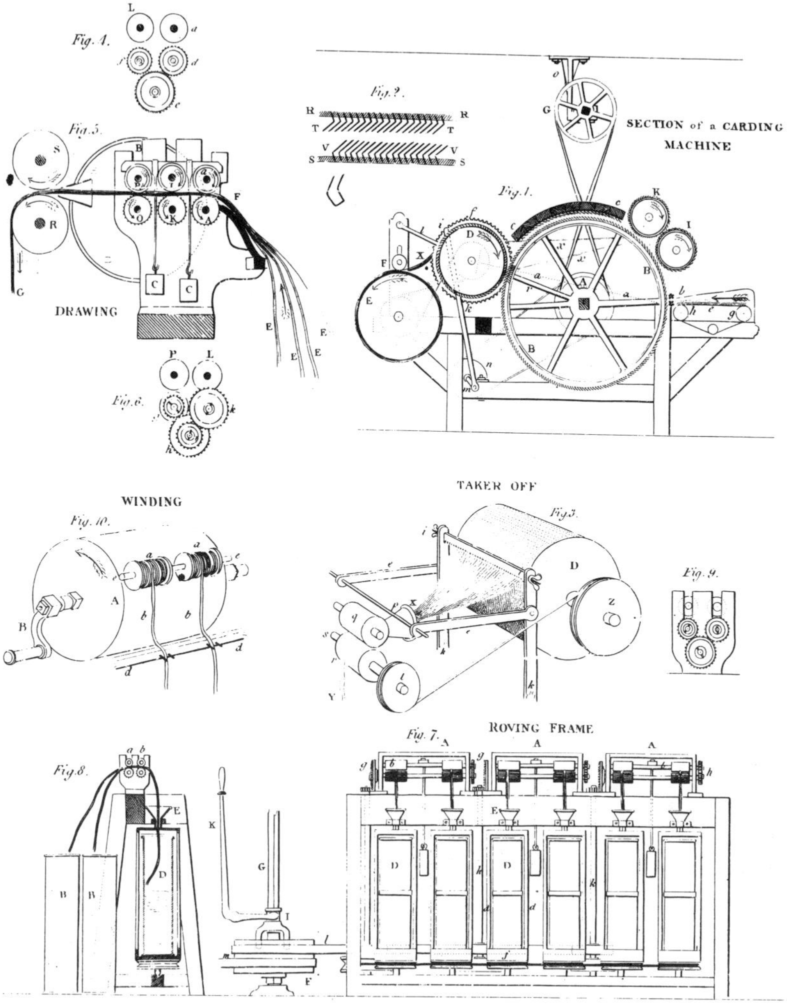
Fig. 4.
Fig. 5.
DRAWING
Fig. 6.
Fig. 2.
SECTION of a CARDING MACHINE
Fig. 1.
WINDING
Fig. 10.
TAKER OFF
Fig. 3.
Fig. 9.
Fig. 8.
ROVING FRAME
Fig. 7.

SPINNING

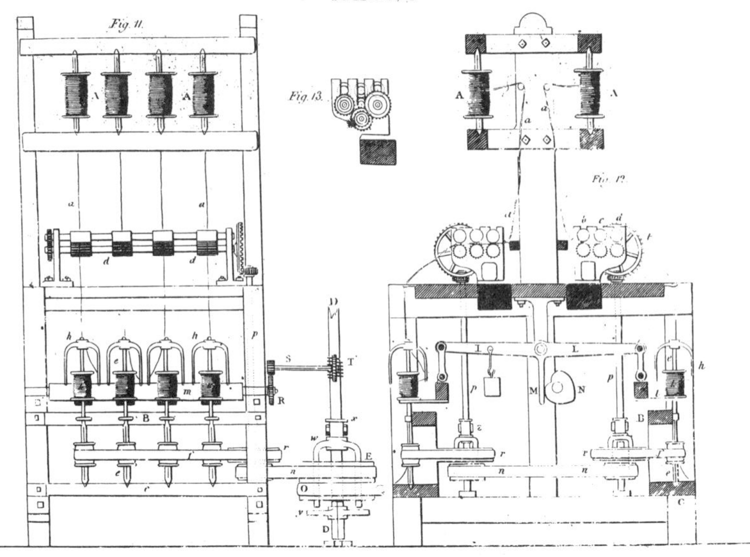
Fig. 11.
Fig. 13.
Fig. 12.

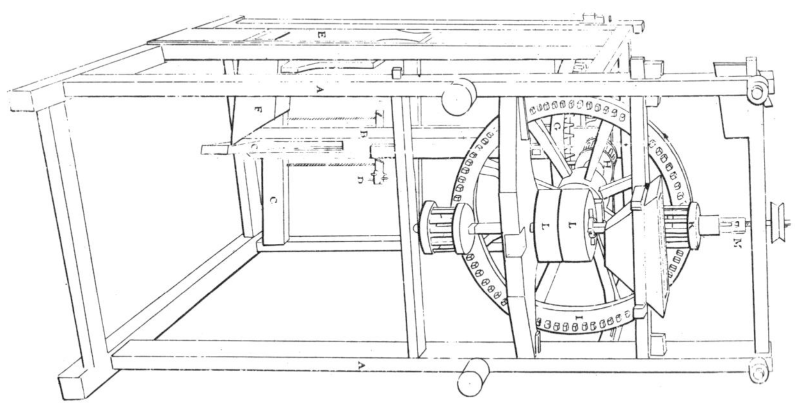

Samuel Slater

English immigrant Samuel Slater was the first to operate power spinning machinery successfully in the United States. Slater's achievement in reconstructing the machinery without the aid of plans has customarily been credited to his solitary feat of memory. The actual story is far more complicated. Slater, in fact, was working with Arkwright spinning frames that were purchased and operated by the wealthy Providence manufacturer Moses Brown. These machines had been built by Rhode Island artisans from models commissioned in Massachusetts in 1787. The spinning frames, however, worked poorly, and the carding machine did not incorporate Arkwright's principle of continuous operation. In the period from January to December 1790, Slater, with the critical assistance of local and area artisans, made the necessary alterations to the spinning frame and built two carding machines, a drawing frame, a roving frame, and another spinning frame. Slater's success was not his alone. It rested on three years of previous experimentation, the skill of local artisans like Sylvanus Brown and Oziel Wilkinson, and the continued financial backing of Moses Brown. The following selections throw light on Slater's English background, his work on the Arkwright machines at Pawtucket, and his reputation among contemporaries. (GK)

Painting of Pawtucket Falls by J. Rubens Smith, ca. 1802–1807. The tall narrow building to the left of the bridge is the clothier's shop where Samuel Slater first put his machines in operation. The Smithsonian Institution, Negative #48109.

Samuel Slater
Moses Brown

Correspondence (1789)

These letters were the first exchanged between Samuel Slater and Moses Brown. They provide the first terms of the agreement between the two men. Brown was a wealthy Providence Quaker with an extensive background in colonial manufacture and trade. Along with his kinsmen William Almy and Smith Brown, he had begun textile manufacture in the late spring of 1789 but was still searching for an English artisan capable of building and operating Arkwright machinery. A chance encounter with a Providence ship captain who knew something of Moses Brown's search prompted Slater to write to Brown in December 1789. The "New York manufactury" mentioned in Slater's letter was a textile workshop employing jennies, which opened in the summer of 1789. Contrary to Brown's offer, Slater never worked under an agreement guaranteeing him all the profits. His first written agreement in 1790 with William Almy, Moses's son-in-law, and Smith Brown, Moses's cousin, both of whom initially ran the business, gave him one-half of the spinning mill's profits. (GK)

George S. White, *Memoir of Samuel Slater* (Philadelphia, 1836), pp. 72–73. The location of the original letters is unknown. They are not a part of the Almy, Brown and Slater Papers at the Rhode Island Historical Society. For additional reading on Moses Brown, see Mack Thompson, *Moses Brown: Reluctant Reformer* (Chapel Hill: University of North Carolina Press, 1962); and James B. Hedges, *The Browns of Providence Plantations,* volume 1, *Colonial Years* (Cambridge, Massachusetts: Harvard University Press, 1952), volume 2, *The Nineteenth Century* (Providence, Rhode Island: Brown University Press, 1968).

NEW YORK, December 2d, 1789.

Sir,—A few days ago I was informed that you wanted a manager of *cotton spinning*, &c. in which business I flatter myself that I can give the greatest satisfaction, in making machinery, making good yarn, either for *stockings* or *twist*, as any that is made in England; as I have had opportunity, and an oversight, of Sir Richard Arkwright's works, and in Mr. Strutt's mill upwards of eight years. If you are not provided for, should be glad to serve you; though I am in the New York manufactory, and have been for three weeks since I arrived from England. But we have but *one card, two machines*, two spinning jennies, which I think are not worth using. My encouragement is pretty good, but should much rather have the care of the perpetual carding and spinning. *My intention* is to erect a *perpetual card and spinning*. (Meaning the Arkwright patents.) If you please to drop a line respecting the amount of encouragement you wish to give, by favour of Captain Brown, you will much oblige, sir, your most obedient humble servant, SAMUEL SLATER.

N. B.—Please to direct to me at No. 37, Golden Hill, New York.

Mr. Brown, Providence.

PROVIDENCE, 10th 12th month, 1789.

Friend,—I received thine of 2d inst. and observe its contents. I, or rather Almy & Brown, who has the business in the cotton line, which I began, one being my son-in-law, and the other a kinsman, want the assistance of a person skilled in the frame or water spinning. An experiment has been made, which has failed, no person being acquainted with the business, and the frames imperfect.

We are destitute of a person acquainted with water-frame spinning; thy being already engaged in a factory with many able proprietors, we can hardly suppose we can give the encouragement adequate to leaving thy present employ. As the frame we have is the first attempt of the kind that has been made in America, it is too imperfect to afford much encouragement; we hardly know what to say to thee, but if thou thought thou couldst perfect and conduct them to profit, if thou wilt come and do it, thou shalt have all the profits made of them over and above the interest of the money they cost, and the wear and tear of them. We will find stock and be repaid in yarn as we may agree, for six months. And this we do for the information thou can give, if fully acquainted with the business. After this, if we find the business profitable, we can enlarge it, or before, if sufficient proof of it be had on trial, and can make any further agreement that may appear best or agreeable on all sides. We have secured only a temporary water convenience, but if we find the business profitable, can perpetuate one that is convenient. If thy prospects should be better, and thou should know of any other person unengaged, should be obliged to thee to mention us to him. In the mean time, shall be glad to be informed whether thou come or not. If thy present situation does not come up to what thou wishest, and, from thy knowledge of the business, can be ascertained of the advantages of the mills, so as to induce thee to come and work ours, and have the *credit* as well as advantage of perfecting the first water-mill in America, we should be glad to engage thy care so long as they can be made profitable to both, and we can agree. I am, for myself and Almy & Brown, thy friend,

Moses Brown.

Samuel Slater, at 37, *Golden Hill, New York.*

Engraving of Samuel Slater, from J. D. Van Slyck, *New England Manufacturers and Manufactories*, volume II, Boston, 1879.

Samuel Slater

Autobiography (1834)

In 1834, the Rhode Island Historical Society requested that Samuel Slater "draw up and present to this society, a history of the first introduction of cotton spinning into this country." The following brief document is Slater's effort to comply. It was written in the third person shortly before his death on April 20, 1835. His claim that he built the Arkwright machinery "principally with his own hands" is qualified, if not contradicted, by subsequent documents in this collection. (GK)

George S. White, *Memoir of Samuel Slater* (Philadelphia, 1836), pp. 41–42.

Samuel Slater was born in the town of Belper, in the county of Derby, June 9th, 1768. In June 28th, 1782, being about fourteen years of age, he went to live with Jedediah Strutt, Esq., in Milford, near Belper, (the inventor of the Derby ribbed stocking machine, and several years a partner of Sir Richard Arkwright in the cotton spinning business,) as a clerk; who was then building a large factory at Milford, where said Slater continued until August 1789. During four or five of the late years, his time was solely devoted to the factory as general overseer, both as respected making machinery and the manufacturing department. On the 1st day of September 1789, he took his departure from Derbyshire for London, and on the 13th he sailed for New York, where he arrived in November, after a passage of sixty-six days. He left New York in January 1790, for Providence, and there made an arrangement with Messrs. Almy and Brown, to commence preparation for spinning cotton at Pawtucket.

On the 18th day of the same month, the venerable Moses Brown took him out to Pawtucket, where he commenced making the machinery principally with his own hands, and on the 20th of December following, he started three cards; drawing and roving, and seventy-two spindles, which were worked by an old fulling mill water wheel in a clothier's building, in which they continued spinning about twenty months; at the expiration of which time they had several thousand pounds of yarn on hand, notwithstanding every exertion was used to weave it up and sell it.

Early in the year 1793, Almy, Brown and Slater built a small factory in that village, (known and called to this day the old factory,) in which they set in motion, July 12, the *preparation* and seventy-two spindles, and slowly added to that number as the sales of the yarn appeared more promising, which induced the said Slater to be concerned in erecting a new mill, and to increase the machinery in the old mill.

Smith Wilkinson

Letter to George White (1835)

Smith Wilkinson wrote the following letter as an account of what he remembered of Slater's first year in Pawtucket. Smith was the son of Oziel Wilkinson and worked for Slater as a ten-year-old laborer in the first mill. He later was the superintendent of a cotton mill in Pomfret, Connecticut. The Andrew Dexter and Lewis Peck mentioned in Wilkinson's letter were Providence merchants who set up a textile business employing hand spinners and weavers in 1787. It was Dexter & Peck who commissioned the construction of one of the Arkwright spinning frames later sold to Moses Brown. That frame was built by Daniel Anthony, a North Providence artisan. Joseph and Richard Anthony were his sons.

Wilkinson's graphic description of laying the cotton onto the carding machine suggests the hazards of mill work. Commenting on this task, the Reverend David Benedict, a first-hand observer and no friend of labor reform, claimed that "many unfortunate children had their hands terribly lacerated by the operation." (GK)

Letter from Smith Wilkinson to George White, May 30, 1835, from White, p. 76. David Benedict, "Reminiscences No. 14," *Pawtucket Gazette and Chronicle,* June 10, 1853.

POMFRET, May 30th, 1835.

Mr. Samuel Slater came to Pawtucket early in January 1790, in company with Moses Brown, Wm. Almy, Obadiah Brown, and Smith Brown, who did a small business in Providence, at manufacturing on billies and jennies, driven by men, as also were the carding machines. They wove and finished jeans, fustians, thicksetts, velverets, &c.; the work being mostly performed by Irish emigrants. There was a spinning frame in the building, which used to stand on the south-west abutment of Pawtucket bridge, owned by Ezekiel Carpenter, which was started for trial (after it was built for Andrew Dexter and Lewis Peck) by Joseph and Richard Anthony, who are now living at or near Providence. But the machine was very imperfect, and made very uneven yarn. The cotton for this experiment was carded by hand, and roped on a woollen wheel, by a female.

Mr. Slater entered into contract with Wm. Almy and Smith Brown, and commenced building a water frame of 24 spindles, two carding machines, and the drawing and roping frames necessary to prepare for the spinning, and soon after added a frame of 48 spindles. He commenced some time in the fall of 1790, or in the winter of 1791. I was then in my tenth year, and went to work for him, and began at tending the breaker. The mode of laying the cotton was by hand, taking up a handful, and pulling it apart with both hands, and shifting it all into the right hand, to get the staple of the cotton straight, and fix the handful, so as to hold it firm, and then applying it to the surface of the breaker, moving the hand horizontally across the card to and fro, until the cotton was fully prepared.

Pawtucket *Chronicle*

Samuel Slater (1835)

Samuel Slater's death on April 20, 1835, evoked national attention. The *Manufacturers' and Farmers' Journal* of Providence carried a brief note on April 22, eulogizing Slater as "one of our most enterprising and respected citizens. . . . In all the relations of life he maintained a character for probity and integrity seldom equalled." This was the tenor of later pieces, one of which was reprinted in George White's *Memoir*.

The Pawtucket *Chronicle* viewed Slater differently, and its editorial is reprinted here for the first time. Those writers who sought to honor Slater have all overlooked it. The editorial is by no means entirely negative, but the weight of it suggests a man of overweening ambition pursuing wealth with single-minded passion.

In the context of the mid-1830s, the *Chronicle* was not a radical paper. Its primary appeal was to the rising artisan, and its pieties were those of the petty-bourgeoisie. But many in Pawtucket remembered the depression of 1829, when artisan manufacturers like the Wilkinsons and the Greenes were driven out of business. Although we have no definite proof, the bitterness of the *Chronicle* suggests that many believed that Slater could have done more to rescue Pawtucket from a decline that stretched from 1829 well into the 1840s. Whether it was in Slater's power to do so is unclear. As an endorser of the Wilkinsons' notes, Slater himself was saddled with an imposing debt, which he paid off only by selling part of his property, including his share of the Old Slater Mill. George White, in his *Memoir,* suggests that the real culprit was William Almy, but many in Pawtucket knew only that the village economy remained depressed and that Slater had left the village. (GK)

"Samuel Slater," Pawtucket *Chronicle,* May 4, 1835.

The absurd story about Samuel Slater having dreamed of a remedy for some defect, which hindered the machinery of his first factory from operating, has been revived in a New York paper. It is the very height of absurdity; and was never heard of here, the spot of his early enterprise, until spawned from the brains of some wonder-making personage.

To those who were acquainted with Mr. Slater since he sojourned in Pawtucket and especially with the events of his early life, no fiction appears more absurd than this dream. Mr. Slater was not a man to be indebted for his success to midnight fancies and drowsy visions. His dreams were deep calculations and his calculations wealth.

He came to this place before 1790, destitute of pecuniary resources, with no capital but industry and no friend but his mechanical skill. Thus thrown on his own hands, he had ingenuity enough to convince the strong mind of Moses Brown and arouse the interests of Wm. Almy to a correct knowledge of that manufacturing business that had helped elevate Great Britain in the scale of nations, and which finally was to make America independent of her father land for the articles she wore as for the principles of her republican government. By the aid of these men he began life.

We have heard the following anecdote about Slater from a person who had it from his own mouth. Mr. S, when quite a young man, was in the employment of a great Manufacturer in England, (named, we believe Strutt), he passed by some loose Cotton on the floor, without picking it up: his employer called him back and told him to take up the Cotton, for it was by attending to such small things that great fortunes were accumulated. And the gentleman related to his wife that he was afraid that Samuel Slater would never be a rich man, as he was ignorant of the art of acquiring wealth! Samuel Slater ignorant of money-getting! If this old employer could have burst the marble jaws of his tomb, returned to earth, and visited America, he might have learned much of the "art of accumulation," even from the earlier workman, whose improvidence he attempted to chide!

Establishing the first Cotton Factory in the Country, he of course had the whole market to himself; and unable to supply the orders sent him for cotton yarn, he often had the money lodged in his possession for months before he could answer the commands of his correspondents. Everything in his fingers turned into money. His business was profitable, without competition, he alone master of the art, he went on in the path of success, the favorite of good luck and proof against misfortune.

At the time of his death, he owned one third of the manufacturing village of Slatersville, the steam Mill in Providence, the whole town of Webster, and immense tracts of landed property, besides a great deal of bank stock, all of which, we understand, has been estimated at $1,200,000, by Moses Brown of Providence. This estimate is founded on a schedule of his property exhibited by Mr. Slater to Moses Brown in 1829.

Mr. S. was not exactly a generous man. He gave little to public institutions and regarded not the appeals of private individuals. His object was gold. No man was more indefatigable. Buonaparte [*sic*] never pursued schemes of conquest, never followed the phantom of ambition more insistently, than did Samuel Slater his business. With him there was no second object to divide his thoughts. Like a shrewd, worldly man, he never boasted of his riches. When the President [Andrew Jackson] visited him two years ago, he told him he understood that he had made quite a large fortune, "why" said Mr. S. "I have made, I think, a competancy!" An answer that opens an avenue into the very heart of his character.

Cotton carding machine from the 1790s used by Samuel Slater in Rhode Island. The Smithsonian Institution, Negative #45060-A.

George S. White

Memoir of Samuel Slater (1836)

The following passage provides details of Slater's early life, his apprenticeship with the English manufacturer Jedidiah Strutt, and his decision to emigrate to America. They are drawn from the Reverend George S. White's *Memoir of Samuel Slater.* White, an Anglican clergyman from Bath, England, and a fervent admirer of Slater, occasionally lapses into eulogy. Despite this, his work remains the most reliable source on Slater's life, though it was never intended as a full biography. An adequate, modern biography of Slater has yet to be written. (GK)

George S. White, *Memoir of Samuel Slater* (Philadelphia, 1836), pp. 30–37. Of the nineteenth-century studies of Slater, the best is William Bagnall's *Samuel Slater and the Early Development of the Cotton Manufacture in the United States* (Middletown, Connecticut, 1890). Many of the myths that surround Slater's life are the result of Masena Goodrich, *Historical Sketch of the Town of Pawtucket* (Pawtucket, 1876), an unreliable study. The more recent studies have been equally uneven. Frederick L. Lewton's pamphlet, *Samuel Slater and Oldest Cotton Machinery in America,* from the *Smithsonian Report for 1926* (Washington, 1927), is a useful study of the Slater machinery, which continues to reside in the National Museum of American History. The most recent biography, commissioned by H. Nelson Slater, Samuel's great-grandson, is E. H. Cameron's *Samuel Slater, Father of American Manufacturers* (Portland, Maine, 1960). The Cameron study was designed as part of a series on the "achievements of American industry" and as a "corrective to loose economic and social propaganda, and teaching, at both high school and college levels. . . ." A recent reevaluation of Slater's technical achievement is Paul Rivard's pamphlet, also titled *Samuel Slater, Father of American Manufactures* (Pawtucket, 1974). This is an able discussion of Slater's role in the transfer of Arkwright technology that undercuts a number of Slater myths. Unfortunately, Rivard creates new myths through unwarranted and undocumented assertions about Slater's "genius" for business management. Though not a biography, a serious and thorough study of Slater is James Conrad, "The Evolution of Industrial Capitalism in Rhode Island," Ph.D. thesis, University of Connecticut, 1973. On Slater in Webster, Massachusetts, see Jonathon Prude, "The Coming of Industrial Order: A Study of Town and Factory Life in Rural Massachusetts, 1813–1880," Ph.D. dissertation, Harvard University, 1976; see also Barbara M. Tucker, "Samuel Slater and Sons: The Emergence of the Factory System, 1790–1860," Ph.D. dissertation, University of California, Davis, 1974.

I am writing of a man of business; not of a man devoted to literature, or what has been called the liberal arts; whose fame has been spread by means of publications, or who had in any way sought publicity, or made claim to any pretensions, *but of one who all his lifetime avoided it.* It is well known, that the late Samuel Slater, Esq. of Webster, Massachusetts, and for many years a resident citizen in the village of Pawtucket, Rhode Island, was a native of England. I have the most direct information of the place of his birth, and of his parentage. His father, William Slater, inherited the paternal estate, called "Holly House," near Belper, in the county of Derbyshire, England. This estate is now owned and occupied by his son, William Slater.

The father of Samuel Slater was one of those independent yeomanry, who farm their own lands, now almost peculiar to that part of the country, as a distinct class from the tenantry of England. He did not, however, confine himself altogether to the business of agriculture, but added to his estate by the purchase of lands. He did so for the sale of timber, and was in fact a timber merchant.

Being a neighbour of Jedediah Strutt, of whom we shall have occasion to speak, he once made a considerable purchase for him containing a water-privilege, on which there is now a very extensive establishment. He was otherwise engaged with Mr. Strutt in making purchases of consequence, who had a high opinion of his abilities and integrity as a man of business. This acquaintance, and these transactions, led to the connection of Mr. Strutt with Samuel, who was the fifth son, and is said to have resembled his father in his person, and to have inherited his talents. This enterprising son transplanted a branch of the Slater family into the new world, where we trust they will grow and prosper for many generations. The mother of Mr. Slater was a fine looking woman, and lived a short time since with her third husband, whom she survived, and often observed, she had been favoured with "three good husbands." She had by her first husband, William Slater, a large family; William, who now lives on the paternal estate with many children, bids fair to keep up the family name on the other side of the Atlantic. John Slater, son of the subject of this memoir, visited him a few years since, at the Holly House farm, the place of his father's nativity, and viewed the establishment where his honoured parent served his long and important apprenticeship, as he did also the other mills owned by Messrs. Arkwright and Strutt, at Crumford, six miles from Belper. When on my last visit to Mr. Slater at Pawtucket, in 1833, he showed me the prints of Arkwright and Strutt, and pointing to that of Strutt, said, "Here is my old master," and pronounced it a good likeness.

Perhaps nothing could have had more influence on the subject of this memoir, to induce him to leave his business, than the desire to visit his aged mother, of whom he spoke always most affectionately, and corresponded with her. And to have viewed the place and scenes of his early days; his brothers and sisters, and their little ones, to the third generation; his school-fellows, his playmates, his schoolmaster, Jackson, who was then living; the sons and grandsons of his old master, Strutt; the old mill; the meadows and orchards, &c. that surrounded Holly house. He left them all, in the bloom of youth, and retained a vivid recollection of every particular. These early remembrances would cause the tear to escape, even in his old age. But the state of his health, the multiplicity of his concerns, and his *concentrativeness*, bound him to Webster, and forbade the thought of a voyage across the Atlantic. He refrained, denied himself, sent his love by his son, and never returned to his father's land. But he ever retained a strong affection and lively concern in the welfare of his native country.

As is usual, Samuel went on trial to Mr. Strutt, previous to his indenture of apprenticeship, and during this probation his father fell from a load of hay. This fall was the occasion of his death. During his father's sickness, and perceiving that he was dangerously ill, he wished his father to article him to Mr. Strutt, as both parties were satisfied. As a proof that his father had confidence in him, and that there was stability in the boy, he said to him, "You must do that business yourself, Samuel, *I have so much to do, and so little time to do it.*" It is believed that this was his last interview with his beloved parent.

He lost his father in 1782, when he was fourteen years of age, at a time when a father's care and advice are much needed. A boy left without guardianship, or watchful eye to restrain him, is frequently exposed and led into temptation and ruin. Young

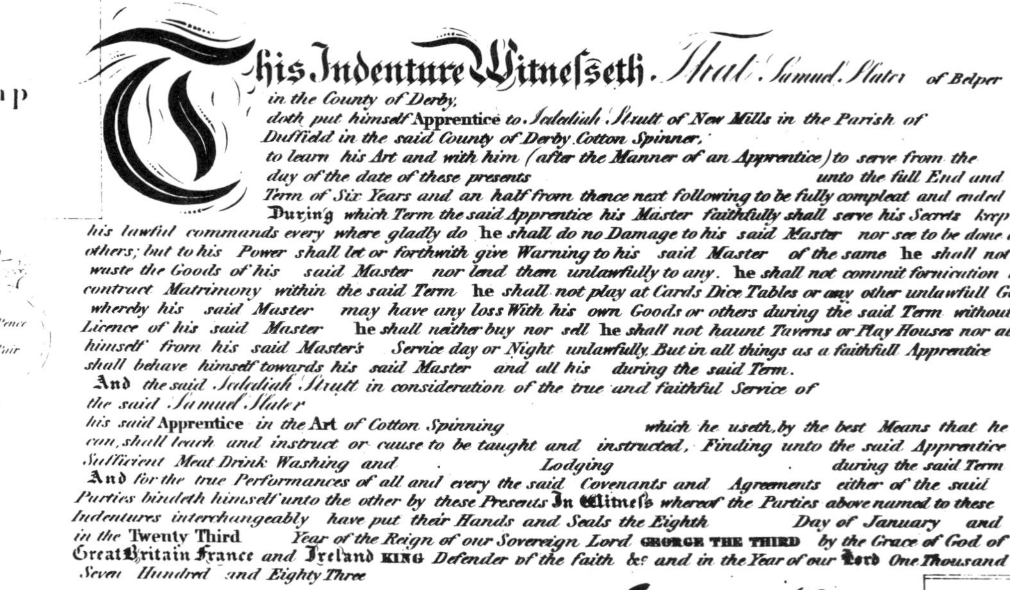

Stamp

This Indenture Witnesseth. That Samuel Slater of Belper
in the County of Derby,
doth put himself Apprentice to Jedediah Strutt of New Mills in the Parish of
Duffield in the said County of Derby Cotton Spinner,
to learn his Art and with him (after the Manner of an Apprentice) to serve from the
day of the date of these presents unto the full End and
Term of Six Years and an half from thence next following to be fully compleat and ended
During which Term the said Apprentice his Master faithfully shall serve his Secrets keep
his lawful commands every where gladly do he shall do no Damage to his said Master nor see to be done of
others; but to his Power shall let or forthwith give Warning to his said Master of the same he shall not
waste the Goods of his said Master nor lend them unlawfully to any. he shall not commit fornication nor
contract Matrimony within the said Term he shall not play at Cards Dice Tables or any other unlawfull Games
whereby his said Master may have any loss With his own Goods or others during the said Term without
Licence of his said Master he shall neither buy nor sell he shall not haunt Taverns or Play Houses nor absent
himself from his said Master's Service day or Night unlawfully. But in all things as a faithfull Apprentice
shall behave himself towards his said Master and all his during the said Term.
And the said Jedediah Strutt in consideration of the true and faithful Service of
the said Samuel Slater
his said Apprentice in the Art of Cotton Spinning which he useth, by the best Means that he
can, shall teach and instruct or cause to be taught and instructed, Finding unto the said Apprentice
Sufficient Meat Drink Washing and Lodging during the said Term
And for the true Performances of all and every the said Covenants and Agreements either of the said
Parties bindeth himself unto the other by these Presents In Witness whereof the Parties abovenamed to these
Indentures interchangeably have put their Hands and Seals the Eighth Day of January and
in the Twenty Third Year of the Reign of our Sovereign Lord GEORGE THE THIRD by the Grace of God of
Great Britain France and Ireland KING Defender of the faith &c and in the Year of our Lord One Thousand
Seven Hundred and Eighty Three

Five Pence Pr. Pair

B

Sealed and delivered being first duly Stamped in the presence of

S. Leeper
Geo. Williams

Samuel Slater — Seal
Jed Strutt — Seal

Slater, however, had an indulgent and faithful mother, and elder brothers, so that he was not left entirely to his own resources. The plate opposite is an engraved copy from the original indenture, which is preserved in the family, as a relic of their father's early fidelity, and as a proof of his favoured means of knowledge.

Mr. Strutt was then building a large cotton factory at Milford, and was a partner with Sir Richard Arkwright, in the cotton spinning business; the latter having been induced to this connection by the prospect which Strutt's machines afforded, of an increased consumption of yarn. Samuel Slater asked Mr. Strutt, before he went into the business, whether he considered it a *permanent* business. Mr. Strutt replied, "It is not probable, Samuel, that it will always be as good as it is now, but I have no doubt it will always be a *fair* business, if it be well managed." It will be recollected, that this was before Mr. Peel invented the printing cylinder. Indeed the whole cotton business of England was, at that time, confined to a small district in Derbyshire, and its whole amount not greater than that which is done at the present day in a single village in New England.

In the early part of our young apprentice's time, he manifested the bent of his mind, for he frequently spent his Sundays alone, making experiments in machinery. He was six months without seeing his mother, or brothers and sisters, though he was short of a mile from home. Not that he lacked in filial or fraternal affections; but he was so intent, and so devoted to the attainment of his business. To show the expertness and the propensity of his mind, the following circumstance is related. Mr. Strutt endeavoured to improve the *heart-motion*, that would enlarge or raise the yarn in the middle, so as to contain more on the bobbin. Jedediah Strutt was unsuccessful in his experiments, and Samuel saw what was wanting, and went to work the next Sunday, (the only time he had to himself,) and formed such a motion, (a diagram of which is given below) to the satisfaction of his master, who presented him with a guinea.

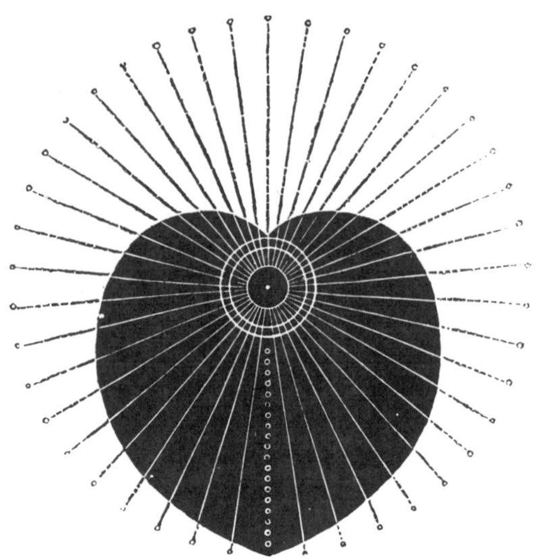

Mr. Strutt was an economist, and enforced his maxims on Samuel, cautioning him against waste, and assuring him that it was by savings that a fortune in business was to be made.* During this time, Samuel became an excellent machinist, as he had an opportunity of seeing the latest improvements. Arkwright and Strutt were in company, and it was at a time when there was much excitement and lawsuits on the patent rights; so that he was initiated into all the crooks and turns of such controversies. This may have prevented him applying for a privilege as the introducer of Arkwright's patents into the United States.

Slater served his indenture with Mr. Strutt, and faithfully performed his part of the contract to the last day of the term, and there was a good understanding between the parties to the last. This accomplishment of his *full time* was characteristic of him, and was praiseworthy and beneficial, as it laid the foundation of

* The following anecdote is told:—"When Mr. Slater was yet a boy, with Mr. Strutt, he passed by some loose cotton on the floor without picking it up; Mr. Strutt called him back and told him to take up the cotton, for it was by attending to such small things that great fortunes were accumulated; and Mr. Strutt observed to his wife, by way of still impressing the subject on the mind of his favourite apprentice, 'that he was afraid that Samuel would never be rich.'"

his adaptation to business, and finally to his perfect knowledge of it. He was different from those restless youths, who think they know every thing before they have cut their eye teeth, and who set up for themselves before their beards are grown, without either knowledge or capital, and who fail and defraud their creditors, during the time they ought to have been serving an apprenticeship. Such boys break their engagements, forfeit all confidence, and follow the example of Franklin, in that particular, though they cannot be compared to him in any thing else. And in this, Franklin was to be blamed; I praise him not. He himself acknowledges it to have been a great error in his life. A conscientious regard to contracts is a principle by which every person ought to be influenced, and without which, there is no hope of their arriving at eminence in their profession. Mr. Slater told me a short time before his death, that after his time was out, he engaged with Mr. Strutt to have the oversight of the erection of some new works, in addition to the mill, and this general employment, with his close observation (for he always saw and heard every thing, nothing could escape his notice,) and retentive memory, was of great service to him in afterwards assisting him to erect his first mill in Pawtucket. If he had been confined to one branch of business, as is usual with an apprentice in England, his knowledge would have been inadequate to perform what he did on his first coming to America. But his residing in Strutt's family, his being the son of his deceased friend and neighbour, as well as his close application to business, his ingenious experiments, and his steady habits, gave him the character of the "industrious apprentice."

He had the confidence of his master, and became his *right-hand man*, and he might have attained the highest eminence by a continuance in England. Mr. Strutt afterwards declared that had he known his intentions, nothing should have induced him to part with him. But Mr. Slater told me that he contemplated trying America for some time; and that his object was, to get a general knowledge of the business, in order to come to this country and introduce the manufacture of cotton, on the Arkwright improvement, and that he remained after the time of his indenture with that special object in view.

There were early indications that he designed embarking in business for himself, and it is said, that he used to enquire of

Arkwright and others, if they thought the business would be overdone in England. Yet it does not appear that he ever made known to any person his intention of leaving England. The father of Samuel Slater must have been a man of considerable property and business for those times, from the fact of his supporting so large a family respectably, and giving them such an education as was equal to any children who were calculated for business, sixty years ago. After making provision for his widow, he left to each of his children what was then a considerable sum for persons in business. There was included, in Samuel's portion, two houses in Belper, a nail store, and another building; all of which sold as they were, under many disadvantages, for nearly two thousand dollars. He did not touch this property when he left home, but probably reserved it for a retreat in case of failure of his object in coming to the United States. He had always that kind of generalship which provides for a retreat in case of accident, or as he would say, "to lay up for a rainy day."

Few persons who are extravagant when apprentices, ever gain in business; and it has been said, that few who saved money then but what succeeded in after life. The following copy of a note* which I have in my possession, shows the early savings of Slater: economy and indefatigable industry were the foundation principle of his fortune. Not by speculation, or by any circumstances peculiarly favourable to the accumulation of wealth, but by the dint of persevering attention to business for half a century.

The motive, or inducement, and first occasion of his thinking of leaving Mr. Strutt, and what finally determined him, was his observing* in a Philadelphia paper, a reward offered by a society

* "*Four-pence Stamp.*

£2 2s.—I promise to pay to Samuel Slater, or order, upon demand, the sum of two pounds two shillings, for value received, with lawful interest for the same, as witness my hand this tenth day of January, 1768.

Signed in the presence of us, WILLIAM ASHMOLE.

Wm. More, J. Pratt."

for a machine to make cotton rollers, &c. This convinced him that America must be very bare of every thing of the kind, and he prepared himself accordingly. He probably knew the risk he should run in attempting to leave England as a *machinist*, and it was characteristic of him, never to talk of his business—where he was going, or when he intended to return. John Slater, a surviving brother, says he remembers his coming home, and telling his mother that he wished his clothes, as he was going by the stage to London; this was the last time his mother, or any of the family, saw him, till his brother John joined him in Pawtucket. He was aware, that there was danger of his being stopped, as the government restrictions were very severe, and very unjust; the officers were very scrupulous in searching every passenger to America. He therefore resolved not to take any pattern, nor have any writing or memorandum about him, but trusted wholly to his acquirements in the business and to his excellent memory. His appearance was also in his favour, it being that of an English farmer's son, rather than that of a mechanic. He told me himself he had nothing about him but his indenture, which he kept concealed, and this was his only introduction and recommendation in the new world.

Patent drawing of David Wilkinson's slide lathe. National Archives.

David Wilkinson

Reminiscences (1846)

David Wilkinson was one of the premier mechanics in the new nation. Born in Smithfield, Rhode Island, on January 5, 1771, he learned the iron trade from his father, Oziel, who was a blacksmith. Moving to Pawtucket with his father in the early 1780s, he eventually played an important role in the beginnings of three separate but interrelated industries: textile machinery, steam engines, and machine tools.

His most noteworthy contribution was the invention of a screw-cutting machine in 1794, which became the basis for the slide-rest industrial lathe he built about 1806 while doing work on textile machinery. His invention antedated that of the great English machinist Henry Maudslay, who has customarily been credited with the development of the industrial lathe. The Wilkinson-type lathe, suitable for heavy, precise turning, soon became indispensable in early nineteenth-century machine shops.

Wilkinson also experimented with steam power generation, and he was one of the country's most important textile machinery makers. He operated a steam boat on the Providence River fourteen years before Fulton and built a vertical steam engine to provide supplementary power for the Wilkinson Mill, constructed in Pawtucket by

"David Wilkinson's Reminiscences," *Transactions of the Rhode Island Society for the Encouragement of Domestic Industry in the Year 1861* (Providence, 1862), pp. 100–111. This text has been lightly edited, and footnotes have been added by the editors. For additional reading on Wilkinson, see Robert S. Woodbury, *Studies in the History of Machine Tools* (Cambridge, Massachusetts: MIT Press, 1972); Jonathan Thayer Lincoln, "The Invention of the Slide Lathe," *American Machinist,* volume 76 (February 7, 1932), pp. 167–171; two museum publications, Edwin A. Battison, *Muskets to Mass Production,* The American Precision Museum (Windsor, Vermont, 1976), and Gary Kulik and Patrick M. Malone, *The Wilkinson Mill,* American Society of Mechanical Engineers and the Slater Mill Historic Site (New York, 1977).

his father in 1810–1811. Later, Wilkinson fabricated the first successful power looms in Rhode Island as part of an expanding business in textile machinery. By the early 1820s, his shops produced a variety of textile machines, for "almost every part of the country" as well as iron shafting, bevel gearing for mill power transmission, and waterwheel governors.

Making the transition from skilled artisan to textile manufacturer, Wilkinson developed extensive interests in southern New England textile mills. Like other artisans who found themselves in a similar position, he possessed more mechanical skill than capital or business cleverness. In the depression of 1829, he failed. He started up again in Cohoes, New York, achieved success, and found it once again ephemeral. He died on February 3, 1852, while working as a common laborer.

His reminiscences, written in 1846 as a personal communication to the Reverend George Taft of Pawtucket and published first in 1862, are reprinted here in their entirety. They demonstrate the important extent of technological convergence in the early nineteenth century, with innovations in one area, such as textile machinery, reinforcing and necessitating innovations in other areas, such as machine tools. They also demonstrate the critical role of skilled artisans in the early decades of industrialism. In particular, Wilkinson's reminiscences once again suggest the importance of Pawtucket artisans to the successful construction of Slater's first textile machinery. Artisans like Wilkinson not only expanded the frontier of technological innovation but, in their mastery of a host of separate mechanical skills, built the first machinery of the industrial revolution and made it work. Relying on a practiced tradition of folk engineering, and in the absence of highly precise tools and measuring devices, this was no small accomplishment. (GK)

Autumn, 1846.

In April, 1776, Eleazer Smith, who had been at work for Jeremiah Wilkinson, junior, a Quaker of Cumberland, came to my father's blacksmith shop, which was making scythes, in the town of Cumberland, Rhode Island, to make a machine to manufacture card teeth, for Daniel Anthony, of Providence, who was going into the card making business. While at work, Smith told my father of Jeremiah Wilkinson's making card tacks of cold iron. In laying the strip of leather around the hand card, he lacked four large tacks to hold the corners in place, while driving the tacks around the outer edge. He took a plate of an old door lock off the floor, cut four points with shears, and made heads in the vice; but afterwards made a steel bow with scores in it, and put it in the vice, and in that way made tacks.

I think in 1777, my father made a small pinch press, with different sized impressions, placed on an oak log, with a stirrup for the foot, and sat me astraddle on the log, to heading nails, which were cut with common shears. He cut the points off of plates drawn by trip hammer. This was the commencement, in the world, of making nails from cold iron.

I think about 1820, I went to Cumberland, with Samuel Greene, my nephew, and purchased of Jeremiah Wilkinson, the old shears, with which he cut the first four nails. He was, I think, ninety years of age at that time. The shears were a pair of tailor's shears, with bows straightened out, and the blades cut off half the length. They were deposited with the Historical Society, in Providence, by Samuel Greene.

My father, Oziel Wilkinson, lived in the town of Smithfield, Rhode Island, in 1775, at the commencement of the war, and owned a black-smith-shop, with a hammer worked by water. It was here Eleazer Smith made the machine for Daniel Anthony. I was then about five years old, and my curiosity was so great to see the work going on, that my father sat me on Mr. Smith's bench, to look on, while he worked. And at this time, seventy years afterwards, I could make a likeness of nearly every piece of that machine,—so durable are the first impressions on the mind of youth. After Smith had finished the machine, so as to make a perfect card tooth, he told the people in the shop that he could make a machine to make the tooth, prick the leather, and set the tooth, at one operation.

Jeremiah Wilkinson carried on the business of making hand cards for carding sheep's wool, and it being difficult to import wire, he drew the wire out by horse power.

In 1784 or '5, my father put the anchor shop in operation, at Pawtucket Falls, on the Blackstone river, in North Providence, Rhode Island.

About this time, I heard of cotton yarn being made in, or near East Greenwich, in which John Reynolds and James Macarris, who employed a Mr. Mackwire, or Maguire, to make yarn on a jenny, for which I forged and ground spindles. I made a small machine to grind with, which had a roller of wood to roll on the stone, which turned the spindle against the stone, and so ground the steel spindles per-
1 fectly. I heard of no machines for carding cotton.

About this time also, a number of gentlemen in the town of Providence, commenced some machinery for working cotton. Andrew Dexter, merchant, the father of S. Newton Dexter, of Orickany, Oneida county, New York; Aaron Mann, father of Samuel F. Mann of Providence; Lewis Peck, merchant; Daniel Anthony, and I think Moses Brown of Providence, were aiding in the work. My father was applied to, to make iron work for a machine for carding cotton, which was done by the help of a carpenter, named Joshua Lindley, and a brass founder, named Daniel Jackson, father of Samuel and John Jackson, of Providence. The card circles, or rims, were made of wrought iron, as there was no furnace near. The card was put in operation in the Market House chamber, in Providence, and was turned by a colored man, named Prince Hopkins, who had lost one leg, and I think one arm, in Sullivan's expedition at Newport, a few years before. The cotton was taken from the card, in rolls about eighteen inches long, and carried one mile from town to Moses Brown's, where it was made into roping, by a young woman in Mr. Brown's employ, named Amey Lawrence.

About this time, too, Daniel Anthony made a trip to Bridgewater, and returning said he had some parts of a machine, called the Arkwright Water Frame, which was commenced by a European, in the employ of Colonel Orr, of Bridgewater, and given up, or the few parts thrown by. He soon had one under way in Providence, which 2
was made and finished in Pawtucket, and put in operation there, by Anthony's two sons, Joseph and Richard, assisted occasionally, by two other sons, Daniel and William. The rollers were made of half inch wrought iron, with swells of brass cast on, and fluted with files. The bobbin which received the yarn from the spindle was made with a score in the bottom, to receive a cross cat-gut twine, with a tightning wooden thumbscrew, like a violin, to regulate the taking up ;—which Mr. Slater performed in his first water frames, by making a wide flat bottom to the bobbin, set on a wooden cloth washer, to regulate the taking up, as the friction would increase by weight as the bobbin filled, and needed more friction. (Mr. Slater run his first machinery by rope bands, for his carding machines, roping and drawing, as the use of belts was not then known in this country. The first leather belts I ever heard of were made by John Blackburn, when he was setting a mule in operation for Mr. Slater. Mr. Slater informed me there had been a new machine for making yarn got up in England, which was a mixture of the Jack and Jenny and the Arkwright Water Frame.)

I assisted the Anthony's in finishing and keeping in order their machine.

Their being no cotton gins at the south, they, (the Providence people above referred to) imported some of the cotton in seed, and picked it off by hand, which being in bad condition, and the machinery imperfect, they made some few tons of yarn, and laid the machinery by Moses Brown bought the machinery, and advertised in New York which brought Mr. Samuel Slater to Providence.

Mr. Slater came out with Moses Brown, to my father's at Pawtucket, to commence an Arkwright Water Frame, and Breaker, and two Finishers, Carding Machines. I forged the iron work, and turned the rollers and spindles, in part. All the turning was done with hand tools, and by hand power, with crank wheels. When the card rims and wheels were wanting, I went with Slater to Mansfield, Massachusetts, to a furnace owned by a French gentlemen, named Dauby, who came I think with Lafayette's army, who has a son and one daughter now living in Utica and Auburn. The card rims broke in cooling. Mr. Slater said the iron shrunk more than the English iron. I told him we would make a crooked arm, that would let the rim move round,—the arms being carried one way, and when the hub cooled would return, and leave the wheel not divided against itself,—which proves a remedy in all cases, if the arms are made the width the right way, to let the curve spring easy, with sufficient strength of iron. I told him cast iron broke more often by division in its own family, than by labor.

About the year 1786–7, my father bought the machinery for cutting iron screws, called the Fly screw, for pressing paper,—of Israel Wilkinson of Smithfield, the son of Israel who built the Hope furnace for the Browns and others,—and with the help of a Mr. Crabb, who was employed by the Browns, John, Joseph, Nicholas and Moses, in building the sperm candle works, on what is now called India Point. They used a screw of cast iron, about seven inches in diameter, and five or six feet long, which was cut by setting it upright, with a wooden guide screw, which was connected with an iron socket, with a mortice to hold
3 the cutter, which was fastened with an iron wedge.

After Wilkinson had finished the Candle works, with Mr. Crabb, he put in operation works for making screws, in Smithfield, and cut in the same manner as the English plan, brought over by Mr. Crabb. The old man (old Israel Wilkinson,) went to different furnaces in Massachusetts, to mould his screws. There were no moulders who would undertake it. My father had once seen old Israel Wilkinson mould one screw, and, after he had bought these old tools of young Israel, as he was called, and at a time when he wanted some moulding done, he took me—then about fifteen years old—into his chaise and carried me to Hope furnace, about fourteen miles from Providence, in Scituate, to mould a paper mill screw, as they had no moulder at their furnace who would undertake to mould one. I had never seen a furnace in operation, or seen a thing moulded, in my life. I moulded three or four screws before I left for home. I stayed there about a month. The screws weighed about five hundred pounds each—were five inch top, with cross holes seven inches diameter, through a lantern head for a lever seven inches diameter. They were cast in dried-clay moulds, hooped and strapped with iron bands. I took the screws home to Pawtucket and cut and finished them there. They were made for Hudson & Goodwin, of New York, and Lazarus Beach, of Danbury, Connecticut. We made many screws of wrought iron for clothiers' presses, and oil mills; but they were imperfect, and I told my father I wanted to make a machine to cut screws on centers, which would make them more perfect. He told me I might commence one. My father, in 1791, built a small air furnace, or reverberatory, for casting iron, in which were cast the first wing-gudgeons known in America, to our knowledge, for Samuel Slater's old factory.

On my way home from Hope furnace, I called at the Ore bed, in Cranston, and found Mr. Ormsbee, (I think Elijah,) of Providence, repairing the large steam engine, which raised the water seventy-two feet from the bottom of the ore pits. The engine was made with the main cylinder open at the top, and the piston raised with a large balance lever, as the news of the cap on the cylinder by Boulton & Watt had not yet come to this country when that engine was built. Mr. 4
Ormsbee told me he had been reading of a boat being put in operation by steam, at the city of Philadelphia, and if I would go home with him and build the engine, he would build a steamboat. I went home and made my patterns, cast and bored the cylinder, and made the wrought iron work, and Ormsbee hired a large boat of John Brown, belonging to one of his large India ships—should think about twelve tons. I told him of two plans of paddles, one I called the flutter wheel and the other, the goose foot paddle. We made the goose foot, to open and shut with hinges, as the driving power could be much cheaper applied than the paddle wheel. After we had got the boat nearly done, Charles Robbins made a pair of paddle wheels, and attached them to a small skiff, and run about with a crank, by hand power. After having the steamboat in operation, we exhibited it near Providence, between the two bridges; I think, while the bridges were being built. After our frolic was over, being short of funds, we hauled the boat up and gave it over.

About this time, a young man called on me, and wished to see the boat, and remained a day or two examining all the works. He told me his name was Daniel French, from Connecticut. I never knew where he came from, nor where he went.

Some three or four years after we laid our boat by, I was at New York and saw some work commenced at Fulton's Works, for steamboat shafts, and saw a small steamboat in North river, built by Col. John

Stevens, of Hoboken. I went over to his place, and saw his boring mill. I thought he was ahead of Fulton, as an inventor.

In the winter of 1814–15, hearing of a trial which was coming on before the Legislature of New Jersey, between Robert Fulton and Col. Ogden, of New Jersey, I had the curiosity to attend — as I always thought it singular that the idea of the paddle-wheel should strike two persons so, at the same time, at such a distance apart; yet I knew so simple a thing might happen. I learned in Trenton, that Fulton had said he made the draft of the wheel, in London. The case in court was managed for Ogden, by Hopkinson and Southard; and for Fulton, by Emmet and Sampson. I, being a stranger there, was in the crowd to learn what I could. After the trial was over,—in company with Emmet, Sampson, Fulton, and others,—I took stage for New York; and, in the midst of an extremely heavy snow-storm, wallowed our way along as far as Jersey City, where we found all the houses full, and no mail had crossed to New York, for two days. Fulton, Emmet and Sampson took a boat, with four oarsmen, and got over by crossing the cakes of floating ice, and launching the boat several times. The boat returned with General Brown and suit. The next boat took me, with several others. Not long after I arrived home, I saw an account of Fulton's death.

About the year 1840, I was on the railroad from Utica to Albany, with an aged gentleman in the cars, and the subject of steam power came up, when I informed him of my early acquaintance with steam power, &c. He was a well informed man, and I think, had been a member of Assembly. He said, he thought more credit had been given to Fulton, than was his due; that Col. John Stevens was more deserving than Fulton. I told him, I never thought Fulton an inventor, but simply a busy collector of other people's inventions. "Well," replied the gentleman, "I always said so, and he would never have succeeded had it not been for Daniel French. "What do you mean by Daniel French?" asked I. "Why a Yankee," said he, "that Fulton kept locked up for six months, making drafts for him."

The name of Daniel French, burst upon my ears for the first time, for forty-nine years, and almost explained some mysteries.

In 1798, when in Philadephia, I called in at the Museum, and saw an old bald-head eagle walking about the yard. The keeper, who I think was named Peal, told me the eagle was ninety-six years old;

that he was taken from the nest, ninety-six years before, at Halifax, or Nova Scotia, and that he would have a new bill in four years,—four years after, I saw mention in a Philadelphia paper, that the old eagle had got a new bill on. I had never seen any other account of the eagle, except in scripture,—of his renewing his age, like the eagle.

In, or about 1794, Col. Noami Baldwin came from Boston to Pawtucket, after machinery for a canal he was going to make, north from Boston. We made the patterns and cast his wheels, racks, &c., and he took them to Charlestown and finished the locks. I was there and saw the operation. It being the first canal in the country, a good deal of curiosity was excited among the people.

About this time, I saw the platform hay scales, at Charlestown Neck, at what was called Page's Tavern. The plan of the scales was brought from Ireland, by a Mr. Cox, of Boston, who built the old Warren Bridge, from Boston to Charlestown, and who was called to Ireland to build a bridge there. On his return to Boston, he brought a three-wheeled carriage, with a Shetland pony, for his son, and the plan of the platform scales, which has been the subject of so many patents in the United States.

We cast at Pawtucket, the iron for the draw for the Cambridge bridge.

A Mr. Mills, who built the South Boston bridge, came to me for the machinery for the bridge. I fixed the patterns, and went to Raynham, got the castings, and carried them to Boston, for the first new bridge.

Jeptha Wilkinson, junior, nephew of Jeremiah Wilkinson, invented a machine for making weavers' steel reeds, by water power.

Gardner Wilkinson invented the rolling axletree in two parts, so useful on rail road curves, &c. He also made the morticing machine, and, I think, he and his brother, made the pivot bridge, used on canals.

About 1794, my father built a rolling and slitting mill, at Pawtucket, On the gudgeon of the wheel of which, I put my new screw machine in operation, which was on the principle of the gauge or sliding lathe now in every workshop almost throughout the world; the perfection of which consists in that most faithful agent *gravity*, making the joint, and that almighty perfect number *three*, which is harmony itself. I was young when I learnt that principle. I had never seen my grandmother putting a chip under a three legged milking stool; but she always had to put a chip under a four legged table, to keep it steady.
I cut screws of all dimensions by this machine, and did them perfectly. 5

I now made a model in miniature, and had thought of trying to procure a patent, but was afraid there might be something somewhere to interfere with me, already in use. So I started off to make inquiries. I went to New York, and found an Englishman, in Greenwich street, on North River, named Barton, making clothiers' screws. He was welding an iron guide on the end of his tap, and forcing it through a socket, with an iron bar, by hand, which was the old imperfection that troubled me always. I could hear of no other in New York. I had heard of one in Canaan, in Connecticut. I went on board a sloop, old Captain Wicks, of Long Island, master, bound for Albany. In five days, I landed at Fishkill, and went ashore, and walked some thirty miles, to Canaan. I found screws made there by Forbes & Adams, by water power, but they welded on, and forced through a socket in the old way. I heard of screws being made in Canaan, from Abram Burt, of Taunton, Massachusetts. He called at Pawtucket, and looking at the old machine, I was at work with by horse power, said he had been making screws, at Canaan, by water power; that he could "set his cutter in the socket, draw the gate, and then it lathered away like the devil," which I fully believed when I saw the machine. I returned to New York, and from there went to Philadelphia, and found no screws made there except after the same mode as in New York. I heard of screws being made on the Brandywine, but my informant assured me they were made the same way as his and Barton's, at New York. I now returned home, and in the year 1797, went again to Philadelphia, when Congress was in session, and made application for a patent; Mr. Joseph Tillinghast, then a senator from Rhode Island, assisting me. On my return home, my father informed me that Jacob Perkins had been there and wanted to see my machine, and that, when he saw it, he laughed out, and remarked that he could do his engraving on cast steel, for Bank Note plates, with that machine,—that he could make a hair stroke with that, for it would never tremble,—that he could put an oval under the end of the rut, and, with an eccentric, make all his oval figures. I suppose Mr. Perkins afterwards derived great benefit from the thing.

Whilst I was at work on Slater's machinery, the owners were unwilling that I should make a slide lathe, on the principle of my screw machine, which was made for large turning; it was too heavy for cotton machinery. Mr. Slater said he had heard of one being made in England since he left, which would turn rollers. He wrote to Derbyshire, to his brother, John Slater, to come over, and bring a man who could build one. John came, and brought a Mr. John Blackburn, who made a slide lathe, which was on the principle of the old fluting machine, with the slide rest grooved in, in four edges, on two edged bars, forced in towards each other, by wedges, in mortises, behind the tenon. They worked this lathe some few weeks, and then threw it out of doors, and afterwards did their work by the old hand tool, as before.

About that time, my father, brothers, brothers-in-law William Wilkinson and Timothy Greene, and James, William, and Christopher
Rhodes, purchased a water power on the Quinnebaug river, Connecticut, at Pomfret, and commenced building a cotton factory. These
owners consented that I might build a gauge lathe, like my large one.
I then went to work, and made my patterns in Sylvanus Brown's shop
in Pawtucket. I left out the three friction rollers from under the rut, 6
as for light work and slow motion, I was willing to risk the friction.

About this time, a Company in Providence got a master machinist
from England, named Samuel Ogden, to build a factory at Hope Furnace. He was a man of great experience and good abilities. He ad- 7
vised me as a friend to abandon my new machine, for said he, "you
can *ner* do it, for we have tried it out and out at *ome*, and given it up;
and don't you think we should have been doing it at *ome*, if it could
have been done?"

Mr. Pitkin, of East Hartford, had an Englishman, named Warburton, with him, building a factory. Warburton told me, "*they* could never make our work in Europe,—that Watt & Bolton gave it to a man for a month's work to finish a piston rod, with hand tools."

When I had finished my patterns for the lathe, and was already to
start next morning, for the furnace, in Foxborough, Sylvanus Brown
took it into his head to put them into the stove, and burn them up. I 8
made others then, and got them cast, and made my lathe, and it worked
to a charm. Mr. Richard Anthony, who was building a factory, in Coventry, with his brother William, paid me ten dollars for the use of my
lathe patterns, to cast after. And this is all I ever received for so valuable an invention.

Captain Benjamin Walcott, father of the Walcotts at York Mills, Oneida county, New York, and of Edward Walcott, of Pawtucket, with Nathan J. Sweetland, put the "live centre" arbour, and the rack, in place of the screw for the feeder, to a lathe they built afterwards. But, on long experience, the screw is found best, and the two "dead centres" will make the truest work,—though they are not quite so convenient perhaps as the "live centre" arbour. But the two great principles of my machine can never be improved upon,—that is, *three bearings* to the rest, and *weight* to hold it down, where you may weigh your friction to an ounce.

The slide lathe has bent sent to all parts of the world. A certain mechanic commenced business in this country, but after using one of my slide lathes awhile, he bought one, and returned to England with it; remarking, that with that lathe in England, he could do better than at any business he could get into in this country.

It was unfortunate for me patenting my machine, when the machine making and manufacturing business in this country was only in its infancy. The patent would run out before it could be brought into very extensive use. It certainly did run out without my deriving that benefit from the invention, I was so justly entitled to. One solitary ten dollar note is surely but small recompense for an improvement that is worth all the other tools in use, in any workshop in the world, for finishing brass and iron work.

The weighted slide, the joint made by gravity, applies to planing, turning, and boring of metals of every kind, and every way, as it needs no watching, and, instead of wearing *out* of repair, it is always wearing *into repair.*

I was always too much engaged in various business to look after and make profit out of my inventions. Other people, I hope, gained something by them.

We built machinery to go to almost every part of the country,—to Pomfret and Killingly, Connecticut; to Hartford, Vermont; to Waltham, Norton, Raynham, Plymouth, Halifax, Plympton, Middleboro, and other places in Massachusetts; for Wall & Wells, Trenton, New Jersey; for Union & Gray, on the Patapsco; for the Warren factories, on the Gunpowder, near Baltimore; to Tarboro' and Martinburgh, North Carolina; to two factories in Georgia; to Louisiana; to Pittsburgh; to Delaware; to Virginia, and other places. Indeed, Pawtucket was doing something for almost every part of the Union, and I had my hands too full of business, and was laboring too much for the *general prosperity*, to take proper care of the details, perhaps, and the advancement of my own individual interests.

In 1829, we all broke down; and although I was sixty years of age, and in very bad health, I thought I would move away, and see if I
9 could not earn my own living. I moved with my family to Cohoes Falls, in the State of New York, and there fixed my new home. I have since recovered my health wonderfully, and, at this moment, being about seventy-six years old, I am hearty and well—enjoy my food

as well as any one, and can bear a good deal of fatigue and exposure. Few men, of my age, enjoy their faculties and health better than I do. Have I not much to be thankful for? I have, and am most sincerely thankful to a merciful God, for the many and great blessings.

The prospects at Cohoes were flattering for a time. But Nullification, Loco-focory, Jacksonism, Free trade, and such abominations, killed the new village just born. Europeans who were applying for water power at Cohoes, at this time, went away, saying, now we were going to have free trade; they could do our work cheaper at "'ome" than they could in this Country, and they would build their factories there.

We were compelled, now, to get our living where we could,—to go abroad, if we could not get work at home. I went to work on the Delaware and Raritan Canal, in New Jersey; then on the St. Lawrence improvements, in Canada; then to Ohio, on the Sandy and Beaver Canal; then to the new Wire bridge, on the Ottawa river, at Bytown, Canada, and Virginia. Wherever I could find anything to do, I went; and it is wonderful how I endured exposure to wet and cold, as I did.

In 1835–6, while engaged on the St. Lawrence river, I met a gentleman at Kingston, who advised me to go back of the Rideau lake, to get what I wanted, about seventy miles north of Kingston, to a village named Perth, which was given to the officers and soldiers who served in the late war with the United States. At the hotel at Perth, the landlord showed me a silver clasp, which was taken from the leg of a large eagle, which was shot in the village. The plate, or clasp, was from some place in Connecticut, I do not remember the town, nor the person's name, but directed to Henry Clay. It was after the war, and the bearer of the express probably thought he might safely take a circuitous route through the British provinces. But these Canadians didn't like the name of Henry Clay; his policy had too anti-British a tendency to suit them, so they took the poor express eagle as a spy, I suppose, and refused to sell the clasp, at any price. Perhaps, they wanted to have the story to tell, that our American eagle had been struck to them, at least.

These are the recollections of an old man, and you will please take them for what they are worth. If they are worth anything to any one, I shall be glad. To yourself, I believe, they will be valuable, and be the means ot recalling many pleasant incidents of olden times, and of an old friend.

DAVID WILKINSON.

Cohoes, Albany County, N. Y.,
December 1, 1846.

Rev. George Taft, Pawtucket, R. I.

1. Robert S. Woodbury has suggested that this may be the first use of the centerless grinder. See Woodbury, "History of the Grinding Machine," in Woodbury, ed. *Studies in the History of Machine Tools* (Cambridge, Massachusetts, 1972), pp. 33–34.

2. On the Arkwright models commissioned by Orr, see William Bagnall, *The Textile Industries of the United States,* volume 1 (Cambridge, Massachusetts, 1893), pp. 84–86.

3. On Israel Wilkinson and the Browns' spermaceti candle works, see James B. Hedges, *The Browns of Providence Plantations, The Colonial Years* (Cambridge, Massachusetts, 1952), pp. 86–122.

4. This was one of the first Newcomen engines installed in the United States. See Hedges, pp. 277–278.

5. This is the forerunner of the industrial lathe. See Woodbury, "History of the Lathe," *Studies in the History of Machine Tools,* pp. 89–93.

6. This is the same Sylvanus Brown who did woodwork on Slater's first machinery and whose account with Almy and Brown is reprinted in this collection.

7. See the headnote to the excerpt from Samuel Ogden's *Thoughts* . . . in this collection.

8. Brown's motive for burning Wilkinson's patterns is unknown. But there is a clue in the claims made by some local historians in the nineteenth century that it was Brown who invented the slide-rest lathe. Zachariah Allen, in an unpublished manuscript, claimed that Brown was the first to build a lathe "with slides adapted to Mill Chisel and made to slide along on a *Gibb* or groove." Brown built the lathe, according to Allen, in 1791 and used it to turn rollers for the Arkwright machinery. Wilkinson improved on this lathe, again according to Allen, by "placing the Slide on top of the parallel slides and held down by a weight," a description which essentially conforms to Wilkinson's own claims. Allen, of course, was not a first-hand observer, but other historians told much the same story. See Allen, "Notes on the Early Introduction of the Cotton Manufacture in New England," (selections reprinted elsewhere in the collection), Zachariah Allen Papers, Rhode Island Historical Society. See also *Dictionary of American Biography,* volume 3, p. 157.

9. The textile depression of 1829 destroyed the Wilkinson business along with that of other mill owners and artisans, and it had a severe effect on Pawtucket. The village did not really recover until the late 1840s.

Oziel Wilkinson Account (1789)

The following account graphically expresses Oziel Wilkinson's importance to the building of Slater's first spinning frame. Wilkinson (1744–1815), a skilled blacksmith, had moved from Smithfield to Pawtucket in the early 1780s to set up an anchor shop. Wilkinson and his sons were later to play a major role in the southern New England textile industry (see David Wilkinson's *Reminiscences* in this volume). The textile mill that Oziel built between 1810 and 1811 is today part of the Slater Mill Historic Site in Pawtucket.

Note the entries for "2 Mo 1" (1 February) 1790: six pair of hinges, six rollers, twenty-four spindles, and six spinning shafts. These are all critical parts for Slater's twenty-four spindle spinning frame, the first machine he built. (GK)

Almy & Brown Papers, Rhode Island Historical Society, Box 32. On Wilkinson's life, see the Rev. Israel Wilkinson, *Memoirs of the Wilkinson Family* (Jacksonville, Illinois, 1869), pp. 468–476. Paul Rivard was first to call attention to the Oziel Wilkinson account Slater machine entries, in his "Textile Experiments in Rhode Island, 1788–1789," *Rhode Island History*, volume 33 (May 1974).

Sundries Bot of

Almy & Brown to M Brown for O Wilkenson, Acct Dr

1789			£
24-3 Mo	To 140 Wool pickers 11/8	Wheel plank 9 —	0-12-5
5 Mo 25	" Loom timber 129 feet	do of Kent 15 feet —	13 6
	" 100 Jenney spindles 10 1/2 3.4.	sett of Irons for Do £4-1 —	8-4-4
	" 3 setts Loom Do £3-3	2 ray wheels for Jenney 10/.	3-13-–
3-7 Mo	4 doz tenter hooks 3/.	1 sett Loom Irons 1-0	2 1-3
6 —	2 Iron Shafts Do 17 6	fixing & turning Grindstone —	18-3
8 Mo 2	1/2 bu Corn dd Brown 1/6	1 pr Cloth Peckers 1/8 —	2-9
22	making Chaise — 6	small Guide — 1 —	1-6
9 Mo 11	making Do covers 2 1/2 days 5/	" bolt & spindle Whipping bar 4/6	19-6
21	Rag wheel & hand — 2	2 pins & dogs 2/3 —	4-3
10 Mo 2	2 long Irons for Singing bar 9/	4 pins & keys 2/5 —	11-5
27	2 pr pickers — 2/6	Sett loom Irons 1 —	1-2-6
11 Mo 12	4 pr Do — 5/6	sett Do — 19/6	1-4-6
24	1 Axle forming Cyllender 3/3.	6 Cyllender frames for Carding Mashene @12	3-12- / 3-3 / 3-12-
1790	18 large screws & 36 nuts @4/		
1 Mo 24	" Nales for Carding Mashine 8	12 Chissels — 4/ —	4-8
25	2d Steel 1/8 & plate 8 — 2-4	screws & nuts for lathe 4/ —	6-4
	small do 2/6 bolt & nuts 2/8	2 dogs & bolts for turning 2/9 —	7-11
	3 ferrels & nails — 1-9	4 doz wood screws 1/4 —	3-1
2 Mo 1	6 pr Hindges — 3-2	6 spinning shafts 17/ —	18-2
	6 rollers — 12	steel plate & Tap 1/ —	13-
	24 spindles — 15-4	4 1/2 feet wood 5/1 —	1-0-5
5	pd Benjn Harris — 1-2	2 drills 1/ 6 sockets 9/ —	11-2
17	2 In pins — 1-2	4 guides for Copper rails 2/8 —	3-10
25	17 pins 6 2 binder Irons 1/6	spikes & gudgeon — 1/2 —	2-3
26	sliding Iron — 1/8	do & Cathes — 3/3 —	2-11
	draw screws & nut 3/4	binder Irons — 3/7 —	6-11
	6 Latches — 1/6	14 washers — 2/6 —	4-
	20 pins 9 2 screws — 1/9	4 Ires — 6 —	7-9
	2 doz roller Irons — 6	12 Stotes &c — 6-6 —	12-6
3 Mo 20	3 Triches & Rings — 5/	7. 2 14 of Iron — 39/	2-4-
	Coal 14/8. 7 days work 42/	rattnut screw — 3/6	2-17-2
	Dog screw & Chisel &c 3/2	pd Prime — 2/0 —	6-
	43 lb Iron — 9	3 screw plates — 1/5 —	10-5
	3 days Work — 18	Coal 5/ pd Prime 2/2 —	1-5-2
	pd Maguire — £1 —	4 Steel Chisels — 2-8	1-2-8

Continued £41-10-6

Sundries Out of

Almy & Brown to M Brown for O Wilkensons Acct D^r [debtor]

1789				
24-3 M°	To	140 Wool pickers 11/8	11 feet plank 9^d	£0-12-5
5 M° 25	"	Loom timber 149 feet	d° of [Kent?] 150 feet	13 6
	"	100 Jenney spindles @ 10d £4-3-4	set of irons for D° £4-1	8- 4-4
	"	3 setts Loom D° £3-3-	2 rag wheels for Tenters 10/	3-13-
3-7M°	"	4 doz tenter books 3/-	1 sett Loom Irons £1-0s	1- 3-
6	"	2 Iron Shafts 17-6	fixing and turning Grindstone 9/-	-18-3
8 M° 2[ ]	"	1/2 bu Corn dd [delivered] Brown 1-6	1 p^r. Cloth Peckers 1/3	- 2-9
22	"	makeing C knife 6	small Guide 1-	- 1-6
9 M° 11	"	makeing 80 covers 2 1/2 days 15/-	bolt & spindle warping bar 4/6^d	-19-6
21	"	Lug weel & han [ ] 2-	2 pins & dogs 2/3	- 4-3
10 M° 2	"	2 long Irons for Singing bars 9/	4 pins & keys 2/5^d	-11-5
27	"	2 pr pickers 2/6	Sett loom Irons £1	1- 2-6
11 M° 12	"	4 p^r D° 5-1	sett D° 19/6	1- 4-6
24	"	1 Axle forming Cyllender 3/3	6 Cyllender Frames for Carding Mashene @ 12	3-12- 3-3
1790		18 large Screws & 36 nuts @ 4/		3-12-
1 M° 24	"	Nales for Carding Mashine 8^d	12 Chessels 4/	4-8
25	"	2^lb Steel 1/8 & plate 8^d 2-4	screws & nuts for lathe 4/-	6-4
		small d° 2/6 bolt & nuts 2/8	2 dogs & bolts for turning 2/9	7-11
		3 ferrels & nails 1-9	4 doz wood screws 1/4	3-1
2 M° 1	"	6 p^r. Hindges 3-2	6 spinning shafts 15/	18-2
		6 rollers 12-	steel plate & Tap 1/	13-
		24 spindles 15-4	4 1/2 feet wood 5/1	1- 0-5
5		p^d Benjn Harris 1-2	2 drills 1/6 socketts 9/	11-2
17		2 In. pins 1-2	4 guides for Copper rails 2/8	3-10
25		17 pins 6^d 2 binder iron 1/6	spikes & gudgeon 1/2	2-3
26		slideing Iron 1/8	d° & [ ] 3/3	4-11
		draw screws & nut 3/4	binder Irons 3/7	6-11
		6 Lattches 1/6	14 washers 2/6	4-
		20 pin 9^d 2 screws 1/9	4 [ ] 6-	7-9
		2 doz roller Irons 6	12 Floters & c 6-6	12-6
3 M° 20		3 [ ]: & Rings 5/	[ ]. 2 14^lbs of Iron 39/	2- 4-
		Coal 11/8 7 days work 42/	[ ] nut screw 3/6	2-17-2
		Dot screw & Chissel & c 3/2	pd Geo. Prince 2/10	6-
		43^lbs Iron 9-	3 screw plate 1/5	10-5
		3 days Work 18-	Coal 5/ pd. Prince 2/2	1- 5-2
		p^d Maguire £1	4 steel chissels 2-8	1- 2-8
			Continued	£41-10-6

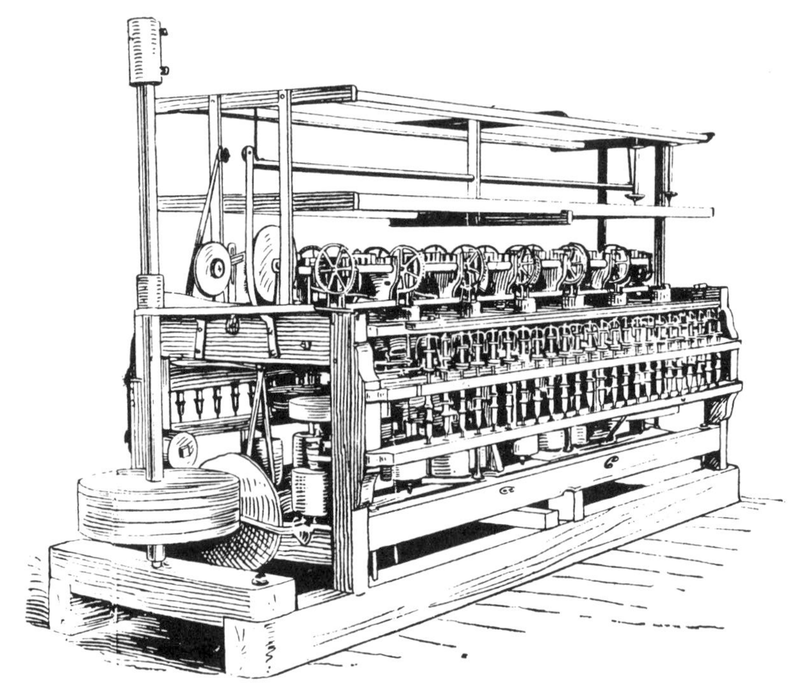

Samuel Slater's spinning frame. Line cut, Slater Mill Historic Site.

Sylvanus Brown Account (1790)

The following excerpt from the account book of the firm of Almy & Brown indicates some of the work that Sylvanus Brown did for Slater during the period when the Arkwright machines were first built. Sylvanus Brown, who was not related to Moses Brown, was a skilled millwright and patternmaker born in Cumberland, Rhode Island, in 1747. He served during the Revolution as Master-at-Arms on the *Alfred,* and he later worked as a stockingmaker in a state-run arms shop in Providence. Subsequently commissioned by the governor of the Eastern British Provinces, Brown supervised the construction of seven sawmills and two gristmills in New Brunswick and Nova Scotia. When he returned to Pawtucket, he resumed his trade as a millwright and began working for Slater in 1790. Because of what Brown considered slow payment, the relationship between the two men became acrimonious and Slater was eventually barred from the use of Brown's woodworking shop. Sylvanus Brown's house is now a part of the Slater Mill Historic Site and is furnished according to a probate inventory taken at Brown's death in 1825.

Note the wood supplied for a "little Frame," the twenty-four spindle spinning frame that Slater first built, and the "Oak for Spindle Rails," necessary parts for both of Slater's frames. Brown, however, did more than supply wood. Other accounts indicate that he worked virtually full time that first year fabricating the woodwork on all of Slater's machines. He also rented his shop to Almy, Brown, and Slater for fourteen months. (GK)

Almy & Brown Papers, Rhode Island Historical Society, Box 31. On Sylvanus Brown's life, see *The Biographical Cyclopedia of Representative Men of Rhode Island* (Providence, 1881), pp. 173–174.

Messrs Almy & Brown

To Silvanus Brown Dr

1790			
Sept 26	96½ feet 2 In. Oak planks	@10/9 pr hd	0-10-5
	37 do Inch Maple	@4/	0-3-0
	14 do 2½ In. Mulberry 17½ ft bound	@8/	0-1-4½
	12 do 3 In. Birch	@3d pr ft	0-3-..
Jan 28	17 do of Oak 2 Inches	@10/9	0-1-9½
	26 do of Pine	@9/	0-2-4
	8 do Birch	@8/4	0-1-8
	40 do 2 In. Oak	@10/9	0-4-3½
For letter frames	22 do 1½ do	@6/	0-1-4
	23 do In. Maple	@8/	0-1-10
	15 do ½ In. Chesnut	@3/	0-..-5½
	12 do 2 In Oak	@10/9	0-1-4
	5 do In Maple	@8/	0-..-5
	Wood for 15 binders		0-2-0
	28 feet of Window Frame Stuff	@9/	0-2-6
	9 ft of Chestnut ½ Inch	@3/	0-0-3
	9 do 1½ Inch Oak planks	@6/	0-0-6½
	37 do Joice	@4/	0-1-6
	10 do 2 In. planks	@10/9	0-1-1
	15 ft 2½ Ash do	@4/	0-2-1
	64 do Oak for Spindle Rails	@5/	0-3-4
	5 ft 3 Inch Birch	@3	0-1-3
	2 old Cyder Mill Screws		0-3-0
	Some Birch for binders &c & some Emmory		0-6-0
	31½ feet of Inch Ash	@7/	0-2-2½
			2-11-0

Errors Excepted

pr Silvanus Brown

Mess[rs] Almy & Brown

To Silvanus Brown D[r] [debtor]

1790			
Apl 26	96 1/2 feet 2 In. Oak plank	@ 10/9 phd	0-10-5
	3 1/2 do Inch Mapple	@ 8/-	0- 3-0
	7 do 2 1/2 In Mulberry 17 1/2 ft board	@ 8/-	0- 1-4 1/2
	12 do 3 In Birch	@ 3[d] pft	0- 3
Jan[y] 20	17 do of Oak 2 Inchs	@ 10/9	0- 1-9 1/2
	26 do of Pine	@ 9/-	0- 2-4
	8 do Birch	@ 8/4	0- 0-8
	40 do 2 In Oak	@ 10/9	0- 4-3 1/2
For little Frame	22 do 1 1/4 do	@ 6/2	0- 1-4
	23 do In. Mapple	@ 8/-	0- 1-10
	15 do 1/2 In. Chesnut	@ 3/-	0- ″-5 1/2
	12 do 2 In. Oak	@ 10/9	0- 1-4
	5 do In. Mapple	@ 8/-	0- ″-5
	Wood for 15 binders		0- 2-0
	28 feet of Window Frame Stuff	@ 9/-	0- 2-6
	9 ft of Chestnut 1/2 Inch	@ 3/-	0- 0-3
	9 do 1/2 Inch Oak plank	@ 6/-	0- 0-6 1/2
	37 do Joice [Joists]	@ 4/-	0- 1-6
	10 do 2 In. plank	@ 10/9	0- 1-1
	15 do 2 1/4 Ash do	@ 14/-	0- 2-1
	67 do Oak for Spindle Rails	@ 5/-	0- 3-4
	5 ft. 3 Inch Birch	@ 0/3	0- 1-3
	2 Old Cyder Mill Screws		0- 3-0
	Some Birch for binders & Some Emmory		0- -6-0
	31 1/2 feet of Inch Ash	@ 7/-	0- 2-2 1/2
			2-18-0

Errors Excepted

p[d] Silvanus Brown

Engraving of Zachariah Allen, from J. D. Van Slyck, *New England Manufacturers and Manufactories*, vol. I, Boston, 1879.

Zachariah Allen

Zachariah Allen, the youngest son of a Providence, Rhode Island, merchant, was born in 1795. At the age of fourteen he entered Brown University from Exeter Academy and devoted himself to the study of medicine, mechanics, chemistry, and physics. Allen hoped to continue his medical studies in Edinburgh, Scotland, but his oldest brother, Philip, opposed his plan and convinced him to remain in Providence and study law. Although he was admitted to the bar in 1815, Allen soon followed in the footsteps of his older brothers, choosing a career in textile manufacturing.

In 1821 Allen inspected the Unity Mill in Cumberland, Rhode Island, with the thought of purchasing it in partnership with his brother Crawford. Within a few months he was considering the purchase of a waterpower site near a cotton factory owned by his brother Philip. After months of negotiations, Allen was able to purchase the site, and on March 6, 1822, the first timber for the mill was cut; the mill was in full production, making satinets, by January 1823.

The woolen business entered a period of decline just as Allen completed his factory and he was forced to curtail production almost immediately. Business conditions remained poor throughout 1824, and in 1825, Allen sailed to Europe with the hope of finding ways to improve the profitability of his factory. Upon his return to the United States his newly acquired knowledge of textile manufacturing and the improved tariff protection for the American woolen industry enabled him to continue to manufacture woolens for another decade.

In 1837, Allen acquired the Phoenix Manufacturing Company of Warwick, Rhode Island, in partnership with David Whitman. The general decline in business conditions may have convinced him to abandon making woolens altogether, for in 1839, he removed the woolen machinery from the Allendale mill, enlarged the factory, and installed the machinery to make cotton cloth. Allen prospered for the next forty years as a cotton manufacturer. His long and full life as a factory owner is documented by the extensive collection of diaries, manuscripts, and published writings he left upon his death in 1882. Allen was a prominent spokesman for the small southern New England textile mill village as well as a prolific writer on technical and historical subjects. The three selections that follow are representative of his contributions. (TZP)

Zachariah Allen

The Science of Mechanics (1829)

The Science of Mechanics was a significant book in the history of American engineering. Most of the technical books available in this country at the time were either mathematical treatises or foreign works republished in the United States. Allen's book did not fit into either category. Instead, it was, as one reviewer said, ". . . the first work of the kind, of domestic origin, that has been put into the hands of our practical mechanics and manufacturers."

One of the main concerns of historians of technology is the means by which technological ideas move from one place to another. *The Science of Mechanics* is a book about mills and millwork that is solidly grounded in Allen's own experience as a Rhode Island mill owner and engineer. Furthermore, it includes his first-hand observations of British textile factories and machinery and incorporates the latest British engineering ideas. In other words, Allen described modern American and British millwrighting and factory engineering as he had observed them and published *The Science of Mechanics* for the express purpose of placing this information in the hands of American mechanics.

The book covers a wide range of topics relating to physics and mechanics and includes discussion of subjects such as the strength of materials, heat, friction, pneumatics, machine design, and hydraulics and hydrodynamics. The pages that follow are excerpted from the sections of *The Science of Mechanics* that focus on the textile factory and its equipment. (TZP)

Zachariah Allen, *The Science of Mechanics* (Providence, 1829) pp. 197–198, 201–209, 211, 246–249.

Water Power.

Projects for erecting mills are usually undertaken in the United States during the Spring of the year, when the streams have abundance of water flowing in their channels. After a hasty view of a stream, and a survey of the fall of its waters, the purchases are too often concluded upon, and costly works are erected. The unfortunate proprietor is only convinced of his precipitancy, when it is too late to remedy the evil consequences of it, after finding that his wheels remain idle for want of the calculated supply of water during the months of the year when the days are longest, and the expenses of lights and fuel are not required. These mills may be seen in different parts of the country abandoned and desolate, standing as monuments of the folly of those who erected them. It is of the utmost importance to the prosperity of manufacturing operations that the moving power should be sufficient for operating the machinery at all seasons of the year. Where the labourers are compelled to remain idle for one or two months, they are during this period thrown upon their own resources for support, being compelled to consume the fruits of former seasons of industry. The best hands will therefore quit a manufacturing village, in which so material a disadvantage exists, in favour of another where the employment is more regular. The improvident and worthless will remain from an inability to remove their families, for whose support during the season of inactivity their employers are under the necessity of making advances, which are rarely repaid. Expensive fixtures for a mill should therefore never be commenced on small streams until the mill site is most carefully surveyed to ascertain the fall of the water, and the quantity of it that may be safely calculated upon during the long droughts of summer. On large streams even the superabundance of the floods of winter are sometimes of serious inconvenience, causing a suspension of the operation of a mill by backwater.

*

On constructing Mill Dams.

After having thus satisfied himself of the sufficiency of the Water Power, the engineer may proceed to select a site for the dam. On large rivers subjected to freshes it is an important consideration to fix upon a spot where the bed of the channel is composed of ledges of rocks, upon which the necessary works of the dam may be more firmly established than upon an earthy bed or channel. When the waters can be made to pour over the top of the milldam upon ledges of rocks below, the continual dashing and attrition of the falling waters are safely sustained. But when the bed of the stream is composed of earth, the force of water soon wears its way to great depths, unless artificial beds of timber and stone are substituted to sustain the immediate action of it. When artificial *aprons* are formed to receive the force of the

the waterfall particular care should be taken to construct it sufficiently broad. A few years since a large mill dam in Rhode-Island was swept away and lost, merely from a fault of this kind in the original construction of it. When the river was swoln by long continued rains the accumulated waters in their descent overshot the apron, and formed an excavation many feet in depth immediately below the dam. The whole structure having been thus undermined, sunk into the abyss formed by the flood, and disappeared. Even the best constructed dams when erected upon a sandy or clayey soil, are subject to injury from the gradual action of the water upon the foundations on which they rest. In all cases of this kind piles should be driven into the ground to sustain the frame of the dam and platform, and heavy stones should be placed in courses arranged like the slates on a roof, with their up-stream ends lowest, for the distance of a few feet below the dam. It is also an excellent plan to cause the water to stand upon the apron by means of a ridge of stones or timber placed across the stream immediately below the dam. The glancing force of water can in no way be so effectually counteracted as by opposing to it a body of the same fluid, which offers an uniform resistance, particle to particle, in every direction of its motion. The durability of the apron is thus promoted by keeping it always immersed. The advantages obtainable from ledges of rocks are, commonly, counterbalanced by the expensive excavations of the rock necessary to form the trenches and wheel pits.

In England the mill dams are usually constructed of hewn freestone, laid in water lime, and secured by iron clamps. This mode of construction is rare in the United States, where the great cost of hewn stone and the comparatively low price of timber contribute to render the use of wooden materials more common In the few instances in which stone dams are erected in this country the junction of the stones is so imperfect as to render the use of a planked covering quite necessary for retaining the water. In constructing dams of every description upon a loose soil, pile-planking should be carefully driven into the solid earth along the face of the whole work, both to prevent waste by leakage and to render the foundations of the dam more secure by cutting off the small streams beneath the ground, which is loosened by them. and rendered more liable to be washed away by freshes. A waste or floodgate, formed in as many sections as may be deemed proper to admit of being easily raised when subjected to the increased pressure against them, should always be made for the purpose of drawing off the mill pond to execute repairs when necessary, as well as for venting the flood waters. The form of the dam, if practicable, should be made somewhat semicircular with the concave facing down the stream In this case an arch is formed, and the pressure of the water cannot carry away the timber work of the frame unless the abutments fail. Another important advantage resulting from this form of construction is the check given to the violent impetuosity of the falling waters, which are thus brought to spend their force in rushing together in counteracting currents.

In laying the foundation of the walls to sustain the lower side of the embankment of the dam, the alluvial loose soil should be removed to the hard bottom or pan of earth. The wall of a dam should also be made much thicker at the bottom than at the top, and the whole slope or batter should be made on the lower face, the upper side, against which the embankment rests, being made perpendicular. A wall thus constructed possesses greater strength to support the pressure of the earth, and of the water of the mill pond resting against it, which under the operation of severe frosts have a tendency to overthrow it.

*

Mill Courses.

The canal to convey water to the mill should be of spacious breadth and depth. When a contracted canal is formed, a rapid current must flow through it to furnish the necessary supply of water, whereby the *head of water* must be reduced, and a consequent loss of power must be the result. In situations where irregular rocks abound, it may be more convenient to make a broad shallow trench, in which case the quantity of water flowing through it will be nearly the same, provided the area of the canal be equal. Thus a canal 12 feet broad and 4 feet deep (48 feet area) will deliver about the same quantity of water as one 8 feet broad and 6 feet (area also 48 feet.) The greatest disadvantage resulting from shallow canals is, that when the water in the mill pond becomes diminished, or drawn down, the canal will not allow a sufficient supply of water to flow through it. If the depth of water in an open canal be less than 4 feet, the passage of the water becomes also greatly obstructed, during the intensely cold nights of the Northern States, by the *anchor frost*, which shoots up in long slender spiculae from every pebble upon the bottom, sometimes uniting in a solid mass with the sheet of ice upon the surface.

The ends of the flooms, and the junction of the abutments of the dam with the timber work should be carefully secured by driving pile planking into the solid earth, leaving the ends of the plank sufficiently long to reach the top of the embankment. Mill dams are more frequently lost by inattention to pile planking than by any other fault of construction, as the water after working its way around the bulwarks intended to resist it soon enlarges the apertures, when the insinuating stream becomes suddenly a torrent, bursting through every impediment to its course, and sweeping away the whole structure, and even the mill itself, should it happen to be erected on the wall of the dam. For this reason it is always desirable to place a mill upon some secure bank, at a short distance from the dam, where the edifice will remain free from accidents by floods.

*

*

It is usual to lay the floor or platform of the wheel pit upon sleepers of sufficient extent for the main walls to rest upon it, in order to prevent the constant agitation of the water from action upon the earth beneath the foundations, or undermining them.

The same observations apply to the mill race as to the canal. If they be made deep and wide, less fall will be required to carry off the water, as it will flow in a gentle full stream, instead of a rippled current. The depth of both the wheel pit and race should be governed by the depth of the bed of the river at the outlet of the race, and also by the usual tendency of the river to rise during a common rain storm, and to cause *back water* to impede the motion of the wheel.

*

WATER WHEELS.

Water Wheels in England are commonly constructed of iron. This material being dearer in the United States, and all the necessary timber for constructing wooden water wheels being comparatively cheaper, the latter material is very generally employed. The iron water wheel possesses several important advantages over a wooden one. It is more durable, and from the small quantity of room occupied by the thin iron floats or buckets an iron wheel will contain more water than one of equal dimensions constructed of wood, and is consequently more effective. Iron shafts are frequently used for the axles of wooden water wheels, but they are liable to become loose, and are frequently broken; or should the wedges allow the wheel to slip upon the shaft, the teeth of the crown wheel or segments, or even the leading shaft may be destroyed. In one water wheel of this description I have known three iron axles or shafts to be inserted, the last being made of wrought iron withstood the stress upon it. It is commonly the practice in England at the present time to make the iron shafts of water wheels hollow and of large diameters. A wooden shaft with the arms of the water wheels properly secured to it, and with short projecting ends that will remain constantly wet by the dripping of the water during the motion of the wheel, may be considered almost as durable as iron I have seen many wooden shafts taken out of old water wheels; but have never observed one of them to be decayed. It is recommended as the very best method of uniting the arms to the axis of the water wheel to have a cast iron centre piece, as used for an iron axis, to prevent weakening the wooden shaft by mortising the arms into it. In the United States shafts of the largest dimensions are obtainable, which will possess sufficient strength after the mortises are made to resist the stress upon them. The best shafts for water wheels now used in Rhode-Island, are of the hard pine imported from N. Carolina. Wood exposed to the action of the water is worn away gradually by the attrition. White pine however becomes covered with a slimy coat which shields it from the injurious action of the water, and is therefore more durable even than oak. In the construction of a wooden water wheel care should be taken to make it sufficiently strong in the first instance to prevent the working of the joints, as the bolt holes in this yielding material soon become enlarged, and the evil is rapidly augmented to an injurious extreme. The joints of water wheels of this description will work and creak like those of an old ship labouring in a heavy sea. If the whole circumference of a wheel be encompassed by strong iron hoops like a cask the strength and durability of it will be greatly increased.

The terms, *Head* and *Fall*, being frequently used in relation to the application of water to the water wheel, may here be defined. By Head is understood the distance from the surface of the water in the floom to the part of the wheel on which the water strikes. The Fall is the perpendicular descent from this part of the wheel to the bottom of it.

Of the great variety of water wheels we shall attempt the description of those only which are in most general use; viz. the *Undershot*, *Breast* and *Overshot* Wheels.

Undershot Wheel.

Upon undershot wheels the water acts merely by its impetus as it shoots against the floats, and not from its weight while resting in buckets, as on the overshot and breast wheel. The undershot wheel is the simplest and cheapest kind of water wheel, but it is employed only in situations where an abundant supply of water is obtainable. It will produce only one half of the effect of an overshot wheel operated on by an equal fall of water.

*

Breast Wheel.

All kinds of hydraulic machines upon which the water cannot descend through a given space, unless the float board or bucket moves therewith, are considered by Mr. Smeaton as of the same nature with overshot wheels, and equal to them in power and effect. All those machines which receive the impulse or shock of the water, whether in a horizontal, perpendicular, or oblique direction, are to be considered of the same nature as undershot wheels. Therefore when the water strikes a wheel at a certain point below the level of the surface of the water in the mill pond, and then descends in the arc of a circle, pressing by its gravity upon the floats of the wheel, the effect will be equal to that of an undershot wheel where the head is equal to the difference of level between the surface of the water and the point at which it strikes the wheel, added to that of an overshot, with a fall equal to the difference of level between the point where the water strikes the wheel and the bottom of it Thus the breast wheel, and also the overshot wheel, is impelled not only by the weight of water, but also by its impetus or momentum. The breast wheel has float boards instead of buckets; nevertheless the water is retained by the sweep of the breast between the floats, so that buckets are virtually formed from which the water cannot escape except the wheel moves. Each of the portions of water contained in these spaces bears partly upon the breast, and partly upon the floats of the wheel. It is, however, impossible in practice to fit the sweep so accurately as entirely to prevent leakage Upon the breast wheel the water is applied against or just below the level of the axis. This kind of wheel is considered by the best engineers as inferior in effect to the overshot wheel, but much superiour to the undershot.

The sweep of the breast is generally constructed of wood in the United States; but in England it is commonly constructed of hewn stone formed with great accuracy to the shape of the circumference of the wheel, in order to prevent as much as possible the waste of water that takes place when it descends in the space left between the floats and the breast, without acting by its gravity upon the wheel.

Breast wheels are employed where the fall is inconsiderable, because the water is retained in such case for so short a time upon the wheel that it could not enter regular buckets, like those of an overshot wheel, and be discharged again, without wasting much of its power.

Overshot Wheel.

This description of water wheel being in most general use, and most effective, our observations will be principally confined to it. To Mr. Smeaton and Mr. Banks, the public are indebted for nearly all the experimental researches which have been made upon this subject. Their rules are adopted in most of the calculations herein made.

The effect of overshot wheels is intended to be derived entirely from the gravity of water, or as little as possible from its impetus.

*

. . . The overshot wheel was formerly used by conducting the water over the top of it to descend in the buckets. This plan is still in common use; but it has been found that although the buckets may be only half filled with the water, yet it soon begins to spill out of them as they descend, or is thrown out of them by the centrifugal force, whereby much of the power of the water is wasted by falling through the air without acting upon the wheel. The principle of the overshot wheel is now very generally carried into effect by constructing the floom to project nearly over the axis of the wheel, giving to the aperture through which the water flows a form calculated to cause the stream to gush in a reverted direction nearly at right angles with the arms of the wheel, and in a line corresponding with the edges of the elbow buckets. By this means the overshot wheel is made to revolve in the same manner as a breast wheel, and the disadvantage of spilling the water is obviated by constructing for it a sweep or breast. The overshot wheel is thus improved by combining in its construction the principal excellence of the breast wheel.

*

Gate or Shuttle of the Water Wheel.

It has been found most advantageous to apply the water to overshot as well as breast wheels by drawing down the gate, that the water may flow over the top of it. The quantity poured upon the wheel is regulated by the shuttle, which is placed in the direction of a tangent to the circumference of the wheel, and is provided with a rack and pinion by which it can be raised or lowered by the action of the governor. The governor is sometimes made to act with less friction upon a *sluice*, which is hung upon pivots and is turned like a throttle valve of a steam engine.

*

ELEMENTS OF MACHINERY, AND THE CONTRIVANCES USED IN THE COMPOSITION OF MACHINES.

It is in the complex wheel work of Mill Gearing, and in all the countless varieties of motion and moving forces produced by the machinery employed in mills that the science of mechanics is remarkably displayed at the present day. Indeed, the interiour of a modern cotton mill exhibits a complete view of the practical application of all the Mechanical Powers, so admirably combined and cooperating in various ways to produce desired effects, that mere machines are made to operate with apparently as much self directed skill as is exhibited in the manual labour of intelligent beings. In the movements of the Steam Engine, power loom, Whitmore's card machine for cutting, bending, and inserting in the leather the wire teeth of cards, and of Wilkinsons' machine for making slaies, admirable mechanical ingenuity is displayed, the operations of all these machines being complete in themselves, and requiring only the inspection of a workman or superintendant. The triumph of mechanical ingenuity, however, appears to be exhibited in the wheel work producing the surprising action of automaton figures, which have been made to resemble man in certain movements, whereby mere matter is made to appear to possess the attributes of mind; almost inducing the beholder to imagine that the modern mechanic, like Prometheus the great inventor of the arts of ancient times, had stolen the sacred fire from heaven to animate the works of his own hands.

It is believed that the general principles advanced in the preceding part of this treatise, cannot be more intelligibly illustrated than by describing the various modes in which they are practically applied in the arrangement of the Gearing of a manufactory, and in the contrivances used in the composition of machines to produce all the variety of movements and moving forces required in the processes of the useful arts.

Before proceeding to treat of the interior mechanism of the mill it may be proper to make a few observations upon the construction of the mill itself, and of the works connected with it. It often happens that many things are neglected, or are deemed of trivial importance in the original plan for the construction of new works upon unimproved water courses, which, as the works become subsequently enlarged, are frequent sources of regret when it is too late to alter or amend them. Having laid out and erected new works upon an unimproved mill site an opportunity has been afforded me of acquiring information upon this subject from actual experience, which has been not unfrequently impressed by disappointments and losses.

The construction of the mill dam and trenches have already been treated of, as well as the important subject of ascertaining the sufficiency of the power of the stream for the purposes required. In locating a mill the general outlines should be fixed upon for the plan of the village, which in most situations in the United States is erected for the accommodation of the manufacturing population, forming a little colony around the waterfall which turns the mill wheel, in order that there may be as far as the nature of the situation will allow an agreeable arrangement of the cottages or dwelling houses. The roads or streets may in the first instance be regularly laid out, without much additional expense, and the buildings placed square with each other, and not diagonally as is observable in almost every manufacturing hamlet. The whole extent of the waterfall should be in the first instance located and improved as far as practicable, as water power is always valuable; and permanent bounds should be erected at the height of the ordinary level of the water in the mill pond to serve as landmarks of possession, should mills be afterwards erected in the same vicinity. Before fixing upon the immediate spot for sinking the wheel pit the earth around it should be carefully sounded by a pointed iron rod to ascertain if there be ledges of rocks which might obstruct the necessary excavations, as by changing the location only a few feet, obstructions of this sort may commonly be avoided; although it is desirable to place the foundations of a mill upon this solid basis, yet a little attention to this subject may save the subsequent expenditure of large sums, which are very frequently lost by the costly excavations in flinty rocks.

In laying out the ground plot for stone or brick mills the trenches should be staked out considerably larger than the intended size of the building to allow of the projection of one or two feet for the foundation stones, which on loose soils should extend considerably beyond the outer face of the main walls. If the lower courses of stone work intended for the foundations beneath the surface of the ground be 3 feet wider than the wall above it, then 2 feet of the projection should extend beyond the outer fronts of the walls and only one foot withinside of them. Walls of buildings have always a tendency to spring off or outwards, but are effectually prevented from falling inward by the floors. Even after the utmost caution has been bestowed in laying the foundations of a mill with large heavy stones, the walls should be secured to the ends of the beams by iron clamps, or screw bolts and plates, to prevent them from springing outwards. Walls sufficiently strong for warehouses have been found to yield at last to the constant tremour produced by the reciprocating motions of machinery, and the violent sudden thrusts occasioned by the irregular action of the teeth of wheels. Power looms, in particular, have a remarkable tendency to rack the walls of a mill if placed in an upper story, as the lathes of an hundred looms may at times have a simultaneous horizontal vibration back and forth. The proprietors of a considerable cotton manufactory in Rhode Island, who had arranged all their power looms in an upper apartment of their building, constructed of wood, found after a short time that the joints of the building were unable to withstand the movements of the looms, and the whole fabric was perceived to acquire a horizontal vibrating motion sufficient to cause water contained in a vessel to oscillate until a portion of it passed over the sides upon the floor. It is almost unnecessary to add that they were compelled to remove all their looms to a lower floor. In one of the finest weaving mills which I visited in Glasgow this difficulty appeared to be most effectually guarded against, by placing each power loom upon four small blocks of stone sunk in the ground floor to a level with the tiles. In this instance the building was only one story high, lighted from the roof by sky-lights to facilitate the operations of weaving the cotton yarn, which was as fine as No. 60. Mr Buchanan observes, that to obtain solidity and steadiness, a mill should not only be sufficiently *strong* and *stiff*, but sufficiently *heavy*, as the greater the mass the less in proportion will it be effected.

The arches above the floom and race of a mill, unless constructed near the centre of the building with each wing to serve as a buttress, are always inclined to yield to the weight pressing upon them, whereby one of the buttresses forming the end wall is commonly crowded off. The tremour of the walls affects the stones of the arch, the least yielding or opening of which allows the key stones to operate in an instant like so many wedges to prevent the span from recovering its former place, whereby the walls soon become seamed with unsightly cracks. It is better to form two small arches, or to support the centre by stone pillars than to form one arch of large span.

When the soil is composed of loose sand or clayey loam, the walls of the wheel pit should be founded upon piles, and in most cases it is common to extend the planked floor of the wheel pit sufficiently for the surrounding walls to be based upon it. Indeed it may be adopted as a general rule that it is true economy to construct all parts of the foundations of mills in the strongest and most solid manner.

The posts which support the beams in the centre of a mill should also rest upon a very solid mass of masonry, as the lines of shafts and other mill gearing are either attached or dependent upon them for being maintained in their proper situations. The settling of a pillar in the basement of a mill merely ½ of an inch will derange all the lines of horizontal shafts in every story above, whereby vast stress is thrown upon the couplings, and all the revolving wheels connected with such shafts immediately begin to wear irregularly, and to produce a clattering noise. If a block of hewn stone be used in any part of the structure, it should not be omitted here. Cast iron pillars or posts are generally used in England, and as they are cast hollow like water pipes they are not very expensive.

Great care is bestowed in laying the most solid foundations of hewn stone, to sustain the working parts of Steam Engines and water wheels in the best English mills. At the present prices of blocks of split granite in most parts of New England, the plummer blocks and other heavy fixtures for water wheels may be formed of granite at an expense which will not prove eventually much greater than if formed of timber, a material which in such situations is very liable to rapid decay. In setting up water wheels and steam engines, particular care should be bestowed in constructing the framing, which sustains the first impulse, or immediate action of the moving force, as independent of the walls and floors of the mill as possible, in order to avoid imparting to the whole building the tremour which is frequently so great as to be communicated in a very perceptible manner to the ground itself upon which the building rests.

Modes of transmitting a Force to a distance from a First Mover.

The most common mode of transmitting a moving force is by lines of revolving shafts.

*

The shafts which serve to transmit the moving power from one part of a mill to a distant part of it should be so secured upon their bearings or pillows, as to be easily moved to restore them to their true situations when displaced by the settling or sagging of the beams to which they are affixed; as the teeth of the wheels revolving upon a long line of shafts all act irregularly the instant the centre of the shaft is deranged from its proper position, in which case not only the wheels but the couplings of the shafts are soon worn away and destroyed. The pillows or framing which supports a line of shafts must therefore be set up in such a manner that they may be moved at pleasure to accommodate any shifting or alteration that may from time to time become necessary in practice to preserve them in a straight line. Indeed every part of the mill gearing and of the machinery contained within the mill, should be put together in such a manner as to admit of being easily repaired or replaced with the least possible derangement to the other parts of the mechanism.

Zachariah Allen

The Practical Tourist (1832)

The Practical Tourist, published in Providence in 1832, three years after the printing of *The Science of Mechanics,* was drawn from the very complete travel diaries kept by Zachariah Allen during his European trip in 1825. *The Practical Tourist* was not designed as an instructional work in engineering and millwrighting, much of the technical information gathered during the trip had already been published in *The Science of Mechanics.* Rather it was intended to give the general reading public a comparative view of the industries of England, France, Holland, and the United States.

The section of *The Practical Tourist* reprinted here is taken from the description of the British textile industry in and around Manchester, England. It was there that Allen first learned about the new British system of high-speed, power transmission technology that he would introduce to America when he rebuilt his mill at Allendale in 1839. These pages indicate not only the profound effect of British textile technology upon the visiting American but also some of the important social and economic differences between the Manchester mills and Allen's experience at home. His description of suffering and turpitude among the British working class leads up to and immediately precedes the idyllic vision of American mill village communities reprinted at the beginning of this volume. (TZP)

Zachariah Allen, *The Practical Tourist,* 2nd ed. (Providence, 1832), pp. 121–124, 128–131, 147–149, 152–153.

In a humid state of the atmosphere, the traveller is apprised of his approach to Manchester, when from the summit of some hill over which the road may wind, he first beholds at a distance the dark mass of smoke, which hovers like a sooty diadem over this queen of manufacturing cities. On approaching nearer, he views the numerous tall chimneys with smoky tops rising high above the roofs of the houses. A remarkable elevation is given to the vents of the furnaces, for the purpose of increasing the draught to render the combustion of the fuel more complete, and also to discharge the smoke into the air far above the windows of the houses. Notwithstanding these precautions, the inhabitants of the region below live amid sulphureous vapors, and the very walls of the houses are stained to a sombre hue by the coal smoke. During the summer, and also in dry and windy weather, Manchester might be deemed a pleasant place for a residence. But, at other times, and particularly on calm mornings in the early part of the spring, whilst a bright sun cheers the adjacent country, it displays to the inhabitants of Manchester its broad red disk, scarcely affecting the feeblest eye which gazes upon it through the dusky vapors, by which it is obscured. During the frequent foggy days in winter, an artificial twilight so completely shrouds the place, that at times the use of the gas lights becomes necessary, even at mid-day, for certain nice operations in manufactures. For the same reason, the lights in the large cotton mills are not extinguished until nine o'clock in the morning, and are rekindled to form a brilliant illumination, as early as about half past three in the afternoon. Most of the labor at such periods is performed by the aid of artificial light. Nearly one half of the surface of the exterior walls of the manufactories is composed of spacious glazed sashes, which are arranged in profusion to admit all the scanty light which a naturally hazy atmosphere, rendered still more obscure by smoke, will transmit. When a slight breeze arises, this dark cloud is put in motion, and is borne away over the country in an unbroken murky volume, perceptible at the distance of twenty or thirty miles, like the long train of smoke which streams from the chimney of a steamboat, and leaves a dusky line extended far over the waters and shores.

*

It has been stated that in 1831, there were nearly 400 Steam Engines in operation in Manchester and in the adjacent suburbs, besides numerous forges, bleacheries, print works, and foundries. Taking the average of these 400 Steam Engines, at 14 horse power each, and the average consumption of coals per hour at 13 pounds for each horse power, it will appear that the quantity of coal consumed by them in each hour exceeds 70,000lbs ; and if the quantity consumed in the foundries, bleacheries, and in other processes of manufactures, and by the inhabitants in their dwelling-houses, be computed at as much more, the whole consumption of coal in Manchester will not fall much short of 140,000lbs during each hour of the day.

The price of this important article of fuel, coal, on the abundant supply of which the prosperity of Manchester is so essentially dependant, varies from eight shillings to ten shillings sterling per ton, delivered on the banks of the canal. With the rate of exchange at 10 per cent premium, this price is equal to nearly two dollars, and two and a half dollars a ton, or about one third of the price of the same fuel on the seaboard of the United States.

Even at this low rate of the price of coals, steam power, with the building and fixtures necessary for operating machinery, has been leased in Manchester at £20 sterling per year, (nearly 100 dollars) as an enterprising American domiciliated here stated to me. An equivalent amount of water power may be obtained in most parts of New-England nearly forty per cent cheaper than this steam power in Manchester, the necessary mill buildings in each case being included. It is, indeed, only after viewing the vast amount of labor expended in mining coal and transporting it to the furnaces of Steam Engines, and the multitude of these costly machines, upon which the engineers are often at work, repacking the pistons and executing frequent small repairs before and after the ordinary hours of labor, and during almost every Sabbath, the only day on which the boilers become cool, that an American can estimate the vast advantages possessed by the United States in the immense water power furnished by their innumerable rivers. The wealth of England could hardly purchase, at the rate of the cost of steam power in Manchester, the water power available within the limits of the United States. To this advantage of a cheap moving power may be attributed the remarkable prosperity of all branches of manufacturing industry in the United States, wherein a great moving force forms a principal part of the ordinary cost. The flour mills of the United States have long been celebrated for superiority in mechanism and effective operation over all other similar mills in Europe; and even the manufacture of coarse cottons and other fabrics which requires the hard twist of the throstle and the violent blows of the power-loom, is greatly indebted to cheap water power for the successful competition maintained with the steam power and cheaper labor of England. The difference in the cost of the transportation of the raw materials to the waterfalls located at remote distances from the seaboard, must, however, be taken into consideration with the comparatively light expense of transportation between Liverpool and Manchester. The first cost of a suitable lot of ground for the erection of a cotton mill in Manchester will be nearly equal to that of some powerful waterfall, with many acres of adjacent lands in the United States.

*

Having a letter of introduction to the proprietor of one of the most extensive cotton-spinning mills in Manchester, containing nearly 90,000 spindles, he very civilly accompanied me in a ramble over his vast works. The buildings are all of brick, in the form of a hollow square, the principal front of which towers to the height of eight stories, and the four outer fronts of the building measure more than 800 feet. The entrance is by a great gate, at which a porter is always in attendance to refuse admission to intruders from without, and to watch lest property should be conveyed furtively from within.

After passing the gate and beneath an arch formed under a side of the building, I entered the open court-yard or square, inclosed within the four interior walls of the manufactory. In the centre of this square is a sheet of navigable water bordered by a quay, on which canal boats may be seen discharging their freights of raw cotton and coals in the heart of the works, and receiving the packages of yarns. A tunnel or arched passage is made beneath the mill, to connect this interior basin with one of the principal canals which traverses a considerable part of England. Every possible facility is thus afforded for transporting the raw material to the very centre of the mill, and for shipping the manufactured goods in return to London or Liverpool.

In the preparatory process of picking over and assorting the sea island cotton, before it enters the machinery, there were more than 60 persons at work in one apartment, beating the flakes of cotton with sticks, in order to open them for more minute inspection. On suddenly entering this apartment, and viewing so many men and women, all simultaneously brandishing rods and beating the cotton, the loose locks of which flutter in every direction from beneath the strokes of the rods, descending with a deafening clatter, you may readily suppose that you are witnessing the disorderly scene of a mad house. The dust and small particles of cotton, floating in the air in this room, are almost suffocating, and must prove most pernicious to the health of the workmen.

When the doors of the various long apartments are successively thrown open, you view the wheels revolving on long lines of shafts, and ranges of machines with the metallic brass bright and glittering, as if polished by some careful housewife. The heads of the numerous busy attendants are visible above the machinery as they move to and fro at their tasks. In going from one apartment to another the spectacle almost produces the bewildering sensations which are sometimes excited by the strange visions of a dream.

The apartments are all warmed by steam from the boilers of the engine, conducted through cast iron pipes, in some cases arranged near the floor, with the design of distributing the heat more uniformly.

*

This manufactory which gives employment, directly and indirectly, to nearly thirteen hundred persons, and rivals in magnitude and importance many national works, was erected by Mr. Murray, who removed to Manchester about forty years ago, and commenced his career as a common mule-spinner. The cotton manufacture was at first, as he stated, "almost all profit." As competition gradually reduced these profits, he continued to enlarge his works; and the result in the aggregate, on a greater amount of production, he observed, has continued nearly the same. Separated from Mr. Murray's mill only by a narrow street, is another cotton mill, of equal, or even of greater magnitude. I was informed that one firm, engaged in spinning coarse yarn, has during the last year manufactured upwards of six thousand bales of cotton. It appears from a published statement, that the number of large cotton factories in the immediate parish or town of Manchester, was in 1820, fifty-four—in 1823, fifty-six—in 1826, seventy-two—in 1828, seventy-three. But the whole neighboring country abounds in them.

*

The rapid increase of the population of Manchester establishes a further proof of the prosperous diffusion of useful employment and consequent facilities for human subsistence. Within the last ten years, the population of this town has been increased by an accession of about 40,000 inhabitants, being a ratio of increase of above 35 per cent. Although the population within the circumscribed limits of the immediate town of Manchester is rated by the census of 1831, at only 142,026 inhabitants, yet including the population of the suburbs within the circuit of two miles, it contains 233,380 inhabitants; and within nine miles of this centre of manufacturing industry, a million of people have concentrated their habitations. The county of Lancashire contained, in 1821, 1,052,200 inhabitants, and in

1831, the returns are given 1,335,000; showing a rate of increase of about 33 per cent in 10 years—which is fully equal to that of some of the most prosperous districts of the United States.

When all the machinery of the cotton mills is simultaneously stopped at the usual hours of intermission, to allow the laborers to withdraw to their meals, the streets of Manchester exhibit a very bustling scene; the side walks at such times being crowded by the population which is poured forth from them, as from the expanded doors of the churches at the termination of services on the Sabbath in the large cities of the United States. On first beholding these multitudes of laborers issuing from the mill doors, I paused to examine their personal appearance, expecting to behold in them the sickly crowd of miserable beings, so vividly described in Espriella's letters, as "keeping up the *laus perennis* of the devil, before furnaces which are never suffered to cool, and breathing in vapors which inevitably produce disease and death." In this respect my anticipations were disappointed; for the females were in general well dressed, and the men in particular displayed countenances which were red and florid from the effects of beer, or of "John Barleycorn," as Robert Burns figuratively called his favorite potation, rather than pale and emaciated by excessive toil in unwholesome employments in "hot task houses." Every branch of business being in a prosperous state when I had an opportunity of noticing them, they may have appeared, perhaps, under favorable circumstances, and in the possession of more than their usual share of comforts and enjoyments.* The children employed in the cotton mills appeared also to be healthy, although not so robust as those employed as farmer's boys in the pure air of the open country.

*

*The following statement of the prices of labor will show that the laboring classes were not in 1825 actually depressed to the lowest degree of wretchedness, as is by many supposed. The rate of wages at present is not materially lower than the sketch here presented.

To this list of prices of labor in England, has been added the prices of similar labor in France and the United States, together with the cost of coals and of wheat, which form the basis of manufacturing capabilities. By this Comparative Table, which has been formed with much care, the relative advantages for producing cheap manufactures, possessed by the three principal nations of the earth, may be estimated by those who are curious in the investigation of this subject.

Comparative Table of Rates of Wages, in England, France, and the United States.

Wages of	England. *s.*	*d.*	*cts.*	France. *cents.*		U. States. *cents.*	
A common day laborer per day, about	3	0	73	37 to	40		100
Do. do. with steady employment	2	6	60	35			80
Carpenter	4	0	97	55	75		145
Mason	4	3	103	62	85		162
Mule Spinners in cotton mills	4	3	103	75	85	90 to	137
" " in woollen mills	3	10	93	45	55	90	137
Weavers on hand looms	3	0	73	37	50	80	95
Boys 11 or 12 years old, per day	1	0	24	14	17	25	30
Women in cotton mills, per week	8	0	192	145	175	250	300
" woollen mills "	8	0	192				
Maid servants in private families, per week, board found	2	6	60			100	133
Machine Makers and Forgers, best, per day	6	6	158			150	167
" " ordinary	3	9	90			100	117
Children, piecers in mills, for mules and billeys	0	7	14			20	30
Overlooker of carding rooms	5	6	135			108	150
Slubbers of woollen roving	4	0	97			80	100
Experienced workmen to attend shearing machines and gig mills for woollens	3	4	82			80	117
Firemen for steam engines	3	9	91			100	125
						N. Y.	Pitsb.
1827. Price of Coals for steam engines, per ton	9*	0	220	700†		700	106
Wheat (per bushel of 60 lbs.)	7	4	179	117		96	49

**Manchester.* *†Louviers.*

When the work is done by the piece or job, and individuals are thus incited to greater efforts, the wages earned commonly exceed those noted in the table.

The abundance of wild and unappropriated lands in the United States, forms the certain resource of the mechanics, and indeed of all other classes of workmen, when thrown out of employment by any of the vicissitudes of business, and serves as a sort of balance wheel to regulate fluctuations in the prices of labor.

There appears to be a greater difference between the quantity of the necessaries of life which a laborer obtains for his day's work in England, and what a similar laborer obtains in the United States, than there is between the nominal pecuniary standard of value. With an equal amount of wages, the mechanic in the United States may purchase nearly double the quantity of bread and other provisions necessary for himself and family that the English mechanic can purchase in England. Beef now sells here at from 13 to 17 cents per pound; bread 5 cents per pound. The day laborer for his three shillings earns 16 pounds of bread and 5 pounds of beef. In the New-England and Middle States, the best pieces of beef sell for about 8 cents to 9, and bread 3¼ cents per pound. Such turkeys as may often be seen in the hand of the American mechanic, on his return from market, would cost here three dollars or more; and of course are beyond the means of most of the laboring classes in England. Provisions of the coarser sort are also much dearer in Manchester than in the United States, from being more generally consumed by the poor. A sheep's head and offal, which may be bought in the United States for 8 or 10 cents, and are there frequently thrown away for the want of a purchaser, will sell here for 30 to 50 cents.

On account of these high prices, the laborers are under the necessity of living here much more economically on a stinted and inferior fare. In consequence of the heavy taxes and tithes, and other exactions, many of the very manufactures of England are retailed in her own shops for home consumption, at higher prices than they are sold for in the shops of the United States. Various fruits, such as apples, melons, &c. which in England constitute the luxury of the rich, and are cultivated in green houses and under the shelter of lofty brick walls constructed expressly for this object, are so abundant in the United States as to be found almost equally on the tables of the poor and of the rich. The sum of the enjoyment within the reach of the mechanics of the two countries appears, therefore, to be greatly in favor of those of the United States.

The most highly colored sketches of the moral depravity and vices of many of the laboring classes of Manchester, fall short of the reality. A stranger, if he walk leisurely through some of the streets during pleasant evenings, is frequently addressed by abandoned females, who press their solicitations with earnestness, and even take uninvited possession of the arm, should the position of the stranger's bended elbow happen to offer a loop favorable to their design. Unless, indeed, peremptorily repulsed at once, they acquire assurance, and press their importunities with a shamelessness, that can only be the result of long practised habits of vice.

Rhode Island Society for the Encouragement of Domestic Industry

Transactions (1861)

By the middle of the nineteenth century, the village manufacturers of southern New England were quite aware that economic and technological events had eclipsed them. Though small "Rhode Island system" factories still produced much of the American cloth, the great manufacturing cities on the Merrimack and Connecticut rivers—Lowell, Massachusetts, especially—had gained national and international attention as the leading sites of industrial progress. In 1858 Nathan Appleton, one of the leading investors in the northern factory cities, sharpened the self-consciousness of the Rhode Island manufacturing community when he published a widely circulated pamphlet called *Introduction of the Power Loom, and Origin of Lowell.* Appleton focused on the technological accomplishments of Francis Cabot Lowell and Paul Moody. Lowell was founder of the Boston Manufacturing Company in Waltham, Massachusetts, where a power loom was successfully operated in the fall of 1814, and Moody was chief mechanic of the firm who added many improvements to early textile machinery. Though cotton manufacturing in America began almost a quarter-century before the experiment at Lowell, Appleton quickly passed over the early years of the industry, remarking that "the manufacture of cotton had greatly increased, especially in Rhode Island, but in a very imperfect manner." He discounted the technological contributions of earlier manufacturers, writing that a "vertical power loom" briefly in use in 1811 in the Taunton, Massachusetts, factory of Silas Shepard "did not promise success." To Francis Lowell, Appleton claimed, belonged "the credit of being the first person who arranged

Transactions of the Rhode Island Society for the Encouragement of Domestic Industry in the Year 1861 (Providence, 1862), pp. 76–85, 98–99.

all the processes for the conversion of cotton into cloth, within the walls of the same building."

The New England textile industry, of course, was first established in Rhode Island and adjoining areas of Massachusetts and Connecticut. The technology first introduced there was of British origin, as were the improvements Lowell brought back to New England at a later date. Appleton's brief history, in addition to neglecting the achievements of New England's mechanics and mill owners previous to 1814, failed to mention the debt owed to British technology by Lowell and his mechanic, Moody. The manufacturing interests whom Appleton slighted were not long to respond. The Rhode Island Society for the Encouragement of Domestic Industry determined to revive its own history and resolved that "the standing subcommittee on Manufactures be requested to procure information and report on the first introduction of the Power Loom and other machines for the manufacture of cotton in this state." A circular was sent out requesting "assistance and the collection of history, anecdote or story." Responses to the circular were collected and published as the society's *Transactions* of 1861. The most cogent and accurate of these are reprinted here. (TZP)

PROVIDENCE, May 28, 1861.

ELISHA DYER, Esq., Chairman of the Committee on Manufactures of the R. I. S. E. D. I.:—

Dear Sir,—Having received a circular from you, requesting information relative to the history of the introduction of Machinery for Spinning and Weaving Cotton, I have thought it might be useful to state some facts from memory, known to me.

My father, Judge Daniel Lyman, Samuel G. Arnold, Edward Manton, Jacob Dunnell, and Dr. Williams Thayer, formed a copartnership in 1807, and purchased a site on which there was a mill privilege, in North Providence, on the Woonasquatucket river, and built the Lyman Mill.

At this time, the cotton was spun only on water-frames, introduced by Mr. Samuel Slater; but soon the mule and throstle frame, from England, were brought over, and highly spoken of. The Lyman Company purchased two mules built by a Mr. Ogden, at his shop near Mill bridge, in this city. They were finished before the mill was ready, and brought out and placed in a barn on Judge Lyman's farm; they run with tin cylinders. The spindles were made in England, and they numbered one hundred and ninety-two in each mule. I believe these were the first pair built in this country. The basement in the factory when built, was used by Peter Cushman, as a machine shop; he having made a contract with the Lyman Company, to build eight throstle frames of one hundred and eight spindles each, in said shop. He also furnished the first cards, drawing and stretcher, for making Roving.

The Company, at the commencement, worked Sea Island cotton into No. 12 to 15 yarn. For several years, the cotton was given out in Providence, to be picked in various houses. Every Saturday, we sent a one-horse wagon and collected this cotton for the mill. Dr. Thayer attended to this branch. The price paid was four cents per lb., the cotton, waste and water making up the original weight sent. Of course, the introduction of the heavy fan and beater machine picker, about 1810, was of great advantage; inducing the Company to build a new building, water wheels, flūme and trench, to accompany this machine. At this time, the yarn spun, both warp and filling, was all reeled by hand; a major part was dyed on the premises, and sent to Joseph S. Cooke, agent, (he having purchased into the concern with S. N. Richmond, the former being chosen agent,) for the purpose, principally, with other duties, to have this yarn woven in the country, into cloth, shirtings, plaids, stripes, ginghams, and shambrays. The next improvement eagerly desired and anxiously waited for, was the loom to weave by power.

The Company was solicited by a Mr. Blydenburgh, to furnish him with means and room to experiment with a loom of his invention. After a full trial, at a heavy expense, it proved to be a failure. About the year 1816, Mr. William Gilmore, from Scotland, called to see Judge Lyman, to interest the Company to try his Scotch loom; bringing plans of the same, with those of a warper and dresser, saying, that the Messrs. Slater, in Smithfield, had given him encouragement to build the same, at their mill. He was disappointed, they having employed him to build a hydrostatic pump, to water their meadows.

His plans and conversation proved satisfactory in the interview; the Judge having visited the Messrs. Slater, at Smithfield, enquired into the character of Mr. Gilmore, and obtained their consent to employ him, and engaged him to move to the Lyman mill. The first operation of Mr. Gilmore, was to draft his warper, dresser and loom on the floor of a vacant room in mill No. 2, (since burnt,) full size; the next, was to make his patterns, and have them cast of iron and brass; in the meantime, the machinists were forging, turning and finishing the wrought iron, taking the measurement of the same from Gilmore's drafts on the floor, for twelve looms, one yard wide. They built the warper first, which run the yarn in sections on large creels, separated by iron wires, in circles about six inches apart. When one section was filled, the bell rung and they commenced another. These creels, when full, were placed at each end of the dresser; the yarn from all the sections taken through holes in a copper, through the lead sizing rollers, then bushed and wound on a loom beam in the centre, run by a diagonal shaft from the roller, and the take up was governed by a friction pulley on the beam. The dressing was a tedious and difficult operation. Mr. Gilmore, a thorough machinist, was entirely unacquainted with the practical operation, and the Company had no one, at first, to start the machinery; they began to grow discouraged; the warper worked badly; the dresser worse, and the loom would not run at all. In this dilemma, an intelligent, though intemperate Englishmen, by trade a hand weaver, came to see the machinery. After observing the miserable operation, he said, the fault was not in the machinery, and thought he could make it work; he was employed. Discouragement ceased; it was an experiment no longer. Manufacturers, from all directions, came to see the wonder. They obtained from Gilmore, for the sum of $10, the privilege to use his pattern for the warper, dresser and twelve looms. Several manufacturers in Pawtucket, made up a purse of $1500, and presented it to Gilmore. To this day, the same loom, with trivial alterations, is in use in all our mills; and now, instead of paying four cents per lb. for picking cotton, six cents for weaving shirting, eight and ten cents for plaid, and twelve and a half to seventeen cents for bed-tick weaving, we can pick the cotton, weave the cloth, and spin the yarn, all at the same time, at a trifling cost.

I am afraid you will consider this communication prolix, but wished to state the above facts which were known to me, having been on the spot while these operations were being made. If you think the above communication of sufficient interest to your Society, you are welcome to it,—otherwise, destroy it.

Yours, respectfully,

H. B. LYMAN.

P. S.—One other fact, occurs to me, since this communication was closed, which may be thought worthy of remembrance.

When we first started the power loom, the filling was prepared for the shuttle on Thorp's winding frame, from the skein, on to a bobbin; this bobbin was put into the shuttle when filled, being as long and of the same diameter as the cop. One of our observing mechanics, Joseph B. Harvey, thinking this tedious—of reeling and winding the filling,—of his own accord, fixed a spindle in the shuttle, and placed a cop from the mule on it, and this is the same process in use at the present time, without alteration.

WOONSOCKET, June 1st. 1861.

ELISHA DYER, Esq., Chairman of the Committee on Manufactures:—

Dear Sir,—I have no recollection of the exact time when Water Looms were introduced into Rhode Island, although I had been engaged for some years in manufacturing, when they were introduced, (as I commenced the manufacture of goods in 1811,) but I think they were introduced about the year 1819 or '20, by whom I cannot say. The upright, or perpendicular loom, was started at Pawtucket, by Mr. Ingraham, but did not answer the purpose, and the horizontal, superseded it entirely.

As I find by your circular that other information is sought for, it may not be improper to say, that the Double Speeder was first introduced into the State, by Mr. Palemon, Walcott, and myself. Mr. Walcott had been down to Waltham, Mass., and worked in the employ of the Waltham Company, until he had obtained the knowledge sufficient to enable him to build those machines, the Double Speeder and Drawing Frames. I took him in company, and we built a number of them fôr myself, and for sale. Knowing that they had patented them as Moody's invention, a Mr. Stephens, of Medway, had gone to making the same machines, when they commenced a suit against him, (or his Company,) and he called on us to make common cause with them, which we agreed to do, and raised among the manufacturers of Rhode Island, who had purchased the machines, several hundred dollars to pay lawyers fees for them. At our first appearing, Mr. Webster called us pirates, and said we had pirated their invention. But we only asked for time to show to the contrary, which was granted as we had discovered that the same machines were patented in England, in 1813. We sent to New York and obtained the volume of that year of the Patented Inventions of Great Britain, and in it was a perfect description of those machines. Mr. Webster was taken by surprise, and was very indignant with his employers.

I think Mr. Edward S. Wilkinson, of Pawtucket, would be as likely to give the information as regards the introduction of the loom, as any one, as he was brought up by his grandfather, Mr. Oziel Wilkinson, and of age at that time sufficient to remember a circumstance of that kind, the whole family being mechanics or manufacturers, and he of business talents; and, perhaps, much other information he might give which might be acceptable. And one other man of about the same age, Mr. James Brown, of Pawtucket, he and his former partner,

Mr. Larned Pitcher, were early in the machine making business, and have themselves made many valuable improvements in machinery.

But the most important machine of that early day, was the picking machine, invented by a Scotchman, by the name of Blair, if my memory serves me, till which time we had to pay four cents per pound for picking our cotton. A clear gain of that amount, as we had to scatter it over the country among the lower orders of Society, the peculation was equal to the expense of the labor of turning the then new machine. Should I hereafter, think of any one who would be likely to give you the desired information, as regards the loom, I will communicate it, or should I fall in with any one having such knowledge, I will inform you. Respectfully Yours,

WILLIAM HARRIS.

WASHINGTON, July 12th, 1861.

EDWARD HARRIS, Esq.

Dear Friend,—I received a circular from the committee of the Rhode Island Society for the Encouragement of Domestic Industry, asking for information in regard to the first introduction of power loom weaving in that State.

I deferred answering it for some days, in hopes of being able to give the exact dates, by referring to my books, which are at East Greenwich. I have been intending to take a trip to Rhode Island before this time, but have been detained from day to day, until I think some apology is due to your committee, for my not having answered more promptly. I well remember having made some of the first looms of T. R. Williams' patent, and some of Shepard & Thorpe's upright looms, also one dozen of W. Gilmore's patent. I can give a chapter of circumstances and events, but a narrative without precise dates would be useless. * * * * * * *

In haste, yours, &c.,

AZA ARNOLD.

WASHINGTON, September 6th, 1861.

Hon. ELISHA DYER, Chairman, &c.—

We are informed that the Differential Speeder, is claimed by Mr. Appleton, as a Waltham invention. But the author of Waltham inventions made no such claim.

No improvement on cotton machinery, appears to have been made at Waltham, up to 1826, except by Paul Moody, he was chief mechanician of Waltham, and claimed to be the inventor of the Waltham Speeder; he claimed eight improvements on the machine, but they proved not to be new.

Jonathan Fisk also built the same kind (Waltham Speeder) at Medway, and took five patents on the machine. William Hines, of Coventry, R. I., had made improvements on the Speeder and patented before them. And it is remarkable that the parts claimed by Moody, are the identical parts which are superseded by my compound motion, and were never used in a Differential Speeder. I shall refer to the case of (Moody *v.* Fisk) in a future page, to show that Moody's claim proves the Waltham Speeder to be essentially different.

Paul Moody took charge of the Lowell establishments, and Jonathan Fisk took charge of the Dover factory. And I shall show that neither Moody or Fisk, knew any method of compounding two different motions, and producing their differential, for four years after I had the machine in operation. I was well acquainted with Moody, saw his machines and considered that he improved the speeder by adopting the long flier; but the long flier was invented by Asa Gilson, at Dorchester.

I have used both Hines' and Fisk's speeders, and well remember the difference. If I exhibit a little egotism in this reminiscence, you will excuse it when you consider the local prejudice that was exercised against my machine as a Rhode Island invention.

I invented the Differential Speeder, and put it in operation in 1822, at South Kingstown, and it was soon in operation at Coventry, Scituate, Pomfret, and a dozen other places, but for three or four years it was discountenanced at Waltham, and Lowell; the Waltham Speeder being exclusively used in both places until I had constructed and put in operation the Great Falls factory, at Somersworth, N. H., which actually produced 30 per cent. more goods per week, than the Waltham or Lowell factories had produced, of equal quality.

This brought down the directors of the Lowell factories to our place at Somersworth, to enquire into the cause of so great a difference. It brought also Mr. Moody, their engineer, and Mr. George Brownell, the foreman of the Lowell machine shop: they also sent the celebrated mathematician, Warren Colburn, to see if our calculation was correct.

I had the pleasure of exhibiting and explaining all the minutiæ of the Rhode Island invention a third time, and the result was that Mr.

Colburn told Moody that it was mathematically correct, and that it was the only plan that he had heard of by which the machine could be made adjustable to all sizes of roping.

We notice the case, Moody *v.* Fisk, (2 Mason, Rep. 112,) tried at Boston, October term, 1820.

In the defence, it was proved that the improvements claimed by Moody, were not new, neither were they invented at Waltham. William Hines, of Coventry, had made improvements on the speeder, and patented in February, 1819, previous to Moody's date. Moody's patent was vacated for want of novelty. The object of referring to it, is to show that Moody's claim proves the Waltham Speeder to be a different machine from the Differential Speeder.

In summing up his claims, he says: " First,—I claim the position of the rolls. Second, the two upper cones. Third, the method of of moving the belt on the two lower cones, and that of communicating motion from the lower driven cone to the spindles, and all the mechanism and method of communicating motion from the upper driven cone to the arbors or axles of the endless screws, and perpendicular racks or screws that raise and lower the spindle rail. Fifth,—I claim the method and machinery by which the said motion communicated to the spindle rail is changed from an ascending to a descending motion, and the manner of connecting the same with the wagon carriage. Sixth,—the wagon, and the wagon carriage, gallows frame, catch wheel, the cycloid cam, slide lever and pulley shaft, which raises the belt on the upper cones, and all the similar parts that raise the belt on the lower cones, (except the cycloid or cycloid cam,) with all the parts, movements, and mechanism connected with the same. Seventh,—the flier tubes, and method of applying and using them. Eighth,—the rotary motion of the cams, and the intermediate gear work.

And further, I claim that these my inventions are applicable, not only to this machine which is adapted to one size of roping, but may be proportioned and applied to the making of any other kind of roping."

So, by his own showing, the Waltham Speeder makes but one size of roping. It is proper to remark that my compounding wheels supersede all the second pair of cones, cycloid cams, the cycloid racks, the second cone belt, and the method of moving the belt, which required to be brought up by ratchets and catches with teeth of different lengths, graduated to suit one size of roping, and which could not be used to make a different grade of roping, finer or coarser, but require another set of parts, graduated differently to suit any other size of roping, and this proportioning and adjusting of the machinery was required at each change from fine to coarse, or from coarse to fine.

The object of my inventing the Differential Speeder, was to do away with the intricate construction, and to simplify and extend the use of the machine, so that one set of gears can be adjusted to each and every size of roping by merely changing the pinions.

When Mr. Moody came to me for an explanation of my invention, we had a free and full discussion of its parts and properties. I remarked to him, that the exact difference between the retarding motion and a certain uniform motion would be always right for the accelerating motion. He seemed not to recognize the fact, and spoke doubtfully of it ; I then remarked, that the same cause that required the graduating of one, requires the graduating of the other, for both depend on the diameter of the roping. Therefore, I take the advantage of using this differential for the accelerated motion, rather than to use another pair of cones and belt fixtures ; but I have another more important advantage by so doing, that is, whenever it is required to alter the one graduation, the other always keeps right along with it, whatever may be the rate of change required, these motions are always reciprocal to each other.

Therefore, I use a rack with equal teeth for moving the belt, and move it by a pinion of any requisite number of teeth, so as to adapt the same machine to any size of roping by merely changing the pinions.

Up to this time, the Differential Speeder had not been seen at Waltham or Lowell, neither had the authors of Waltham inventions, taken the pains to investigate its merits. But after this, I had a cordial and good understanding with both Moody and Fisk.

I have subsequently been informed by Mr. Geo. Brownell, that soon after this interview, they commenced making my kind of gears at Lowell, and not only built my kind of speeders, but also took up their Waltham Speeders, and geared them over, and converted them into Defferential Speeders, by putting in my compound motion. This is a historical fact of some significance ; George Brownell, I think is still living at Lowell, and will confirm these remarks ; James Dennis, Gideon C. Smith, and Daniel Osborn, who were with us at Somersworth may perhaps recollect some of the circumstances.

While on the subject we may remark further, that the speeder, (fly frame,) had been used in England, but the compound motion or differential had never been applied to an English machine, until Charles Richmond carried to England a model of my wheels, (unbeknown to me.) He was there in 1824–5, when Mr. Houldsworth took up the subject of improving the fly frame.

Dr. Ure informs us that Houldsworth applied the differential system and patented it in 1826 ; that is, three years after the date of my patent. It was not requisite for him to claim it as his original invention.

I have been informed through a former partner of Charles Richmond, that the model which he carried to England, was made in Taunton, and was sold in England, and had since been patented there.

We said that J. Fisk did not understand producing and using the differential motion until three years after we had the machine in operation. It happened that J. Cowing, in describing my speeder, told Fisk that it had but one pair of cones and one cone belt. Fisk remarked, then it could not work. Cowing replied,—"but it appears to work right well, and makes more roping than the Waltham Speeder." Mr. Fisk then entered into argument, saying, "it is impossible to produce both graduations by one pair of cones and one belt, because while one is a retarding motion to vibrate the spindle rail, the other requires to be an accelerated motion for the winding up."

So, it was evident that he did not understand it, or he would not have made this assertion. If my Differential Speeder had ever been supposed to have been a Waltham invention, we should have heard of it during my three years contest with six corporations of Lowell, yet not a word of any such claim was offered, but on the contrary they tacitly acknowledged my right to the invention ; and after having the law repealed, thereby defeating my first claims up to that time, they then gave me $3500, for the right to use the same, for the last year of the term of my patent.

And this they did after searching all the evidence that could be found against my claim. Mr. Lyman, of Boston, who acted as their agent, who paid me the money and received the license for them, told me they found no evidence against it.

Few readers will take the trouble to understand the specific difference between two complex machines, but when one mode of operation enables the manufacturer to produce twenty per cent. more goods, (with the same cost of labor) than has before been done, it becomes of national importance.

Dr. Ure well remarks, that since the differential system has been adopted, manufacturers have been able to produce a better article at a less cost, and have thereby increased the trade.

I am, dear sir, most respectfully,

Your friend and servant,

AZA ARNOLD.

Statement of Peleg Wilbur, of Coventry.

The first of March, 1812, I was employed by the Washington Manufacturing Company, in Coventry, as agent for said Company. The cotton had been picked by hand by the families in that and the adjoining towns. About that time, some machines were introduced for picking cotton. The ingenuity of man was taxed, and the variety of productions were many, till they have arrived at something nearly perfect in its operation.

The drawing frames contained single heads, more or less; the drawing passed through four heads, each head on a separate frame in front of each other, and requiring four hands to attend the drawing, more or less.

The roping was made in tin cans; the frames contained twenty tin cans each, about the size and form of an old-fashioned wooden butter churn, and required one hand on the back side to put up the drawing, and generally two in front to attend it; the drawing passed through a trumpet into the can, through a brass nose that formed the top of the can; the can was turned by a pulley on the bottom of the can, with a belt passing horizontally over it. The can had a door on the side, held to when running, with a wire hoop that dropped down over it or around it; the turning of the can gave twist to the drawing as it passed into the can through the nose on the top. When the can was full, the hoop was slipped up, the door opened, the roping pressed down with the hand, taken out and thrown on the floor; from thence it was wound on to spools. Four blocks, as they were called, were required to a frame of twenty cans; these blocks supported a small drum about twelve inches in diameter, placed on a table, and each drum turned with a crank as a grindstone is turned; each drum turned one spool, requiring four hands to fill four spools at the same time.

About this time, Hines, Arnold & Co., got up what they called a bobbin-fly, for making roping; the flyer was not attached to the spindle. They made some improvement on that by gearing, and finally, soon produced what was called the Hines's speeder, that had a good reputation among the many inventions that succeeded it.

The frames for spinning warp yarn, before 1812, were called water frames,—sixty-four or eighty-four spindles in a frame, with separate heads for eight spindles; so that, when doffing the frame, only eight spindles were stopped at the same time, while the other parts of the frame were running. They were weighted very heavy, and required much power to operate them. All the gearing for machinery, at that time, was made of brass. About that time, the throstle frames were built by Hines, Arnold & Co., since which time there have been many inventions and great improvements.

In the year 1813, Hines, Arnold & Co., experimented on looms, without success ; others, tried their skill. I believe the perpendicular web was the most successful in its operation, however awkward it may appear. When Gilmore and Fales introduced their looms, all other inventions were thrown into the shade. Like all other machinery, the improvements on looms have been progressing.

Early among the inventions of power looms, was the bed-tick loom, got up by Job Manchester, now Rev. Job Manchester, of Providence, aided by Perez Peck, of Coventry. I believe Mr. Manchester invented the first bed-tick loom, which may have suggested the many plans since adopted for weaving twilled and figured goods.

The foregoing is a short sketch of my knowledge and experience in the manufacturing business. In describing some of the machinery in use at my commencement, I have stated the number of hands required in their operation, showing the saving of labor by improvements in machinery.

PELEG WILBUR.

Zachariah Allen

History of Cotton Manufacture in America (1861)

Zachariah Allen joined the effort of Rhode Island manufacturers to correct the historical distortions fostered by Boston investors like Nathan Appleton. While a number of his colleagues contributed the brief reminiscences reprinted in the *Transactions of the Rhode Island Society for the Encouragement of Domestic Industry* (RISEDI), Allen prepared a full-blown historical treatise that he offered to the RISEDI for publication in 1861. The manuscript was never published, for reasons which remain unknown, though the fact that Allen's study is largely derivative and ill-organized may explain why it was never seen in print. The section on power loom weaving, here edited and printed for the first time, is valuable, however, for it is based on Allen's intimate familiarity with the early textile industry of southern New England. In framing his account, Allen had available to him the reminiscences published in the RISEDI *Transactions* (1861), the original manuscripts of which survive in Allen's papers. (TZP)

The manuscript of Allen's "Historical, Theoretical, and Practical Account of Cotton Manufacture" is among the Zachariah Allen Papers, Rhode Island Historical Society; excerpts printed here by permission. The manuscript has been heavily edited. Redundancies have been silently deleted, paragraphing introduced, and punctuation regularized. Allen's spelling has been retained.

The delays, difficulties and great cost of weaving by hand looms delayed the progress of the cotton manufacture to such a degree, that it became a subject of general complaint among the Manufacturers in New England as it had previously been in Old England. Numerous experiments were continually made up to the year 1815, by ingenious mechanics in various parts of the United States, to *invent power looms* to supercede hand looms. In now looking back upon the past history of the successful introduction of Power Looms in England previously to the year 1809, it exhibits an extraordinary lack of vigorous enterprise among the cotton manufacturers of that time in the United States that they should have used no exertions to procure from that country either models of power looms, or drawings and workmen for constructing them. It was well known that such looms were successfully in operation there, from the reports of hand loom weavers, who immigrated in considerable numbers to America. Whilst the utmost powers of contrivance were excited to invent new ways of constructing power looms, from the rude descriptions gleaned from English hand loom weavers, an unaccountable apathy prevailed in regard to obtaining reliable information from publications within their reach. An ample and accurate description of the newly invented power looms with both horizontally and vertically arranged warps
1 was published in the year 1807 by John Duncan of Glasgow, with illustrative drawings of all the working parts, as well as perspective
2 views; including also an account of the new dressing machine.

On examining these drawings and explanations contained in this old but excellent "Treatise on Weaving". . . , it is truly surprising to find how complete in all its working parts the power loom was made early as the year 1807. The *protector* springs for stopping the loom whenever the shuttle failed to pass through the sheds of the warp, so as to preserve the yarns from being broken by jamming the entangled shuttle against them;—the ratchet for taking up the cloth as fast as woven;—the movements of the lay by both the plan of cranks as well as cams, the operation of the heddles, and of the other parts,—were all substantially as they are now found in common use, excepting the modern contrivance of a picker staff at each end of the lay, and of the finger Catch for stopping the loom when the filling yarn breaks. The speed of these looms was arranged at 80 or 90 beats of the lay per minute, but the author states that in some instances over one hundred beats were obtained.

In the plans of the dressers, therein described, are the fans for drying the yarn,—the vibrating brushes for brushing the warp yarns, the reeds or coppers for keeping them separated, and the sizing rollers and troughs heated by steam, all complete, so that a skillful mechanist could have constructed both the power loom and the dressing machine.

The copy of Mr. John Duncan's book on Weaving now before me, containing all the drawings and specifications necessary for constructing the different kinds of newly invented power looms and the dressers, was published in the year 1808, and was purchased in New York in the year 1816 of a bookseller who said he was glad to sell the book, for it had been on his shelves ever since its publication. During all this time the American Manufacturers were groping their way blindly in costly, experimental attempts to invent what had already been successfully accomplished in England, and actually published by an intelligent weaver for the benefit of the world, as stated in the well written preface to the work, wherein the author remarks that "an objection had been urged against his publication, because the Communication of this information might impart knowledge to persons out of the Kingdom whereby our manufactures might be rendered less productive."

In answer to this objection, the Author manifests a magnanimity and intelligence in regard to political economy, as well as skill in his profession, which has anticipated alike the results of the repeal by English statesmen of all their stringent laws against the exportation of machinery, and their cruel penalties of fines and imprisonment, for preventing the emigration of English workmen from the kingdom.

The reason for this indifference to adopting improved machinery in the United States during the period of the Nonintercourse Act, and of the subsequent war with England [1807–1814], is ascribable to the extraordinary profits of all manufacturing enterprises resulting from the exclusion of foreign fabrics. The manufacturers were too impatient of delays in waiting for new inventions or improvements. Notwithstanding, several poor and unaided mechanics struggled to bring into the world power looms of various forms.

It appears by the Reports of the Patent office that in March 1812, before the declaration of war against Great Britain, Mr. John Thorpe of Rhode Island secured his invention for an improvement in power loom weaving. Mr. Samuel Blydenburgh in the same year made several power looms for the Lyman Company, and Thomas R. Williams followed him the next year (1813). Mr. Elijah Ormsbee constructed power Looms near Providence. These looms were put in operation by cranks. The loom used by Thorpe was constructed with a vertical movement of the lay, the shuttle traversing over the surface of the reed, as had been practised previously in some of the England plans of power looms. All of these kinds of power looms appear to have been put into immediate operation practically in Rhode Island. Mr. Silas Shepard of Taunton, in reply to my enquiry on this subject, states, that in 1811 he constructed a power loom with a crank motion for the lay & cams for the heddles. The shuttle being thrown by a movement derived from the lay; but in the winter of 1812 Mr. John Thorpe came to Taunton, and in connection with him commenced making power looms for sale as well as fifteen or twenty for use in their own factory.

Several power looms were caused to be constructed in the year 1812 by the Lyman Manufacturing Company in North Providence; R.I., under the direction of Mr. Samuel Blydenburgh; but for want of the proper dressing of the yarn for these power looms, it appears that they failed of giving satisfaction. The cost of the labor for making these looms appears to have been fifty dollars each, as entered on the books of the Lyman Company, May 6th 1813.

In the arrangement for the operation of the first power looms, it appears that whenever the shuttle did not enter the box adapted to receive it, but became arrested among the warp threads, thus endangering the breakage of them, termed by weavers "a smash," resulting from the blow of the lay against the side of the shuttles whilst entangled among the warp threads, the loom was caused to be instantaneously stopped by a spring suddenly detached to throw out the single driving loom pulley from its connection by friction with the internal surface of a fixed pulley. This plan of detachment, which was the original one exhibited in Mr. John Duncan's drawing of the scotch loom published in 1808 is quicker and therefore better than that of the fast & loose pulley now in general use [1861], although less advantageous in starting the loom by its sudden impulse. The writer has noticed that in a new loom made at Manchester, N.H., this superior ancient plan is advantageously restored.

To lessen the risk of this occurrence of a "smash," the shuttle was thrown with great velocity, which caused it frequently to fly out of the loom, and to become a somewhat dangerous missile. It is narrated that a neighboring farmer, after selling his load of wood at Mr. Franklin's mill in Olneyville, where some power looms were used in 1813, indulged his curiosity to see them operate, by stepping into the Weaving room. Whilst he stood gazing with wonder at the novel spectacle before him the shuttle suddenly flew out and hit him in the head causing him to drop his whip in his haste to escape out of the door. Fearful of again entering into the room, he begged one of the weavers to hand it out to him as he said he was not used to power loom weaving.

Next in progress of time were introduced at Waltham by Francis C. Lowell power looms without cranks, which were operated by a cam to force back the lay, and to raise a weight of iron attached thereto by a strap of leather passing over a pulley within the frame of the loom. After the cam had driven back the lay, its form allowed the weight to descend and pull it forward to press the reed against the filling, shot in between the warp threads at each passage of the shuttle across them. The frames of these looms were of wood, and the velocity of the lay was limited by the gravitation of the weight; so that the number of beats per minute could not be increased beyond a certain limit, much less than that available from the use of the crank. This cam with the appended weights is exhibited in Plate XIV of John Duncan's Treatise on Weaving, published six years previous to the construction of the first power looms put in operation at Waltham, at the close of the year 1814.

Mr. Nathan Appleton in his statements relative to the "Introduction of the Power Loom," recently published by him [1858] gives the following enthusiastic account as an historical sketch of this interesting event in Massachusetts. "I well recollect the state of admiration and satisfaction with which we sat by the hour, watching the beautiful movement of this new and wonderful machine, destined, as it evidently was, to change the character of all textile industry."

As the statements published by Mr. Nathan Appleton were made at the formal request of some of his friends professedly to serve as a history of "the Introduction of the Power Loom," it cannot be deemed disrespectful to that worthy gentleman to review his statements before they are allowed to pass down to posterity as veritable historical details showing how much was accomplished by his own manufacturing company and how little was done by others.

The part which Mr. Nathan Appleton performed toward accomplishing the important enterprise of the Manufacture of cotton at Waltham, from his own account appears to have been confined to the mercantile department. "In the first instance I found an interesting and agreeable occupation in paying attention to the sales." For this reason his statements in regard to the mechanical department contain some erroneous representations of facts, that deprive them of the reliable character of an authentic history of the introduction of the Power Loom. He claims unqualifiedly for Mr. Lowell "the credit of being the first person who arranged all the processes for the conversion of cotton into cloth within the walls of the same building." Without extending the inquiry to broad manufacturing regions of Europe, it may be affirmed that the arrangements of all the processes for the conversion of cotton into cloth were made in the mill of Mr. Franklin near Providence, whereas Mr. John Waterman the manager of that mill has stated to the writer "some power looms were put in operation in the year 1813, and operated there several years. One girl attended two looms and made into good cloth of No. 15 yarn 240 to 275 yards per week, at the wages of three dollars per week." These looms the writer saw in actual operation at that time, and before the power looms at Mr. Appleton's Waltham mill were used in the autumn of the year 1814. By means of the unprecedentedly great capital of four hundred thousand dollars invested by the wealthy merchants of Boston in the Waltham cotton mill, a perfection as well as magnitude of operations was available, which rendered that establishment distinguished above all others for an unembarrassed course of success. This manufacturing enterprise appears to have been the earliest attempt made in the United States with ample funds, and the projectors of it merited the greatest praise for demonstrating what could be effected by the combination of Capital and skill. At the same time the merit of the first introduction of the power loom, and of the earnest and difficult struggles of manufacturers of humble means should not be overlooked, for the use of a single power loom, or even of a hand loom in any other mill where the yarn was spun and dressed, involved the necessity of "arranging all the processes for the conversion of cotton into cloth within the walls of the same building," to use the words of Mr. Appleton, the credit of which arrangement he emphatically claims as having originated from his factory at Waltham. In the large factory in Warwick, Rhode Island, owned by the Providence Manufacturing Company there were hand looms with fly shuttles in use in the year 1812, whereby the cloth was completed in the same establishment.

The kind of loom selected for operation by waterpower at Waltham was the most inefficient of those described and exhibited in the drawings contained in the work of Mr. John Duncan published five years previously to the commencement of the Waltham Mill and which was doubtlessly brought from England by Mr. Lowell on his return from his residence there for the purpose of obtaining information. The lay of their loom was drawn back by a cam and then pulled forward by the descent of a weight over a pulley to make the blow to beat in the filling; and was so much inferior to the crank loom, that it was speedily superseded [by] the latter, as acknowledged by Mr. Appleton. With the advantages derived by Mr. Francis C. Lowell from his observations in England, and from books and other sources, in relation to the manufacture of cotton, it cannot be doubted that his superior intelligence enabled him to profit by them practically in constructing the Waltham Mill.

Mr. Appleton states that "their greatest improvement was in the double speeder." This claim to originality of invention was enforced at the time by applications for patents, that were contested by other manufacturers as being, like their power loom, an English invention, and their claim was disallowed by a decision of Court after trials at law. In alluding to this claim for originality of invention, Mr. Samuel Slater states in writing to one of his correspondents (p. 243, *Life of Slater* [George S. White, *Memoir of Samuel Slater*]), "A certain cotton manufacturing company, who have been in the cotton business a few years only, have pretended to be the inventors of almost everything, and have taken out patents accordingly; but it is well known, that before they commenced business one of their brightest partners was in England for some time (cloaked as a merchant) obtaining information and workmen."

Mr. Appleton, after expressing his "feelings of satisfaction on the part I have performed in the introduction of the manufacture so important in every point of view to the interests of the whole country," concluded "to go to Rhode Island with a view of seeing the actual state of the cotton manufacture there, after the sudden conclusion of the peace with Great Britain in 1816," which he said "was ruinous to the manufactures generally." Accordingly he states that "on visiting the machine shop of Mr. [David] Wilkinson, a maker of cotton machinery in Pawtucket, RI, [he] took us into his establishment—a large one; all was silent. Not a wheel in motion, not a man to be seen, not a spindle running in Pawtucket except in Slater's old Mill. All was dead and still. He (Mr. Wilkinson) stated that during the war the profits of manufacturing were so great that the question never was asked, whether any improvement could be made in machinery, but how soon it could be turned out of his machine shop." "Mr. Lowell endeavored to assure the manufacturers that the introduction of the power loom would put a new face on the manufacture. They were incredulous;—it might be so, but they were not disposed to believe it."

Thus after complacently depicting his feelings of satisfaction in his concern at Waltham, he travelled out of his way to sneer at the less fortunate position of Mr. Wilkinson and of his poorer fellow manufacturers in Rhode Island, and to depict them all as lacking sagacity to appreciate the advantages of improved machinery and the energy to help themselves and, like the man in the fable, all praying to Hercules, instead of putting their shoulders to the wheel.

The visitors did not reflect that they came to see men whose plans had all been baffled by a sudden transition from a state of war and restrictions of importations, to one of peace and freedom and importations of foreign fabrics, and whose thoughts were only devoted to weathering the present storm like tempest tossed mariners on a troubled sea. Mr. Wilkinson was not a man lacking in ingenuity and enterprise, as proved by his original improvement of the slide engine for turning iron, which was before the gentlemen in his Machine shop without attracting their notice. By means of this engine, the steel chisel is held with the firm untrembling grip to cut off large chips of iron as readily as those of wood are pared off by a chisel held by the hand on a turning lathe. The effectiveness of this engine led to the subsequent adaptation of steel chisels to plane the flat surfaces of huge masses of iron. Mr. Wilkinson's ingenuity had then actually contrived a machine of such masterly power for putting refractory metals in desired forms, that it has to this day proved the most effective tool placed within the control of mankind for shaping refractory metals and for accomplishing the triumph of mind over matter. The slide engine is employed in the great machine shops of America and Europe.

Without the aid of the slide engine for cutting iron modern machinists would lack the ability of perfecting the mighty steam Engines for propelling huge ships of national armaments. As a recognition of the advantages of the use of his invention in the Navy yards and Armories of our country, after the lapse of half a century from the date of his original patent, a grant of ten thousand dollars was made by Congress in 1848 to Mr. Wilkinson to relieve him in his old age and poverty, and as a rare testimonial of National gratitude.

Nor was Mr. Wilkinson lacking in enterprise for modern improvements in machinery, as was verified by the fact that he speedily commenced making power looms with cranks which entirely superseded the Waltham loom by the perfection of their operation to this day.

In the progress of introducing the power loom into general use in New England, considerable delay occurred from the hesitation of the manufacturers in deciding upon the best kind of loom to be adopted. The plan of loom adopted at Waltham with its lay operated by the cam and weight, was not deemed desirable in comparison with the use of a crank loom. In this state of doubt and perplexity, the advent of Mr. William Gilmore from Scotland in the year 1816 with plans of the improved looms and dressing machines in use there, appeared to have effected as sudden an improvement in the finishing processes of dressing and weaving cotton cloth, as the advent of Mr. Slater produced in the preparatory processes of carding and spinning. Mr. Gilmore was first employed by the Lyman Manufacturing Company in North Providence, R.I., to build twelve looms—the following account of which is furnished by Mr. Henry B. Lyman; at the time coversant with the details. "At the Lyman Mill in North Providence, built in 1807, Sea Island cotton was at first used for making yarn No. 12 to No. 15, the picking of which was done by hand in families distributed over the country, at the regular price of four cents per pound. The cotton picker with beaters and fan was introduced in 1810, which greatly reduced the cost of this process. A new building with a water wheel was constructed for this picker."

"The yarn was all reeled by hand, the greater part dyed on the premises, and sent to the agent in Providence who put it out to be woven by hand loom weavers, in various parts of the Country, into plaids, ginghams, stripes, chambrays, and shirtings."

"The Lyman Company were early anxious to weave their yarn by power looms, and employed Mr. Blydenburgh to experiment with a loom of his invention, which failed after a heavy expenditure in the years 1812 and 1813."

"About the year 1816, Mr. William Gilmore from Scotland brought plans of the Scotch loom, Warper, and Dresser. He was employed at the Lyman Mill to build the machines; and drew the plans of full size on the floor of a vacant room. He next had patterns made for casting the frames of iron and the bearings of brass for 12 looms of 1 yard width. On the warper the yarn was wound on sections and the beams, when full, were placed at each end of the Dresser. The ends of the yarn from each section were drawn through holes in a sheet of copper and passed under the pewter rollers in the sizing trough, then cleared by the brushes, and wound on a loom beam placed in the Centre of the dresser, and turned by a diagonal shaft from the roller. The 'let off' was governed by friction of a rope on the head of each beam."

The dressing at first proved a tedious operation, as Mr. Gilmore was not acquainted with the practical operation of dressing yarn. An intelligent, but intemperate, Englishman was employed to make it work. Discouragement ceased. It was an experiment no longer. Manufacturers came from all quarters to see the wonder. Mr. Wilkinson of Pawtucket obtained from Mr. Gilmore for the sum of ten dollars the privilege of using his patterns for the warper, Dresser and twelve looms, and several manufacturers of Providence made up a purse [of] $1500 and presented it to Mr. Gilmore.

"Some of this identical pattern of looms with iron frames are still in use at the present day, with trivial alterations, in mills in Rhode Island. And now, instead of paying 4 cents per pound for picking cotton, 6 cents per yard for weaving shirtings, and 8 to 10 cents per yard for weaving plaids, and 9^{d} to 1/ per yard for weaving bedticks, principally by means of this power loom the cloth is made for about the former cost of weaving alone, including the cotton also."

The protracted delay in successfully starting the operation of power looms in the United States appears to be ascribable to the same cause—the want of a proper Dressing machine, which so long impeded the successful use of the power loom in England.

Various other additions have subsequently been made to the original power loom both in England and in the United States, for the purpose of improving the evenness of the weaving of the cloth and the speed of the process. . . . By means of all these improvements the speed of operation of the power looms has been gradually increased from about 100 beats per minute to 160 or 180 beats; but after a course of experimental trials in several cotton mills, it was practically found that about 180 beats per minute is a maximum for an economical result of economy in cost of production and of repairs; for the wear and resistance by friction, with loss of motive power, increases in the ratio of the squares of the increased velocities.

1. John Duncan, *Practical and Descriptive Essays on the Art of Weaving* (Glasgow, 1807–1808), 2 volumes.

2. "Dressers" or "slashers" are machines which coat woven thread with starch "sizing," to make it both stiff and smooth. This step was not required in hand weaving but proved necessary to enable cotton fiber to tolerate the vigorous and exact motion of power weaving machinery. Thus, the development of power weaving was the story of two machines, both the power loom *and* the dressing machine. Hence, Allen's emphasis on the "dresser."

III

Organization of the Mill Village

In order to exploit the new technology of power spinning and weaving, the first American industrialists had to alter the landscape and create a new kind and shape of community. The documents in this section explore the dimensions and patterns of these new communities and the entrepreneurial logic and cultural considerations that underlay their organization.

The section opens with excerpts from the most important early compilations of data about American industry: the federal censuses of manufactures conducted in 1810 and 1831. The documents suggest the geographical spread of the textile mill villages and also the kinds of information to be gleaned from such sources. Another group of documents—advertisements for mill villages—suggests the widespread turnover of management that occurred even as the number of mill villages was increasing. There follow clusters of documents—historical accounts, maps, ledgers, public records, company correspondence, and personal memoirs—relating to four villages, chosen both to indicate the variety of sources available for studying the process of organizing a mill village and to suggest some of the variety in their patterns of organization. Most striking are some of the differences in method and approach between the villages that grew up in the orbit of Providence, Rhode Island, and the village of Ware, Massachusetts, which was built by a Boston-based corporation. The section concludes with documents touching on some of the conflicts that arose in the organization of these new communities. Two legal texts deal with litigation over the regulation of water rights. A third document reports a struggle between townspeople and factory owners over the regulation of time. (MBF)

View of Slater Mill as it may have looked ca. 1840. From Richard Grieve and J. F. Fernal, *Cotton Centennial,* Providence, 1891.

Albert Gallatin

Census of Manufactures (1810)

Albert Gallatin, as secretary of the treasury, published the results of the first census of American manufactures in the midst of the industrial expansion occasioned by the Embargo of 1807, which kept English textiles out of American markets and stimulated American textile manufacture. Gallatin's report of mills in operation and projected for completion in 1811 indicates the spirited, if shallow, optimism of the period. The character of the industry, which was still primitive even by the standards of the next decade, is indicated by the number of diminutive mills with as few as ninety-six spindles or powered by horses and the absence of industrial weaving and the emphasis on domestic manufacture. Gallatin's statistics, important as they are, are incomplete and unreliable. For instance, the unnamed cotton manufactory tabulated in table (D) cannot be readily identified among the Rhode Island mills listed in lesser detail in table (C). Note in table (B) the preponderance of mills in New England and, within the region, the dominance of Rhode Island mills (including "Sundry towns adjoining" and the Connecticut towns of Pomfret and Sterling): Out of 87,000 spindles in sixteen states, 66,900 were attributed to New England and 56,500 of those were in the orbit of Providence. (MBF)

[Albert Gallatin], *Report from the Secretary of the Treasury, on the Subject of American manufactures, prepared in obedience to a resolution of the House of Representatives, April 19, 1810* . . . (Boston, 1810), pp. 9–15, tables (B)–(E).

COTTON, WOOL AND FLAX.

I. *Spinning Mills and Manufacturing Establishments.*

The first cotton mill was erected in the state of Rhode Island, in the year 1791 ; another in the same state in the year 1795, and two more in the state of Massachusetts, in the years 1803 and 1804. During the three succeeding years ten more were erected or commenced in Rhode Island, and one in Connecticut, making altogether fifteen mills erected before the year 1808, working at that time about eight thousand spindles, and producing about three hundred thousand pounds of yarn a year.

Returns have been received of eighty seven mills which were erected at the end of the year 1809; sixty two of which (48 water and 14 horse mills) were in operation, and worked at that time thirty one thousand spindles. The other twenty five will all be in operation in the course of this year, and together with the former ones (almost all of which are increasing their machinery) will, by the estimate received, work more than eighty thousand spindles at the commencement of the year 1811.

The capital required to carry on the manufacture on the best terms, is estimated at the rate of one hundred dollars for each spindle ; including both the fixed capital applied to the purchase of the mill-seats, and to the construction of the mills and machinery, and that employed in wages, repairs, raw materials, goods on hand and contingencies. But it is believed that no more than at the rate of sixty dollars for each spindle is generally actually employed. Forty five pounds of cotton worth about 20 cents a pound, are on an average annually used for each spindle ; and these produce about thirty six pounds of yarn of different qualities, worth on an average one dollar and 12 1-2 cents a pound. Eight hundred spindles employ forty persons, viz. five men and thirty five women and children. On those data, the general results for the year 1811, are estimated in the following table :

Mills.	Spindles.	Capital employed.	Cotton used.		Yarn spun.		Persons employed.		
No.	No.	Dollars.	Pounds.	Value.	Pounds.	Value.	Men.	Women & Children.	Total.
87	80,000	4,800,000	3,600,000	720,000	2,880,000	3,240,000	500	3,500	4,000

The increase of carding and spinning of cotton by machinery, in establishments for that purpose, and exclusively of that done in private families, has therefore been fourfold during the two last years, and will have been tenfold in three years. The table (B.) shews the situation and extent of those several mills, and that although the greater number is in the vicinity of Providence, in Rhode Island, they are scattered and extending throughout all the states. Those situated within thirty miles of Providence, are exhibited in the table (C.) and the statement marked (D.) gives the details of one of the establishments, as furnished by the proprietors.

The seventeen mills in the state of Rhode Island, included in the table (C.) which were in operation, and worked 14,290 spindles in the year 1809, are also stated to have used during that year 640,000 pounds of cotton, which produced 510,000 lbs. of yarn ; of which, 124,000 lbs. were sold for thread and knitting; 200,000 lbs. were used in manufactures attached to, or in the vicinity of the mills; and the residue was either sold for wick, and for the use of family manufactures, or exported to other parts. Eleven hundred looms are said to be employed in weaving the yarn spun by those mills into goods, principally of the following descriptions, viz.

Bed ticking,	sold at	55 to 90	cents p.	yard.
Stripes and checks,		30 to 42	do.	do.
Ginghams,		40 to 50	do.	do.
Cloth for shirts and sheeting,		35 to 75	do.	do.

Counterpanes, at 8 dollars each.

Those several goods are already equal in appearance to the English imported articles of the same description, and superior in durability ; and the *finishing* is still improving. The proportion of fine yarns is also increasing.

The same articles are manufactured in several other places, and particularly at Philadelphia, where are also made from the same material, webbing and coach laces, (which articles have also excluded, or will soon exclude similar foreign importations) table and other diaper cloth, jeans, vest patterns, cotton kerseymeres and blankets. The manufacture of fustians, cords and velvet, has also been commenced in the interior and western parts of Pennsylvania, and in Kentucky.

Some of the mills above mentioned are also employed in carding and spinning wool, though not to a considerable amount. But almost the whole of that material is spun and wove in private families; and there are yet but few establishments for the manufacture of woollen cloths. Some information has, however, been received respecting fourteen of these, as stated in table (E.), manufacturing each, on an average, ten thousand yards of cloth a year, worth from one to ten dollars a yard. It is believed that there are others from which no information has been obtained; and it is known that several establishments, on a smaller scale, exist in Philadelphia, Baltimore, and some other places.—All those cloths, as well as those manufactured in private families, are generally superior in quality, though somewhat inferior in appearance to imported cloths, of the same price. The principal obstacle to the extension of the manufacture is the want of wool, which is still deficient both in quality and quantity. But those defects are daily and rapidly lessened, by the introduction of sheep of the merino and other superior breeds, by the great demand for the article, and by the attention now every where paid by farmers to the increase and improvement of their flocks.

Manufacturing establishments for spinning and weaving flax, are yet but few. In the state of New

York, there is one which employs a capital of 18,000 dollars, and twenty six persons, and in which about ninety thousand pounds of flax are annually spun and wove into canvas and other coarse linen. Information has been received respecting two in the vicinity of Philadelphia, one of which produces annually 72,000 yards of canvas, made of flax and cotton; in the other, the flax is both hackled and spun by machinery, thirty looms are employed; and it is said, that 500,000 yards of cotton bagging, sail cloth and coarse linen, may be made annually.

Hosiery may also be considered as almost exclusively a household manufacture. That of Germantown has declined, and it does not appear to have been attempted on a large scale in other places. There are however some exceptions; and it is stated that the island of Martha's Vineyard exports annually nine thousand pair of stockings.

II. *Household Manufactures.*

But by far the greater part of the goods made of those materials (cotton, flax and wool) are manufactured in private families, mostly for their own use, and partly for sale. They consist principally of coarse cloth, flannel, cotton stuffs, and stripes of every description, linen and mixtures of wool with flax or cotton. The information received from every state, and from more than sixty different places, concurs in establishing the fact of an extraordinary increase during the two last years, and in rendering it probable that about two thirds of the clothing, including hosiery, and of the house and table linen worn and used by the inhabitants of the United States, who do not reside in cities, is the product of family manufactures.

In the eastern and middle states, carding machines, worked by water, are every where established, and they are rapidly extending southwardly and westwardly. Jennies, other family spinning machines, and flying shuttles, are also introduced in many places; and as many fulling mills are erected, as are required for finishing all the cloth which is woven in private families. (See note F. and statement G.)

Difficult as it is to form an estimate, it is inferred from a comparison of all the facts which have been communicated, with the population of the United States, (estimated at six millions of white and twelve hundred thousand black persons), that the value of all the goods made of cotton, wool and flax, which are annually manufactured in the United States, exceeds forty millions of dollars.

The manufacture of cards and wire is intimately connected with this part of the subject. Whittemore's machine for making cards has completely excluded foreign importations of that article. It will appear by the communication (H.) that the capital employed in that branch may be estimated at 200,000 dollars; and that the annual consumption amounted till lately to twenty thousand dozen pair of hand cards and twenty thousand square feet of cards for machines, worth together about 200,000 dollars. The demand of last year was double that of 1808, and is still rapidly increasing. But the wire itself is altogether imported, and a very serious inconveniency might arise from any regulation which would check or prevent the exportation from foreign countries. It appears however, by the communication (I.) that the manufacture may and would be immediately established, so as to supply the demand both for cards and other objects, provided the same duty was imposed on wire, now imported duty free, which is laid on other articles made of the same material. The whole amount of wire annually used for cards, does not at present exceed twenty five tons, worth about 40,000 dollars.

(B)

Statement of Mills for spinning Cotton, of which an account has been received.

State or District.	Town or Situation.	NUMBER.		SPINDLES.	
		In operation.	Erecting.	In 1809.	In 1810.
Maine,	Waldoborough,	1	-	150	300
New Hampshire,	New Ipswich,	2	-	1,200	2,000
..	Other Towns,		4	-	2,000
Massachusetts,	Near Newburyport,	1	-	200	200
..	Dedham,	1	-	192	900
..	Sundry towns adjoining the state of Rhode Island,	(c.) 8	-	4,820	7,500
..	Ditto,	(c.) -	5	-	13,000
Rhode Island,	Providence and its vicinity,	(c.) 17	-	14,296	22,900
..	Ditto,	(c.) -	7	-	7,600
..	East Greenwich,	1	-	500	1,000
Connecticut,	Pomfret and Stirling,	2	-	1,390	4,500
..	New Haven and Derby,	2	-	(e) 700	700
..	Killingly and Plainfield.	-	2	- -	3,600
Vermont,		2	-	260	350
do.		-	2	- -	350
New York,	Washington county,	1	-	608	700
..	Hudson,	1	-	(e.) 500	500
..	Whitestown,	1	-	200	300
..	Washington county,		1	- -	500
..	Dutchess county,		2	- -	1,000
New Jersey,	Patterson and Belleville	2	-	(e.) 500	500
Pennsylvania,	Near Philadelphia,	2	-	(e.) 500	500
..	Shippensburgh,	† 1	-	548	600
..	Pittsburgh,	† 1	-	300	300
Deleware,	Near Wilmington,	1	-	480	500
..	Ditto,	† 1	-	(e.) 200	200
Maryland,	Near Baltimore,	2	-	1,100	6,000
..	Ditto,		1	- -	5,000
..	Patuxent,		1	- -	300
..	Washington county,	† 1	-	300	300
Virginia,	Petersburgh,	1	-	96	500
South Carolina,	Charleston,	† 1	-	252	350
Georgia,	Louisville,	† 1	-	200	300
Ohio,	Cincinnati,	† 1	-	576	600
Kentucky,	Six several places,	† 6	-	700	3,000
Tennessee,	Nashville,	† 1	-	220	250
Total,	Spindles,			30,500	87,000

NOTES.

(e.) These are on estimate: the residue of the spindles for 1809, are from actual returns.

† All are water mills, except those marked thus †, which are impelled by horses.

(c.) For a detail of these, see statement (C.)

(C)

List of the Cotton Mills within thirty miles of the Town of Providence, November 14, 1809.

Year of Establishment.	FIRM.	WHERE SITUATED. State.	WHERE SITUATED. Town.	AGENT, OR PRINCIPAL OWNER.	Number of Spindles now in operation.		Number of Spindles which might be employed.	
1791,	Almy, Brown & Slater,	Rhode Island,	North Providence,	Almy & Brown,	1,150		1,500	
1795,	Warwick Spinning Mill Owners,	..	Warwick,	Ditto,	600		1,200	
1805,	Coventry Manufacturing Company,	..	Coventry,	Samuel Arnold,	1,692		2,000	
..	Union Cotton ditto,	..	Johnston,	Henry P. Franklin,	896		1,200	
1805--6,		..	Cumberland,	Walcott,	200		400	
1806,		..	..	Richard Waterman,	288		400	
..		..	North Providence,	Tiffans & Reed,	400		500	
..		..	..	Hosea Humphrey,	96		200	
1807,	Providence Manufacturing Company,	..	Warwick,	John K. Pitman,	2,190		3,200	
..	Warwick ditto,	..	..	Almy & Brown,	1,260		2,50C	
..	Hope ditto,	..	Scituate,	Thomas S. Webb,	1,584		2,500	
..		..	Smithfield,	Almy & Brown,	1,700		3,000	
..	Natick Manufacturing Company,	..	Warwick,	Adams & Lothrop,	1, 84		1,500	
1808,	Potowomut ditto,	..	..	Blodget & Power,	564		1,000	
..		..	South Kingston,	Cyrus French,	200		500	
..		..	Coventry,	Theodore A. Foster,	96		300	
1809,	Manchesier Manufacturing Company,	..	Warwick,	Caleb Greene, junr.	96		1,000	
						14,196		22,900
1803,	Pawtucket C. and C. ditto.	Massachusetts,	Rehoboth,	Ingraham,	1,240		2,000	
1804,	Samuel Slater & Co.	..	..	Samuel Slater,	1,344		1,500	
Within two years.		..	Canton,		500		500	
		..	Franklin,		400		400	
		..	Medway,	Blackburn,	528		600	
		..	Swanzey,	Dexter Wheeler,	180		500	
		..	Taunton,	Leonard,	500		1,500	
1808,		..	Rehoboth,	John Pitman & Co.	128		500	
						4,820		7,500
1805--6	Pomfret Manufacturing Company,	Connecticut,	Pomfret,	Smith Wilkinson,	1,200		2,500	
1808,		..	Sterling,	John Dorrance,	190		2,000	
						1,390		4,500
		Rhode Island,	Scituate.	Caleb Fisk,			2,500	
1809,		..	Johnston,	Daniel Lyman,			1,200	
..		..	Cranston,	Roger Williams,			1,000	
..		..	Smithfield,	Richard Buffum,			1,000	
..		..	Ditto,	Oliver Bartlett,			1,000	
..		..	Coventry,	Dutu Arnold,			500	
..		..	Cranston,	William Potter,			400	
								7,600
..		Massachusetts,	Mendon.	Butler & Wheaton,			10,000	
..		..	Attleborough,	Ebenr. Tyler,			1,000	
..		..	North Bridge,				800	
..		..	Mendon,				700	
..		..	Swanzey,				500	
								13,000
		Connecticut,	Killingly,	Walter Paine.			1,500	
		..	Plainfield,	Tyler.			1,500	
								3,000
						20,406		58,500

In addition to this list of mills, there are several intended to be erected the ensuing spring; for some the mill seats are already purchased. I am well informed that by next April, there will be upwards of 40,000 spindles in operation, and it is expected that the whole number above mentioned will be in operation in the couse of one year.

6 (Signed) THOS. COLES, *Collector*.

(D.)

Statement of a Cotton Manufactory, owned by   ***near Providence.***

Year of establishment.	Capital employed.		*No. of Spindles employed.*	No. of Looms employed.		Number of persons employed.				*Number of pounds of cotton annually consumed.*	Quantity and quality of yarn spun and cloth wove annually.					*Greatest number of spindles which might be employed.*	REMARKS.
						In the manufactory.		In neighboring private families.			YARN.		CLOTH.				
	The first year.	*At the present time.*		*In the manufactory.*	*In neighboring private families.*	*Males.*	*Females.*	*Males.*	*Females.*		*Number of pounds.*	*Quality.*	*Yards wove in the manufactory.*	*Yards wove in neighboring private families.*	*Quality.*		
	DOLLS.	DOLLS.								lbs.							
1806	20,000	56,000	960	9	70	24	29	50	75	40,000	35,000	1 and 2 for weaving, knitting, sewing, &c. &c.	10,000	58,000	Tickings, sheetings, shirtings, stripes, checks, ginghams, duck, coverlid, bagging, diaper, &c.	2,000	This establishment suffered much in the outset in being put to much expense by English workmen, who pretended to much more knowledge in the business than they really possessed. At present only two are employed, and Americans as apprentices, &c. are getting the art very fast. No dividend of profits is expected for a considerable time, and from want of experience in the durability of the machinery, &c. are at a loss to calculate what they may expect; they however calculate, under a proper care of government and the growing disposition of the American citizens to consume the fabrics, to make it a good business. *Providence, Aug. 31, 1809.*

(E.)

Statement of Manufactnres of Woollen Cloth, of which an account has been received.

STATES.	TOWNS.		CAPITAL.	*Number of workmen.*	*Number of yards annually made.*	*Species of goods.*	REMARKS.
New Hampshire,	New Ipswich,	1	10,000	8 to 10	not stated,		
Massachusetts,	Byfield, -	1	not stated	20	15,000	cloth, baize, flannel,	declining—good wool scarce.
Rhode Island,	Warwick, - -	1	9,000	28	10,000	cloth ¾ to ⅞ wide, sold for 83 cents to 3 dollars,	only 900 yards of first quality can be made, employs 26 looms—profitable.
ditto,	Portsmouth, -	1	3,000	not stated,	10,000	cloth worth 75 cents to 2,25,	profits 12 to 15 per cent.
Connecticut, - -	Humphreysville,	1	not stated,	ditto,	not stated	broad cloths and others,	wool of merino breed.
New York, - -	Poughkeepsie	1	do.	ditto,	ditto,	ditto,	ditto.
Delaware, - -	Brandywine,	1	11,000	12	8,000	ditto,	ditto.
ditto, - -	ditto,	1	not stated,	not stated,	5,000	coarse cloth,	
Maryland, - -	Baltimore,	1	20,000	20	15,000	some broad cloth,	
ditto, - -	ditto,			56 looms,		various descriptions,	all small establishments.
ditto, - -	Elkton,	1	16,000	20	8,000	cloth worth 1,50 to 5 dollars,	good wool scarce.
ditto, - -	Frederick,	1	not stated,	29	not stated,	woollen cloths and mixtures,	
Pennsylvania, -	Philadelphia,	2	do.	not stated,	27,000	cazinet. a mixture of wool and cotton,	
ditto, -	ditto,	1	sundry other small	establishments,			
ditto, -	Germantown.	1	not stated,			coarse cloth and cazinet,	wool purchased from skin dressers.

Documents Relative to the Manufactures of the United States (1833)

A federal census of manufacturing establishments was taken in 1810 (see the Gallatin documents in this collection) as well as in 1820 (see the Phoenixville documents). None was undertaken at the time of the decennial population census in 1830, in spite of the fact that the previous decade had been by far the greatest period of industrial expansion in the nation's brief history. To rectify the oversight, a special census of manufactures was authorized by Congress in 1832. The results were compiled under the direction of Louis McLane, secretary of the treasury, and published the following year in two volumes as *Documents Relative to the Manufactures in the United States.*

This work, informally referred to as "McLane's Report," is one of the most important sources of American industrial statistics before the mid-1800s. Valuable as they are, however, McLane's volumes are not without flaws and omissions. As the tabulation at the end of Caroline Ware's chapter (reprinted in this collection) indicates, McLane's figures do not square with those compiled by other interested parties. The note to the Sturbridge/Southbridge returns that follow indicates other problems.

From almost 2,000 pages of tiny print, the excerpts reprinted here only give a suggestion of kinds of information available. We include the questionnaire circulated, several examples of returns from individual proprietors, several examples of returns from separate towns,

[Louis McLane], *Documents Relative to the Manufactures in the United States, Collected and Transmitted to the House of Representatives,* volume 1 (Washington, 1833), pp. 933–934, 950–952, 1045–1046, 536–537, 540–541, 970–971, 976. Reprinted by Augustus M. Kelley, 1969. See also Samuel Slater's preface to the Returns from Rhode Island reprinted in this collection.

one portion of a summary of state statistics by town, and a summary of state statistics by industry. Since this census was conducted voluntarily, it is far from systematic, and the returns from some states are very sketchy. The most highly industrialized areas of southern New England are the most fully represented. (MBF/TZP)

Queries

1. County and town in which your manufactory is situated, and the *name* of the establishment?
2. Kind or description of the manufactory, and whether water, steam, or other power?
3. When established, and whether a joint stock concern; also, the *number* of spindles, looms, and other machines, respectively, and, if practicable, increase of machinery and operative capacity from time to time, designating when?
4. Capital invested in ground and buildings, and water power, and in machinery?
5. Average amount in materials, and in cash for the purchase of materials, and payment of wages?
6. Annual rate of profit on the capital invested since the establishment of the manufactory, distinguishing between the rate of profit upon that portion of the capital which is borrowed, after providing for the interest upon it, and the rate of profit upon that which is not borrowed?
7. Cause of the increase, [or decrease, as the case may be] of profit?
8. Rates of profit on capital otherwise employed, in the same State and county?
9. Amount of articles annually manufactured since the establishment of the manufactory; description, quality, and value of each kind?
10. Quantity and value of different kinds of materials used, distinguishing between foreign products and domestic products?
11. Cost in the United States of similar articles of manufacture imported from abroad, and from what countries?
12. Number of men, women, and children employed, and average wages of each class?
13. How many hours a day employed, and what portion of the year?
14. Rate of wages of similar classes otherwise employed, in the same State and county, in other States, and in foreign countries?
15. Number of horses or other animals employed?
16. Whether the manufactures find a market at the manufactory? If not, how far they are sent to market?
17. Whether foreign articles of the like kinds enter into competition with them at such place of sale? and to what extent?
18. Where are the manufactures consumed?
19. Whether any of the manufactures are exported to foreign countries? and if so, where?
20. Whether the manufacture is sold by the manufacturer for cash? and if on credit, at what credit? if bartered, for what?
21. Whether the cost of the manufactured article (to the manufactory) has increased or decreased; and how much in each year, from the establishment of the manufactory; and whether the increase has been in the materials or the labor, and at what rate?
22. The prices at which the manufactures have been sold by the manufacturer since the establishment?
23. What rate of duty is necessary to enable the manufacturer to enter into competition in the home market with similar articles imported?
24. Is any change necessary in levying or collecting the duty on such articles, to prevent fraud?
25. What has been the rate of your profits, annually, for the last three years? and if it be a joint stock company, what dividends have been received, and what portion of the income of the company has been converted into fixed capital, or retained as a fund for contingent or other objects, and, therefore, not divided out annually?
26. What portion of the cost of your manufactures consists of the price of the raw material, what portion of the wages of labor, and what portion of the profits of capital?
27. What amount of the agricultural productions of the country is consumed in your establishment, and what amount of other domestic productions?
28. What quantity or amount of manufactures, such as you make, are produced in the United States, and what amount in your own State?
29. If the duty upon the foreign manufacture of the kind of goods which you make were reduced to 12½ per cent., with a corresponding reduction on all the imports, would it cause you to abandon your business, or would you continue to manufacture at reduced prices?
30. If it would cause you to abandon your business, in what way would you employ your capital?
31. Is there any pursuit in which you could engage from which you could derive greater profits, even after a reduction of the import duties to 12½ per cent.?

32. Are not the manufactures of salt and iron, remote from the points of importation, out of foreign competition within a certain circle around them, and what is the extent of that circle?

33. Amount of capital, and what proportion the borrowed capital bears to that which is real?

34. What amount of reduction in the duties would enable the actual or real capital employed to yield an interest of six per cent.? and how gradual the reduction should be?

35. If minimums should be abolished, and the duty assessed upon the actual value of the imported article in the American port, what rate of ad valorem duty would be equivalent to the present with the minimum?

36. What would be the operation of this change upon the frauds at present supposed to be practised?

37. Proportion which the production by the American manufacturers bears to the consumption?

38. Extent of individual and household manufacture in the United States, and how much it has increased since the tariff of 1824?

39. Average profit of money or capital in the United States?

40. Average rate of wages?

Individual Returns

DOCUMENT 8.—No. 17.

Mill Merino, Johnston, Providence county.—Answers to Questions.

1. Mill Merino, situated in the town of Johnston, county of Providence, and State of Rhode Island.

2. A stone building, water power, used for manufacturing cotton shirting from the raw material.

3. Erected in 1812, added to in 1828; not a joint stock company; 2000 spindles, 60 looms, and all other necessary machinery; commenced with 600 spindles, and added to from time to time till last year.

4. Forty thousand dollars.

5. Twenty-five thousand dollars per annum.

6. No regular per centage can be calculated on. When the mill was built cotton was low, say 10 cents per pound, and goods high; and it was profitable. During the war cotton rose to 40 cents per pound, and many mills having on hand at peace large stocks of goods suffered severely. Since peace, the business has been good and bad from period to period; 1828 and 1829 bad, 1830 good. Now goods are very low and profits small.

8. Various, according to circumstances.

9. 1813, yarn sold, 1,500 dollars; 1814 and 1815, cloth, 10,000 dollars; 1816 to 1826, cloth averaging from 15 to 20,000 dollars; 1828 to 1831, 30 to 33,000 dollars. Cloth consisting of $\frac{7}{8}$ fine bleached shirtings, and $\frac{3}{4}$ negro shirtings.

10. 300 bales cotton of the growth of the United States.

11. Such goods as the mill now makes are not *now* imported to any extent. Many other coarse articles might now be made of cotton and sold, was it not for the *very heavy* importations, *yearly*, from *Dundee* and other places, of cotton bagging, burlaps, Osnaburghs, &c., all made from Russian materials and imported at a very low rate of duty. (See New York importations for a year or two past.)

12. About one hundred; 10 men from 3 dollars to 75 cents per day, 60 young women at 42 cents per day, 10 boys at 50 cents per day, and 20 boys and girls at 25 cents per day.

13. 300 days, averaging about eleven hours.

14. Laborers about 10 dollars per month, mechanics from 75 to 150 per day and boarded. Do not know price in foreign countries; they are known to differ, but all very low.

15. About 6.

15. Some sold at home; principal sales in the United States; and when sales have been dull have sold considerable quantities at Madeira, South America, West Indies, Africa, Brazil, and Canton.

17. They do, but the quality of our goods, being all made of American cotton, is preferred.

18. In all parts of the world where they are admitted.
19. See answer to query No. 16.
20. Cash, credit, or barter; but principally by agents, on credit, in the United States.
21. Goods are now sold for about the price paid for weaving, by hand, when the mill was built; power looms, &c., have reduced the cost very materially.
22. Prices have gradually been declining. In 1814, '15, and '16, goods that were selling for 30 cents and upwards, can now be sold from 6 to 11 cents; I am thinking they are about as low as they ever will be.
23. See answer to query No. 29.
24. No doubt extensive frauds are committed, particularly in importing woollen goods, I am thinking. The goods imported must finally be valued, by competent judges, when landed in the United States.
25. See answer to query No. 6.
26. Materials and labor about equal; profits little or nothing at this time.
27. From five to seven thousand dollars worth per year.
28. Am not exactly informed; many mills alter from time to time to suit the market.
29. Do not consider the duty of any consequence to the kind of goods we are now making, but, should the finer mills not be protected, many would no doubt make coarser goods, and reduce the business so low that many would be obliged to stop.
30. When, by the ruin of your business your capital is lost, it matters but little to think what you would do with it if you had not lost it.
32. Very much depending on cost of transportation.
33. The moving capital varies from time to time, depending on the ready sale and quick return for goods; we borrow when obliged to, and pay generally 6 per cent. interest.
34. See answer to query No. 29.
35. Should minimums be abolished, and ad valorem duties be substituted, probably invoices would be made to suit the times.
36. See answer to query No. 35.
37. A very large proportion of the coarse cottons.
38. Should suppose household manufacturing had not increased since the introduction of labor-saving machinery.
39. 40. Am not informed.

H. P. FRANKLIN, *for Mill Merino.*

PROVIDENCE, *March* 9, 1832.

DOCUMENT 8.—No. 18.

Steam C. Manufacturing Company, Providence, Providence county.—Answers to Questions.

1. City of Providence, county of Providence, Steam C. Manufacturing Company, Samuel Slater, owner.
2. Cotton; steam power.
3. Established in 1828, Samuel Slater sole proprietor, containing 1 picker, 1 spreader, 1 lapper and doubler, 32 cards, 2 drawing frames, 4 speeders, 2 stretchers; 18 mules containing 4,344 spindles, 4 dressers, spoolers, warpers, and 100 looms, will contain 12,000 spindles and 300 looms.
4. Capital invested in buildings and machinery, including iron foundry and machine shop, 130,000 dollars; floating capital 50,000 dollars.
5. 52,000 dollars in the cotton manufacturing department.
6. Answer in general report.
9. Manufacture ⅞ and 4-4 fine Sea Island shirting, say 370,000 yards of number 45 annually.
10. 100,000 pounds cotton, 700 tons anthracite coal, 1,000 gallons oil, leather, &c., &c.; wages paid in cash.
12. Men, 26, $1 12½ per day; women, 45, 50 cents per day; children, 68, 29 cents per day.
13. 12 horses.
15. None.
16. Goods are sold in New York, Philadelphia, Baltimore, and other markets.
18. In all parts of the United States.
20. Sold on a credit of 6 to 9 months.

Document 8.—No. 19.

Allen's Woollen Mill, Providence, Providence county.—Answers to Queries.

1. Providence, Rhode Island; owned by Zachariah Allen.
2. Woollen mill, operated by water power.
3. In the year 1822; sole proprietor ; 600 spindles; 21 looms for broadcloth. The water power capable of more than double the present work.
4. Ninety-four thousand dollars.
5. See answer No. 4.
6. Average profit has not been, on the fixed capital, two per cent.
7. The scarcity of stock forming the raw material of woollen fabrics.
9. About eighty thousand dollars the ensuing year.
10. About 40,000 pounds American wool, and 20,000 pounds foreign wool and yarn.
12. 27 men, $1 16 per day; 30 women, 3 dollars per week; 10 children, 30 cents per day.
13. About 11 hours; all the year.
15. 5 animals.
16. To New York and Philadelphia.
17. English broadcloths and a few French cloths.
18. Principally sent to the New Orleans market.
19. None exported.
20. On a credit of 8 months.
21. The manufacture of broadcloth has increased about one fifth in 5 years.
22. Various prices, from 6 dollars per yard to 1 50 per yard.
23. The present rate, if the raw material, wool, should be raised to supply the domestic demand.
24. If the cloths imported at the 1 dollar minimum were stamped, it would, probably, in a great measure check frauds.
25. It has not paid 3 per cent.
26. About two-thirds of the value in raw materials, and the remaining one-third for labor, &c.
27. About 9,000 dollars per annum domestic.
28. This is for you to find out.
29. It would break up the woollen manufactures throughout the United States.
30 Go to farming.
31 Might get a living by agriculture and starve by manufacturing.
32. Do not know.
34. This would depend upon the price of the raw material.
36. Probably greatly increased.
37. Probably as 7 domestic to 1 part foreign.
39. Probably less than 6 per cent. compound interest, for a given term of years.

Document 8.—No. 20.

Randal's Mills, North Providence, Providence county.—Answers to Questions.

1. County of Providence and town of North Providence; Randal's mill.
2. Candlewick is manufactured by water power.
3. Established in 1814; contains 468 spindles with all the necessary machinery for manufacturing candlewick.
4. About 6,000 dollars.
5. 600 pounds of New Orleans cotton per week, at 13 cents, is - - $78
Amount of wages per week - - 32
Per week - - - - - - $110
9. 26,000 pounds of candlewick at 24 cents is 6,240 dollars.
12. 3 men, 3 women, and 7 children.
13. 12 hours each day the year through.
16. In different parts of the country, from Maine to Georgia.
17. I never knew that any candlewick was ever imported.
18. In different parts of the United States.
19. I once made a lot of 8,000 pounds to go to South America.
20. Sold on a credit of 4 months.

21. In 1814, '15, and '16 sold all I could make at 75 cents per pound, but now it is difficult to sell at 24 cents.
22. Answered above.
27. About 4,500 dollars worth of agricultural productions annually.
28. I do not know.
29. My business must be abandoned if the duties are reduced, as it is with difficulty that I now support myself with all the economy I can use.
30. If my business is abandoned I shall consider my capital totally destroyed.
31. Yes; beggary.
32. I do not know.

STEPHEN RANDAL, Jr.

DOCUMENT 9.—No. 41.

Pomfret Manufacturing Company, Maull & Co., at Pomfret, Windham county, Connecticut.

HUMPHREYSVILLE, Con., *15th November*, 1831.

SIR: The Committee appointed by the Convention held in the city of New York, on the 26th of October last, "to report on the production and manufacture of cotton," solicit your early answer, addressed to the subscriber, to the following inquiries:

1. Where is your establishment situated? Pomfret, Windham county, Connecticut; firm, "Pomfret Manufacturing Company."
2. What is the amount of your whole capital employed? 100,000 dollars.
3. How many mills, and when were they built? Two; one erected 1806 and one 1824.
4. What number of spindles, mules, and throstles have you in operation? 3,072 spindles, viz. 8 mules, 1,440, and 24 throstles, 1,632.
5 What number of looms have you in operation? Seventy-six.
6. What is the average number of the yarn you spin? No. 15.
7 How many pounds of yarn do you sell annually, and what number is it? None.
7. How many yards and pounds of cloth do you make annually? 549,186 yards, 167,870 pounds.
9. How many male persons do you employ, and what is the average of their wages per week? Fifty-six; average 5 dollars each per week.
10. How many females do you employ, and their average wages? Seventy-five; average, $1 80 each per week.
11. How many of those employed are under twelve years of age, what is the average of their wages, and what portion of the year are they allowed for schooling? Twenty; 90 cents per week; one third of the year.
12. How many persons live on your premises, and are dependent on your establishment for their support? Three hundred.
13. How many pounds of cotton do you manufacture annually? 186,522 lbs.; 530 bales from account.
14. How much starch and flour do you use for sizing? 10,000 lbs starch; no flour.
15. How much wood and coal for fuel? 150 cords wood; no coal.
16. How much oil do you consume? 1500 gallons.
17. Give the value of articles used in your manufactory, not herein enumerated. 1,700 dollars.*

* *Memorandum of articles not enumerated in the questions, consumed annually.*

Article	$	Article	$
Iron	$200	Clothing for cards	150
Steel	30	Hemp cords	7
Coal	100	Tin boxes, oil cups, and baskets	25
Baling cloth, twine, and paper	344	Files	10
Reeds	50	Spools and bobbins	50
Pickers	200	Lamps	10
Shuttles	125	Wire	6
Brooms and brushes	30	Wood screws	7
Picker. string, lacing, belting, and roller leather	220	Lumber for repairs	25
Glue, sand paper, and tacks	30	Castings for do	25
Cloth for sizing, rolls, &c.	20	Nails, &c. for do	30
Hand cards	6		$1,700

In answer to question 17th. In this estimate we have included materials and stock necessary to keep the machinary and utensils in constant operative repair, extraordinaries excepted. Three workmen for repairs being included among our hands, being one man to about 1,000 spindles.

18. If you employ hand weavers, please to state the number employed, the quantity and kinds of cloth woven, and such other particulars respecting them as may be of use to the committee. None

19. If you make cotton machinery, please to state the amount of capital and number of hands employed; also, the quantity you can build annually, designated by the number of spindles; with such other information as you may judge will be useful to the committee. We have help to repair our machinery, but do not build any.

20. If you have bleach, print, or dyeing establishments, give such particulars respecting them as you may think may be useful, taking the preceding queries for your guide. In these it may be well to designate the quantity of foreign articles used. We had formerly dyeing and bleaching establishments, but have abandoned both.

Please to give any information you may possess which you shall think will be useful to the committee, although not included in these queries. If you have any mills contracted for, but unfinished, you will please to include them in your return, stating the extent of them, and the time when you expect to get them in operation. Not any.

The object of these inquiries is to enable the committee to exhibit to the Government and people, an accurate statement of the cotton manufacture of the United States, so that the magnitude and importance thereof may be justly appreciated. With this object in view, they are anxious to make their report to the Central Committee by the first of January, in season to have it published during the next session of Congress.

Please have the kindness to fill the foregoing blanks and return this letter to me by an early mail.

Very respectfully,
JOHN H. DE FOREST.

SMITH WILKINSON, Esq., *Pomfret.*

POMFRET, *November* 26, 1831.

DEAR SIR: It fell to my lot to become acquainted with the cotton business at a very early period, having commenced work at ten years old, in a small mill of only 72 spindles, erected by Mr. Samuel Slater, at Pawtucket, R. I., in 1791, under the patronage of Messrs Brown and Almy, of Providence, and worked one year at four shillings per week. At that time American cotton was scarcely known at that place; Mr. Slater used Cayenne and Surinam cottons chiefly. Captain James Brown, of Warren, R. I., employed in trading to North and South Carolina, brought my father, Mr. Oziel Wilkinson, late of Pawtucket, about 40 lbs. picked out of the boll, and in the seed, which he got ginned at Providence by a man named Joshua Langley. Mr. Slater thought at that time that American cotton was not fit to work. I have but one item more to add, by way of opinion, which is the result of twenty-five years experience, during which time I have had the entire care of this establishment, that, so far from manufacturing business lessening the consumption of foreign goods, *it increases it,* so far as respects all who derive their support from the business, *more than three-fold,* as their *wants* keep pace with their *means;* which was adverted to in Mr. Rush's report, when Secretary of the Treasury, when he observed, the way to increase the *consumption* was to increase the *means.* We usually hire poor families from the farming business, of from four to six children, and from a knowledge of their former income being only the *labor of the man,* say 180 to 200 dollars, the wages of the family is usually increased, by the addition of the children, to from 450 to 600 dollars. They all live better than before; some spend all, while others lay up from 75 to 200 dollars per year, to a single family. Single weavers, females of steady, frugal habits, save from 40 to 60 dollars per year. We are holding the savings of from 10 to 15 such, on interest, say 50, 75, 200, 250, and some 300 dollars each. Our greatest difficulty at present is, a want of females, women and children; and from the great number of factories now building, have my fears that we shall not all be able to operate all our machinery another year.

Yours, respectfully,
SMITH WILKINSON, *Agent P. M'g Co.*

JOHN H. DE FOREST, Esq.,
Humphreysville, Con.

Returns by Town

The pages of McLane's Report for the Massachusetts towns of Sturbridge and Southbridge indicate some of the problems in interpreting these documents. Southbridge, which was established in 1816 from parts of the towns of Sturbridge, Charlton, and Dudley, was one of the most heavily industrialized communities in Massachusetts in the early nineteenth century. The switching of town boundaries and the rapid change in ownership of the factories listed by McLane make it difficult to understand their history. According to McLane's Report, Sturbridge contained a single manufacturing village listed simply as, "Cotton Manufacturing Company, Josiah J. Fisk, agent." A second village, however, was on the border between Sturbridge and Southbridge and is listed here in the latter town under the name of S. D. Plympton & Co. See the 1831 map of Sturbridge and Southbridge for clarification.

Samuel Luther Newell, whose personal inventory is included elsewhere in this volume, was a major partner in the "Columbian Manufacturing Company" of Southbridge. (See also the regulations of that company reprinted in this volume.) One of the mills that became the "Hamilton M'g Co." appears in Edward Alexander's Globe Village painting. And it was "Hamilton," of course, not "Mamilton" as the census reports it. (MBF/TZP)

Branch of business. When begun. Names of proprietors.	Value of real estate, buildings and fixtures occupied and used for the business. What kind of power, water, steam, or animal.	Value of tools, machinery and apparatus other than the fixtures. Number and value of horses & other animals used in the business.	Value of average stock on hand, and in the process of manufacture.	The different kinds, with the quantity and value of each kind of materials used per annum, which are produced in the United States, and of agricultural and other domestic productions consumed per annum.		At what places, or in what States such materials and products are principally produced.
Columbian Manufacturing Company; cotton; begun 1821; E. D. Ammidown, Moses Plympton, S'l Hartwell, Stillman Plympton, Wm. Healy, & Joseph Congden	$12,000	$24,000	$3,000	Cott'n, 128,000 lbs.	$13,375	Sou'rn States
				Starch, 4,200 lbs.	266	Philadelphia
				Oil, 261 gallons	240	New England
				Wood, 200 cords	350	In town
				Leather	75	Vicinity
				Cast iron, 100 lbs	50	Connecticut
				Other materials	500	
S. D. Plympton & Co ; cotton; Jas. Walcott, Samuel A. Groves, Sam'l H. Judson, E D. Plympton; begun about 1814, but by other owners	10,000	15,000	2,000	Cotton, 80,000 lbs.	83,000	Same as above
				Starch, 4,653 lbs.	275	
				Oil, 175 gallons	150	
				Wood, 125 cords	188	
				Leather	50	
				Cast iron	50	
				Other materials	200	
*Harvey Dresser& Silas H. Kimball; cotton; begun about 1814, but by other owners	10,000	15,000	2,000	-		-
Mamilton M'g Co ; woolen; begun in 1815, by different owners, and for cotton manufacture; Wm. Sayles and Austin Hitchcock	15,000	2,000	50,000	Wool, 120,000 lbs.	90,000	New England & Mid. States
				Pot & perl ash 1,500 lbs.	75	
				Woad, 2,400 lbs	120	Connecticut
				Allum, 500 lbs.	25	
				Sperm. oil, 1,000 gls.	1,000	N. E'd States
				Soap, 13,000 lbs.	1,400	Massachusetts
				S'g, 600 bls.	450	In town & vi'y
				Teazles, 800,000	1,000	Connecticut
				Wood, 900 cords	1,000	In town
				Glue, 5,540 lbs.	825	Massachusetts
2 establishments for manufacture of cot'n Batting, owned by Larkin Ammidown, and Sam'l C. Fisk	2,000	1,200	600	Waste, 92,000 lbs.	2,800	From c. fact's
				Oil, 110 galls.	92	New England
				Wood, 50 cords	75	In town
Shoe making	1,600	200	500	Leather, thread, and pegs,	4,000	2-3 Mass. 1-3 N. York
Cutlery, by Henry Harrington	1,000	150	-	-		-

EXPLANATIONS, REMARKS,

* This establishment has heretofore produced about the amount of the last above, (S. D. P. & Co ,) but is calculated the present year to be enlarged to do the amount of the Columbian Manufacturing Company; and in the statement previously made has, under the differ-

Buildings, &c., Materials used, Manufactures, Markets, Workmen, Wages.

The different kinds, and the quantity and value of each kind of foreign materials.	The different kinds, and quantity and value of each kind, of articles annually manufactured.	What proportions are disposed of in this State. What proportion in other States, and where. What proportion in foreign countries, and where.	Average number of males over 16 years old employed, and rate of wages per day, they boarding themselves.	Average number of boys under 16 years of age, and rate of wages per day, they boarding themselves.	Average number of women and girls employed, and wages per day, they boarding themselves.
-	Sheetings 366,886 yds. $34,150 Yarn, 480 lbs. 126 Batting, 4520 lbs. 340 Wicking, 506 lbs. 10 Twine, 150 lbs. 47 Waste, 5,360 lbs. 149	2-3 in Mass., 1-3 in N.Y., Ph'a &c. In this State	11 at 95c. $3,239 50	20 at 25c. $1,550	43 at 34c. $4222 20
-	Sheeting, 225,000 yds. 20,800 Yarn, 600 lbs. 150 Waste, 6,000 lbs. 160	As above In this State	3 at 1 00 930	11 at 40c. 1,364	36 at 32c. 3,565
-	Sheeting, 36,600 yds. 34,100 Waste, 6,000 lbs. 160	-	11 at 1 00 3,410	20 at 30c. 1,860	40 at 36c. 3,464
Indigo, 3,600 lbs. $4,500 Dye Woods, 30,000 lbs. 3,750 Madder, 4,000 lbs. 1,000 Vitriol, 100 lbs. 10 Olive-oil, 3,000 galls. 3,000	Broadcloths, 50,000 yds 150,000	3-4 in this State, 1-4 in N. York, Ph'a, & Hartford	36 at 1 00 11,160	8 at 25c. 320	60 at 37½ 6,975
-	Batting, 69,000 lbs. 4,800	New York	2 at 75c. 465	5 at 30c. 465	3 at 25c. 232 50
•					
-	Boots and shoes, 8,000	Massachusetts	10 at 1 00 3,100		
-	Surgical instruments, Rogers' Knives, &c. 1,500	-	2 at 1 25 930		

AND ADDITIONAL FACTS.

ent heads, not here stated, been returned the same as the Columbian Manufacturing Company.

68

SAMUEL WARD.

Branch of business. When begun. Names of proprietors.	Value of real estate, buildings, and fixtures, occupied and used for the business. What kind of power, water, steam, or animal?	Value of tools, machinery, and apparatus, other than the fixtures: Number and value of horses and other animals used in the business.	Value of average stock on hand, and in the process of manufacture.	The different kinds, with the quantity and value of each kind, of materials used per annum, which are produced in the United States, and of agricultural and other domestic productions consumed per annum.	At what places, or in what States, such materials and products are principally produced.
STURBRIDGE.					
Cotton manufacturing company; Josiah J. Fisk, agent	$27,000	$30,000	$2,500	80,000 lbs. cotton, at 12½ $10,000	New Orleans
				Starch, 400	Philadelphia
				Oil, coal, wood, leather, brushes, brooms, &c. &c. 875	Massachusetts
Josiah Hobbs, tanner and currier	1,500	150	2,000	Hides and skins, 1,550	Massachusetts
				60 cords bark, 300	
				5 bbls. oil, 80	
				200 lbs. tallow, 20	
				8 hhds. lime, 20	Rhode Island
CHARLTON.					
Prentiss Whiting	1,500	75	500	Leather, 1,000	Massachusetts
				Woollen cloth 200	
				Hardware, 200	
				Oil, paints, and varnish, 300	
				Laces and worsted, 75	
				Fuel and oil, 50	
				Curled hair, tacks, screws, thread, &c. 25	
Asa Brown, tanner and currier	2,000	200	2,000	Hides and skins, 2,500	Massachusetts
				60 cords bark, 360	
				6 bbls. oil, 96	
				400 lbs. tallow, 40	
				10 casks lime, 25	Rhode Island
Boot & shoemaking	1,000	50	300	Leather, thread, and pegs, &c. 1,750	Massachusetts

The different kinds, and the quantity and value of each kind, of foreign materials.	The different kinds, and quantity and value of each kind, of articles annually manufactured.	What proportions are disposed of in this State. What proportion in other States, and where. What proportion in foreign countries, and where.	Average number of males over 16 years old employed, and rate of wages per day, they boarding themselves.	Average number of boys under 16 years of age, and rate of wages per day, they boarding themselves.	Average number of women and girls employed, and wages per day, they boarding themselves.
- -	445,000 ys. cotton cloth at 6½ cts. per yard, $28,925	Sold in Taunton, Mass. for calico printing	11 at 6s. is $3,410	10 at 2s. is $1,033 33	50 at 42c. $6,510
- -	Leather manufactured, 3,500	Mass. & Con.	3 at 6s. 930		
Wrought iron, $150 English saddlery, 50	Manufactures, 5,500	Mass. and Rhode Island	4 at 6s. 1,240		
- -	Leather manufactured, 5,000	¾ in Mass. ¼ in N. York	4 at 6s. 1,240		
- -	Manufactures 3,000	New York	4 at 5s. 1,033 33		

SAMUEL WARD.

DOCUMENT 8.—No. 41.—*Schedule of Manufactures, &c., in Rhode Island, taken*

Location.	Name.	Description of manufactory.	When established.	Joint stock company or not.	Water or steam power.	Number of spindles.	Number of looms.	Capital.
Providence county.								
Scituate -	Hope mill	Cotton	1806	Joint	Water	3400	84	90000
Johnson -	Mill Merino	do	1813	Not	do	2000	60	40000
Scituate -	Fiskville mill	do	1812	Joint	do	1712	40	40000
Cranston -	Bellfonte	do	1812	do	do	600	22	30000
Cumberland -	White Stone mill	do	1822 / 1828	do	do	2500	60	60000
Smithfield -	H. A. Scott & Co.	do	1827	do	do	624		9000
North Providence	Randall's mill	do	1814	Not	do	468	-	6000
Cumberland -	Carrington mill	do	1831	Joint	do	4500	108	100000
Smithfield -	Philip Allen	do	1813	Not	do	4300	98	8 680
Scituate -	Richmond Manu'ng Co.	do	1814	Joint	do	2292	48	70000
Johnson -	Union factory	do	1805	do	do	1600	40	65000
Smithfield -	Jenks's mill	do	1823	Not	do	2800	54	25000
Do -	Central Falls mill	do	1825 / 1829	do	do	2520	58	55000
Do -	Almy, Brown, & Slater	do	1817	Joint	do	9500	225	240000
Do -	Georgia Manuf'ing Co.	do	1815	do	do	3880	104	100000
Do -	Thomas Sprague	do	1826	Not	do	1834	36	40000
Do -	Israel Arnold & Co.	do	-	Joint	do	656	16	15000
Do -	Albion mills	do	1823	do	do	7100	160	142000
Do -	Valley Falls mills	do	1823	do	do	3442	70	64840
Do -	Olney Manufact'ing Co.	do	1830	do	do	672	10	11000
Do -	Crook Falls Co.	do	-	do	do	316	4	6320
Do -	Branch factory	do	-	do	do	950	20	20000
Do -	Kelley mill	do	1812	Not	do	1200	34	30000
Do -	Lansdale mill	do	1831	Joint	do	4000	108	80000
Do -	Thread Manufa'ing Co.	do	1824	do	do	2500	40	66500
Do -	R. Richards & Co.	do	1830	do	do	1992	56	45000
Do -	Central Falls, W. Room	do	1830	do	do	-	30	18000
Do -	Gardner & Marchant	do	1831	do	do	6500	183	130000
Kent county.								
Warwick -	Natick mills	do	1820	Not	do	13000	326	250000
Do -	Rhodes Natick	do	1807	Joint	do	4000	100	100000
Do -	Lippitt Manufac'ng Co.	do	1810	do	do	4200	108	110000
Do -	John H. Clark	do	1827 / 1830	Not	do	1262	28	19000
Do -	Crompton mills	do	1823	Joint	do	9396	263	2/4300
Coventry -	Arkwright mills	do	1810 / 1820	Not	do	4500	150	125000
Washington co'ty.								
North Kingston -	Bellville mill	do	1814	Joint	do	2000	50	40000
Newport county.								
Portsmouth -	Enterprise	Cotton & wool	1815	do	do	240	none	14550
Newport -	Newport steam factory	Cotton	1831	do	Steam	-	-	48000
Providence county.								
Providence -	Steam Cotton Manu. Co.	do	1828	Not	do	4344	100	130000
Do -	United Manufac'ing Co.	do	1812	Joint	Water	1152	36	60000
Burrellville -	Tar Kiln	do	1813	do	do	408	none	8000
Kent county.								
Coventry -	C. A. Whitman	do	1828	Not	do	840	24	20000
Do -	Andrus Harris	do	-	do	do	912	20	25000

by order of the Secretary of the Treasury of the United States, April, 1832.

Amount materials used and amount of cash paid for labor.	Quantity and value of articles used.	No. of men, women, & children employed.			Average wages per diem.			Hours employed.	Aggregate wages.	Animals employ'd.	
		Men.	Women.	Children.	Men.	Women.	Children.			Horses.	Oxen.
41,000	See rep.	20	50	60	-	-	-	12	17,000	4	4
25,000	do	10	60	30	75c. to $3	42 c.	25 to 50c.	11	7,800	6	
21,500	do	14	19	30	$1	43	23	12	9,042	3	
15,000	15,000	15	16	13	1	50	25	-	4.000	2	2
42,045	20 825	28	55	54	1	42	33	12	14,040	4	4
See report	See rep.	4	3	4	1	42	25	12	1,750		
5,720	do	3	3	7	-	-	-	12	1,664		
See report	do										
do	do	35	60	35	-	-	-	-	20,000		
21,325	6,000	19	16	28	1	37½	21	12	9,000	3	
18,000	See rep.	12	32	18	-	-	-	12	9,000	1	
See report	do	14	34	25	1	37½	21	12	10,000		
25,000	12,000	12	35	13	1 12½	50	29	12	13,000	1	
See abstract	See abs't	66	109	169	-	-	-	12	57,000		
See report	See rep.	20	50	70	-	-	-	12	20,000		
See abstra't	See abs't	27	18	20	-	-	-	12	10,000		
do	do	3	11	10	-	-	-	12	3,500		
do	do	30	100	120	-	-	-	12	32,200		
do	do	28	54	53	-	-	-	12	15,600		
do	do	16	6	8	-	-	-	12	5,500		
do	do	3	4	5	-	-	-	12	1,900		
do	do	6	15	19	-	-	-	12	4,700		
do	do	5	17	8	-	-	-	12	7 000		
do	do	28	54	55	-	-	-	12	20,323		
do	do	12	37	11	-	-	-	12	15,000		
do	do	14	36	16	-	-	-	12	12.000		
do	do	3	12	-	-	-	-	12	2,800		
do	do	70	110	40	-	-	-	12	35,500		
133,750	133,750	100	150	200	-	-	-	12	60,000		
50,000	28,000	30	40	40	1	50	25	12	22,000		
69,662	See rep.	28	54	55	-	-	-	12	22,000	5	
15,000	14,000	10	12	19	1 25	50	30	12	12,500	1	
100,000	45,967	117	131	164	1	35	23	12	50,000	6	
50,000	24,000	36	72	69	-	-	-	10 to 12	26,000	1	4
25,000	23,500	25	30	36	90	45	22	11	13,770	2	4
2,500	See rep.	2	3	7	-	-	-	12	1,250	1	
See report	do										
52,000	28,000	26	45	68	1 12½	50	30	12	24,000		
See report	See rep.	11	33	18	1 12½	49	23	12	7,500	1	
See abstra't	See abs't	2	7	7	-	-	-	12	2,000		
10,000	See rep.	8	10	12	90	50	20	-	6,200	4	
See abstra't	See abs't	30	21	3	Also man	ufacture	machi'ry.	12	10,200		

DOCUMENT 8.—No. 41.—Continued.—*Abstract of Manufactories in Rhode Island*, 1832.

Kind of manufactory.	Number of establishments.	Number of spindles.	Number of looms, &c.	Persons employed.			Aggregate wages.	Amo't invested in ground, buildings, and machinery.	Value of materials used in manufacturi'g.	Value of artic's consumed at the establishment.
				Men.	Women.	Children.				
Cotton - -	119	238,877	5,856	1,744	3,301	3,550	$1,214,515	$5,139,190	$1,627,134	$853,072
Wo llen - -	22	39 sets double cards		150	124	106	78,400	335,000	284,435	37,716
Bleacheries -	5	-	-	200	40	60	69,500	228,000	89,000	
Print works -	2	-	-	120	30	36	40,000	212,000	75,875	
Iron and steel -	10 foundries	30 mach. shops	-	1,242	-	-	453,203	802,666	339,973	
Leather - -	40 tanneries	-	-	135	-	-	13,500	76,800	140,200	
Jewelry - -	27 shops	-	-	160	122	-	67,680	-	100,200	
Comb manufact'ies	2	-	-	132	21	-	31,128	35,000	37,500	
			*5,856	3,883	3,638	3,752	$1,967,926	$6,828,656	$2,694,316	$890,788

* Looms for weaving cotton cloth.

Besides the above enumerated establishments, there are several hundred blacksmith and other shops where a variety of articles are manufactured for the use of cotton and woollen mills, such as brass and tin manufactories, reeds, bobbins, spools, pickers, shuttles, brushes, temples, &c.

Mill Villages for Sale (1821–1835)

Groups of small investors built many of the textile mills and villages begun about the time of the War of 1812. By the 1820s and 1830s, when Lowell and other large manufacturing centers were being developed by well-financed corporations, a number of these earlier villages had come under control of single owners and small groups of partners, many of whom were personally involved in the manufacturing operations or in the wholesaling of textiles. Changes of ownership, resulting from business failures, dissolutions of partnerships, and deaths, were common, and some mills passed through a number of hands in rapid succession.

Newspaper advertisements were one means by which prospective buyers learned which mills were for sale. Papers printed in areas where there were large concentrations of mills ran occasional ads for local properties that were on the market. The most comprehensive and widely read source of information, though, was the *Manufacturers' and Farmers' Journal,* published in Providence beginning in 1820. It carried ads describing mills and mill villages for sale throughout New England and sometimes in other parts of the United States.

These advertisements, together with mill owners' probate inventories and detailed descriptions that occasionally are to be found in property deeds, provide some of the best documentary evidence available concerning the buildings and land included in mill villages, the sizes of mills, and types and quantities of machinery in use. Written in characteristic styles of the time, the ads were intended, of course, to show manufacturing and mill properties in a highly favorable light. Sometimes they explicitly portrayed factories and factory villages as attractive, integral parts of the rural communities in which they were located. This point of view needs to be balanced by some of the more critical opinions expressed elsewhere in this volume. But it was a way of looking at industrialization that probably was more widely held then than now and one that must be taken into account in understanding the functions that mill villages filled in early nineteenth-century New England.

All of the following advertisements are from the *Manufacturers' and Farmers' Journal*. (RP)

[March 15, 1821]

TO MANUFACTURERS, FARMERS, MERCHANTS, AND MECHANICKS

To be sold at Public vendue in April next, on the premises, the whole of that valuable Real Estate, owned by the heirs of the late Col. Thomas Denny, deceased, and to Mr. Alpheus Demmond, situate on Ware river, in the town of Ware, and county of Hampshire, [Massachusetts]; consisting of a large brick FACTORY, four stories high, 75 feet long and 37 wide, built upon the margin of the aforesaid river, where there is water sufficient to keep extensive works in operation the whole time, perfectly secure from the danger of ice and floods.

Also—a Gristmill, with two runs of stones, bolting apparatus, &c. This Mill at all seasons of the year commands an excellent custom, and in the dry seasons, the inhabitants of all the adjacent towns resort to it for grinding.

Also—a Sawmill, which is essential to the establishment, and a profitable appendage to the estate.

Also—A Machine Shop, under the same roof with the Gristmill, and well calculated to do all kinds of work by the power of water.

Also—three hundred and twenty acres of excellent interval, plane and hill land, with valuable wood and timber lots on the same; a large new Dwelling-House two stories high completely finished, except painting—also—a spacious and convenient Store—also—One other old Dwelling-House two stories high—also—One new and convenient Dwelling-House one story high—also—Two Bars and a Blacksmith Shop. All the buildings are compact, and beautifully situated on the Bank of the River.

The manufacturer will not only find an elegant building already suited to his purposes, but numerous and valuable water sites on the premises, sufficient to extend his works as far as he pleases, in the midst of an agricultural County abounding in wool and Provisions for his workmen. In this way the Farmer will find a home-market for his produce, the Merchant for his merchandise, and the Mechanick for his skill; all of whom are invited to view for themselves, and embrace the opportunity of becoming rich and useful members of the community.

The aforesaid Factory is calculated for cotton or wool as the purchaser may choose, and the estate will be sold in lots or all together as will best accommodate the bidders.

At the same time will be sold a quantity of cotton machinery, entirely new. . . .

[March 25, 1824]

MANUFACTURERS ATTEND!
A RARE CHANCE

The well known Sturbridge Cotton Manufacturing Establishment will be sold at a GREAT BARGAIN. Gentlemen that wish to invest capital in Cotton Mills would do well to call and view the premises here offered before their money is invested. The establishment is situated in Sturbridge, county of Worcester, and state of Massachusetts. Two miles from the village of Southbridge; forty miles from Providence, sixty miles from Boston, and forty miles from Hartford. The situation of this establishment is pleasant, and in a flourishing part of the country. The water privilege is excellent, being supplied with water from the never failing Quinebaug River. The establishment consists of one cotton mill, fifty by thirty-four, containing nine hundred and fifty-four spindles with all necessary preparations; also sixteen water-looms all in successful operation; also, one other building designed for manufacturing, seventy-eight by seventy-five feet, ready to receive machinery; also, one good Saw Mill, one excellent Grist Mill, one Blacksmith Shop, three dwelling Houses, sufficient for the accommodation of seven families; one convenient Store, one Picking House, and other conveniences with five acres of Land. The above will be sold entire or in parts as may best accommodate purchasers, and a reasonable credit given to the purchaser or purchasers. For further particulars inquire of MR. JOHN MASON, living in Southbridge or COL. WILLIAM FOSTER and WILLIAM DWIGHT, JR. of Sturbridge, or JAMES WOLCOTT, JR. of Southbridge.

[November 11, 1824]

COTTON MANUFACTORY FOR SALE OR TO LET

The Bozrah Manufacturing Company offer for sale at a reduced price, their establishment with its appertenances. It is situated in the town of Bozrah, Connecticut, six miles west of Norwich Court House, on the post road from Norwich to Hartford—The Mill privilege is on the Yantic river, with a permanent Stone Dam. The water is carried about 28 rods in a raceway, to a secure and pleasant site, where there is about 24 feet head and fall, with another situation of about 9 feet fall, 15 rods below the factory where the water can be used a second time, and is considered among the best water privileges in the state. The Buildings are near the center of an excellent farm of about 84 acres, belonging to the establishment, well proportioned with wood, pasturing, plough land and mowing, and is well fenced—about 200 rods of heavy stone wall has been recently built: not far from 30 tons of the first quality of Hay is cut on the farm annually. The buildings are all well made of good material—a superior stone building 68 by 38 feet, 3½ stories high, with well slated roof, and plastered entire, the floors interlined with mortar, and *every* way guarded against fire, containing 1440 spindles and 26 Power looms, with all the necessary preparations, in a good state of repair, and is now in full operation; together with a set of tools, and engines for repairing and building machinery.—A second building 44 by 30 feet, three stories high, lower story of stone, containing a Grain mill, picker, calender, Press, &c. for finishing and packing goods. A Blacksmith Shop, very complete, with a Trip Hammer, water Grindstone, three Forges, a Furnace for small casting, with a variety of tools for all kinds of work and large Coal house, nearby. A small Bleach House, with steam Vats, 3 copper Kettles, wash stocks, &c.—Also, two large barns, store house, wood house, sheds, &c. Five large double Houses with three smaller ones, which are plastered and finished within and painted outside, containing good tenements for more than twenty families. A large well finished store, where the Company have kept a general assortment of Goods, and have found it an excellent stand for retailing, being every way surrounded by wealthy and reputable farmers. This situation with the appendages, perhaps is not excelled by any establishment of the kind in the country. An unquestionable title will be given. If not sold by the 10th of January next it may be rented on reasonable terms [to] all persons of good character and responsibility.

For particular information application may be made to DAVID L. DODGE on the premises, or to *Jonathan Little & Co. New-York.*

[November 15, 1827]

ASSIGNEES SALE

There will be sold at Public Auction, on the 12th of May next, all the real and personal property, assigned to the subscribers by the Kent Manufacturing Company, in Coventry [Rhode Island], viz: The Factory, together with the privilege thereto belonging: said Factory has formerly operated 372 Spindles, and is as good a stand for a Machinist as there is in the state. A building nearly as large as the factory, formerly occupied as a spooling and dye shop, a House built for the convenience of the Factory, with the lot on which it stands. The factory house, (so called). The Fish House, (so called) built for and occupied as a public house [tavern], until within a few years: it is considered to be one of the best stands in that part of the town, as there is no public house kept in the village: the building is three stories high. The Morse Farm, (so called,) containing about 40 acres of excellent land, with two dwelling houses, and outbuildings thereon. Likewise, Sheds and Barns and about five or six acres of land, will be sold in lots to accommodate purchasers. Also, 1 mule, containing 204 spindles, 2 throstles containing 84 spindles each, 1 double speeder of 18 spindles, on the Asa Arnold [plan], and nearly new; 5 cards, 3 heads tacks drawing; 1 picker; 2 spooling and warping machines; 1 large copper boiler; 2 large iron Kettles, formerly used for colouring, and some other articles, such as hollow ware, oil canisters, patent balances, shop furniture, &c. The above property is situated in Coventry, 12 miles from Providence, within the village of the

Coventry Manufacturing Company's establishment and half a mile from the Washington Factory, pleasantly situated in the flourishing part of the town and as good a stand for business as there is in the state. CORNELIUS G. FENNER, Surviving Partner of the late firm of Hunt and Fenner, Assignees.

[January 5, 1835]

COTTON FACTORY AND FARM FOR SALE

The subscribers offer for sale, the Cotton Factory and Farm assigned to them by Ariel Cook, Esq. for the benefit of his Creditors. The Factory is situated in Smithfield [Rhode Island] about one mile West of Mannville and consists of a factory building, of wood, about 30 by 40 feet, 3 stories high, containing four hundred and eighty spindles, 4 looms for weaving duck, 10 cards, and all other necessary machinery for operating the same, in good repair: connected with the establishment are about 8 acres of land; 2 dwelling houses, wood house; barn, work shop, &c. . . .

Smith Wilkinson cotton mill, Pomfret, Connecticut. Vignette from map of Windham County, Connecticut, published by E. M. Woodford, Philadelphia, 1856. Mill Village Collection, Research Department, Old Sturbridge Village. Photo by Donald F. Eaton.

William R. Bagnall

Pomfret Manufacturing Company, Pomfret, Connecticut (1893)

The frame for the Pomfret Manufacturing Company's factory was raised on July 4, 1806. The holiday occasion drew more than 2,000 people from miles around and aroused "popular interest and cooperation" in the first of many textile mills in the Quinebaug valley of eastern Connecticut. The site, now part of the city of Putnam, had been known until that time as "Cargill's Mills" and had included a sawmill, gristmill, distillery, and other facilities that served local farmers. The agent and part owner, Smith Wilkinson, was the brother-in-law of Samuel Slater and member of one of the prominent families in the early years of the textile industry in New England. Wilkinson had gone to work as a child in Slater's first mill at Pawtucket, and he was still in his early twenties when he was placed in charge of the Pomfret mill by his fellow shareholders, among them his father, brothers, and a brother-in-law. Wilkinson continued in charge until his death forty-six years later, after which the company, like many others, went through frequent changes of ownership. William R. Bagnall's sketch suggests the transformation of one significant waterpower site as nineteenth-century industry became a part of the life of a rural community.

Bagnall was an early historian of the textile industry who compiled a number of short company histories similar to this one. Some of these were collected in a volume entitled *The Textile Industries of the United States* (1893; reprint, New York: Augustus M. Kelley, 1971). The text rerprinted here is from the 1893 volume, pp. 416–423. A more complete version of Bagnall's work exists in typescript: "Sketches of Manufacturing Establishments in New York [ca. 1880] and of Textile Establishments in the Eastern States [1780–1890]," ed. Victor S. Clark, 4 vols. (Washington, D.C.: Carnegie Institution of Washington, Contributions to American History, 1908), a copy of which is in the Baker Library, Harvard University. A microfiche edition was published by Merrimack Valley Textile Museum, North Andover, Massachusetts, 1977.

See also Smith Wilkinson's letter to G. S. White reprinted in this collection ("The Moral Influence of Manufacturing Establishments"). (RP)

POMFRET MANUFACTURING COMPANY, Pomfret, Conn.

The territory on both sides of the Quinebaug River, near the railroad station in Putnam, Conn., was the seat of various industries from an early date. The first settlement of the territory, afterward included in the township of Pomfret, was made about the year 1700, the first conveyance of land having been made by James Fitch to John Sabin, June 22, 1691. Prior to 1696, Sabin had taken possession and built a house. Three or four other persons had purchased land, and probably occupied it, before 1700. In 1713, the first efforts were made to secure a town organization, and a meeting of the men in the settlement was held, May 3, 1713, and a petition was prepared, which was presented to the General Assembly, May 14th, and at once granted, and, on the 27th of May, the first town meeting was held.

In 1730, on the 31st of December, the lands around the falls of the Quinebaug River, in the northeast corner of the town, were sold by John and Noah Sabin to David Howe, a clothier of Mendon, Mass., who immediately removed to his new purchase, and erected a dwelling house, grist-mill, malt-house, and dye-house, with machinery for pressing, shearing, and finishing cloth. These facilities attracted customers from the vicinity to a considerable distance, and "Howe's Mills" became a noted resort.

In 1742, on the 17th of March, the property was purchased by John Daniels, and the name was changed to "Daniels' Mills." He sold the mills to Nathaniel Daniels, February 6, 1760. On the 13th of the same month, Nathaniel Daniels, probably a son of John, sold the mills, and other real estate, to Benjamin Cargill, of South Kingstown, R. I. Mr. Cargill was a man of remarkable energy and enterprise, and, during the period of more than thirty years that he was engaged in business at the Falls of the Quinebaug, greatly enlarged it, introducing new industries, and "Cargill's Mills" became one of the most important centers of trade in Eastern Connecticut. On the 26th of September, 1793, he sent an advertisement to the Providence Gazette, beginning with the following words: —

> "Being stricken in years and past labor, and having a desire to lead a more peaceable and retired life, is now to be sold and entered upon, the ensuing spring, the noted inheritance of Benjamin Cargill, of Pomfret, situated on Quinebaug River, containing five hundred acres of land, . . . together with houses and barns, a smith-shop, with two trip-hammers for scythe-making, a saw-mill, fulling-mill, malt-house and gin-distillery; also a grist-mill, having three pairs of stones, under one roof. . . . The above works are built in the best manner."

The property included land on both sides of the Quinebaug River, now occupied by cotton and woolen factories, from the Ballou and Powhatan Mills to below the Monohansett Mill.

No purchaser of the property was secured till 1798. On the 26th of June of that year, Benjamin Cargill sold to Major Moses Arnold, of Warwick, R. I., and John Harris of Cranston, R. I., seventy-five acres of land in Pomfret. This conveyance included the site of the old fulling-mill and of the other mills and shops, which had been known for more than thirty years as "Cargill's Mills." On the same date, Mr. Cargill sold to the same parties four hundred and twenty-five acres of land in Thompson and Killingly, on the eastern side of the Quinebaug River. These two conveyances covered the whole property referred to in the advertisement of 1793.

On the first of April, 1800, Moses Arnold conveyed his interest in both purchases to Jeremiah and Nehemiah Knight, of Cranston, R. I.; Nehemiah receiving two thirds of the interest, or one third of the whole property, and Jeremiah, one third of the interest, or one sixth of the whole property. On the third of March, 1801, John Harris sold to his son, George, one third of his interest, or one sixth of the whole. On the first of June, 1801, Nehemiah Knight conveyed to his sons, Daniel and Nehemiah, Jr., one half of the Cargill property. This deed refers to his "late father" Jeremiah Knight, and indicates that the latter had died, and that his son, Nehemiah, had inherited his interest of one sixth. On the 14th of April, 1802, Daniel and Nehemiah Knight, Jr., sold their interest in the property to John Harris, who thereby, with his son, George, acquired the entire Cargill estate. On the 12th of April, 1805, they sold the seventy-five acres in Pomfret, together with all the mills and other buildings, included in the first conveyance from Cargill to Arnold and Harris, to James Rhodes of Warwick, R. I. There are no indications in the deeds, and no other proofs, either from records or tradition, that any change had been made in the industries at the Quinebaug Falls between 1793 and 1805. There is little doubt, also, that it was the object of Mr. Rhodes, in the first purchase, just spoken of, to use, either alone or in partnership with his brothers Christopher and William, the splendid water-power at the Falls in the cotton manufacture, which, under the stimulus of the success attained by Samuel Slater and his partners, was then attracting the attention, so largely, of the enterprising business men of Rhode Island.

During the summer of 1805, Smith Wilkinson, having in view the establishment, in partnership with his father, brothers, and brothers-in-law, of a cotton-mill in some suitable location in eastern Connecticut, visited the village of Jewett City, and negotiated with Jedediah Barstow for the purchase of the mill-privilege owned by the latter on the Pachaug River, near its entrance into the Quinebaug River, forming, at the present time, a part of the water-power of the mills of William A. Slater. Mr. Barstow, after having agreed to sell the mill-privilege and adjoining land, at a price satisfactory to Mr. Wilkinson, receded from his agreement, and the latter was compelled to seek another location. Going to Pomfret, and inspecting the location which had been secured by James Rhodes, and being much pleased with its manifest adaptation to his purpose, he proposed to the latter that they should unite their interests. An agreement was at once made to form a company, and to make such additional purchases as should secure the whole water-power of the Quinebaug River, within the limits of the present town of Putnam. With this view, Mr. Rhodes purchased, ostensibly on his own account, but really on account of the proposed company, two tracts of land, one of fifty acres in Pomfret and Thompson, the other of four hundred and twenty-five acres in Thompson and Killingly. The former purchase was made December 24, 1805, and embraced land on both sides of the Quinebaug River, extending about three quarters of a mile north of the old Cargill Mills, and included a dam and mill-privilege, then known as Bundy's Falls. This tract of land had been sold, September 6, 1796, by Benjamin Cargill, Jonathan and Joseph Sabin, and Ithamar May, to Ebenezer Bundy, who erected a dam and a grist-mill, which he operated till the close of 1798, and, on the 1st of January, 1799, sold the property to his sons, Silas and James, who operated the mill about seven years, and then, as has been stated, sold the estate to James Rhodes. On the 26th of December, 1805, Mr. Rhodes made his third purchase of land in that vicinity, four hundred and twenty-five acres on the east side of the Quinebaug River, then within the limits of Thompson and Killingly.

There were five cotton factories in operation in the State of Connecticut in 1805, the first having been started in New Haven, in 1793, by John R. Livingston, of New York, and in operation till 1837; the second in 1794, by the Pitkin brothers in East Hartford, now North Manchester, and operated for seventy years, from 1819 to 1889, by the Union Manufacturing Company; the third in 1795, by Richard Crosby at Suffield, in operation till 1815; the fourth in 1802, by John Warburton, in North Bolton, now Vernon, on the site of the present mills of Talcott Brothers, and operated by Mr. Warburton till 1815, when it was sold to Thomas Bull of Hartford, and changed to a satinet mill; and the fifth in

1803, by Frank Franklin in Middletown, in that part of the town now known as Cromwell, being operated as a cotton-mill by different owners till 1834.

The new company was organized January 1, 1806, under the name of the "Pomfret Manufacturing Company." The formal transfer, by Mr. Rhodes to the company, of the five hundred and fifty acres of land, with all the buildings and mill-privileges, was made February 13, 1806. The stock of the company was then held in sixteen shares by the following persons: James Rhodes, four shares; Oziel Wilkinson, one share; his son, Smith, two shares; his three sons, Abraham, Isaac, and David, jointly, four shares; his son, Daniel, one share; his son-in-law, Timothy Greene, two shares; his son-in-law, William Wilkinson, one share; and the two brothers of James Rhodes, Christopher and William, jointly, one share.

Smith Wilkinson was appointed agent, and immediately entered upon the preparations for building the new factory, preparing the land, procuring stone for the foundations, and timber for the frame. It being ready for raising about the 1st of July, 1806, the national holiday, July 4th, was selected for the purpose. This was shrewd policy on the part of Mr. Wilkinson, not only getting a hard job done quickly, by making it a part of the celebration of the day, but also enlisting for the enterprise popular interest and coöperation. It is said that two thousand persons came together, many of them from a considerable distance, either to assist or to look on, and that free punch was furnished to all. The building was of wood, one hundred feet long, thirty-two feet wide, and four stories high. Work on the building, and its equipment with machinery, the best which could then be obtained, was pushed forward, and on the 1st of April, 1807, the first cotton factory in Windham County, Conn., and the pioneer of many similar enterprises in that county, started within the next seven years, went into operation.

Like all the cotton factories of the period, it was for the manufacture of cotton yarns, which were sold to parties in the vicinity, to be woven into cloth for their own use, or was given out to be woven on account of the company. Under the able administration of Smith Wilkinson, the Pomfret Cotton Factory, as it was called, was an exceptionally successful enterprise from the beginning, passing, without embarrassment, through the financial crises of 1816, 1829, and 1837, which affected so seriously, and in many cases ruinously, the cotton industries of New England.

A large additional purchase of land in the vicinity of Bundy's Falls was made by James Rhodes, December 18, 1813, and on the 3d of the ensuing March, he transferred an interest of three fourths of it to his partners in the company, in the proportions which they severally held of the stock. Various changes were made in the ownership prior to October 30, 1835, at which time Smith Wilkinson had become the owner of eleven twentieths, or a controlling interest, William Wilkinson of four twentieths, and James Rhodes retained his original interest of one fourth or five twentieths.

In 1823, a new mill was erected of stone, ninety feet long, thirty feet wide, with four main stories, a basement, and attic. A deed of September 8, 1832, showed that the machinery of the two mills then consisted of nine mules, having sixteen hundred and twenty spindles, twenty-four throstle-frames (sixteen hundred and thirty-two spindles), three Taunton speeders, six grand and belted speeders (one hundred and fifty-two spindles), seventy-six power-looms, three warpers, three dressers, three spoolers, one picker, etc.

On the 30th of October, 1835, a division of the property was made, Mr. Rhodes receiving the original mill of wood, and the Messrs. Wilkinson receiving the stone mill, an equitable division of the remaining property being also made. The two establishments were operated by separate proprietors for forty years, being reunited in 1875, as the property of the Putnam Woolen Company.

The stone mill, having become by the division of October 30, 1835, the property of Smith and William Wilkinson, was operated by the former, as agent, on their joint account till 1845. On the 28th of June, 1845, William Wilkinson sold his interest, of four fifteenths, to Smith Wilkinson, who thus became the sole proprietor, and so continued till his death, November 5, 1852. The mill was then operated by his son Edmund, in the interest of the heirs, till 1864. On the 20th of September, 1864, Edmund Wilkinson sold his own interest of two fifths of the property, and on the same date, as trustee for certain other heirs of his father, sold two fifths to Thomas Harris of Providence, R. I. On the 19th of September, 1864, Eliza P. Perkins, guardian of the minor grandchildren of Smith Wilkinson, sold their interest of one fifth, also, to Thomas Harris, who by the three conveyances became the sole proprietor, and at once proceeded to equip the mill with woolen machinery. During the latter part of the same year, he organized the Harris Woolen Company, and, on the 31st of December, 1864, conveyed the property to that company. It operated the mills till 1870, and, on the 3d of August, of that year, sold the property to the Putnam Woolen Company, a corporation organized July 18, 1870, having been chartered by the Legislature of Connecticut at the May session of that year. The mills have since been owned and operated by the Putnam Woolen Company.

Soon after the death of Mr. Rhodes, Tully Dorrance, who, with his wife, a daughter of James Rhodes, had come into possession of the old Pomfret Cotton Factory, in the division of the estate, received into partnership his son, James Rhodes Dorrance, and Daniel S. Anthony, of Pomfret, who for some years had been the resident agent of the mill, under the style of the Cromford Manufacturing Company, which operated the mill till the beginning of 1849, Mr. Anthony continuing the management at the factory, and the Messrs. Dorrance being the mercantile agents at Providence, R. I. Mr. Anthony sold his interest in the company, January 11, 1849, to his partners, and removed to Jewett City, where he was actively engaged in the cotton manufacture, as narrated in our sketch of the Ashland Manufacturing Company. The Messrs. Dorrance continued running the mill till the early autumn of 1855. The old mill of wood had been burned in the mean time, and a larger mill of brick had been erected.

On the 19th of September, 1855, Tully Dorrance and his wife sold the property to Joseph F. Eaton of Putnam. It next passed, by legal process, into the hands of Asa Cutler, of Putnam, who sold it, June 22, 1858, to Andrew J. Currier, of Norwich, Conn., who on the 8th of April, 1862, sold it to Frederic W. G. Jones, of Boston, Mass. Mr. Jones sold the property to Earl P. Mason, of Providence, R. I., July 11, 1864. On the 29th of March, 1869, Michael Moriarty purchased the mill, and operated it nearly six years, under the name of the Saxon Woolen Company. On the 15th of February, 1875, Mr. Moriarty sold it to Mr. Mason, who immediately leased it to Mr. Moriarty and Frank D. Sayles, son of Sabin L. Sayles, the well-known manufacturer of Killingly, Conn. They operated it till after the death of Earl P. Mason, when, on the 11th of November, 1879, the heirs of Mr. Mason sold it to the Putnam Woolen Company.

THE SCHOLFIELD MANUFACTORY, Stonington, Conn.

The early record of John Scholfield and his brother Arthur, as the pioneers of the woolen manufacture according to the improved English methods in this country, first at Byfield, Mass., and afterward at Montville, Conn., has been given. In our sketch of the industry, established by them in the latter town, it is stated that John Scholfield, leaving his sons, James and Thomas, in charge of his business there, went in 1806, to the town of Stonington, Conn., and there established a similar industry. In the town clerk's office of Stonington a deed is recorded by which Jonathan Nichols, of Warwick, R. I., conveyed to John Scholfield, of Montville, Conn., for $1,440, January 25, 1806, a lot of land on Pawcatuck River, containing nine acres. This land was in the eastern part of Stonington, in what is now known as Stillmanville.

Map (1835)

The map of Pomfret, Connecticut, shows a developmental pattern similar to that of Phoenixville in that the initial wooden mill, erected in 1806 (1813 in Phoenixville) was followed by the building of a stone mill in 1823 (also 1823 in Phoenixville). The mills at Pomfret, unlike their counterparts at Phoenixville, were added to an established community of sawmills and gristmills and dwellings that had been in place for some time. The natural falls of the Quinebaug at Pomfret had been used to power small mills since the middle of the eighteenth century; the new factories took shape around this previous development and incorporated parts of it into the growing village. (TZP)

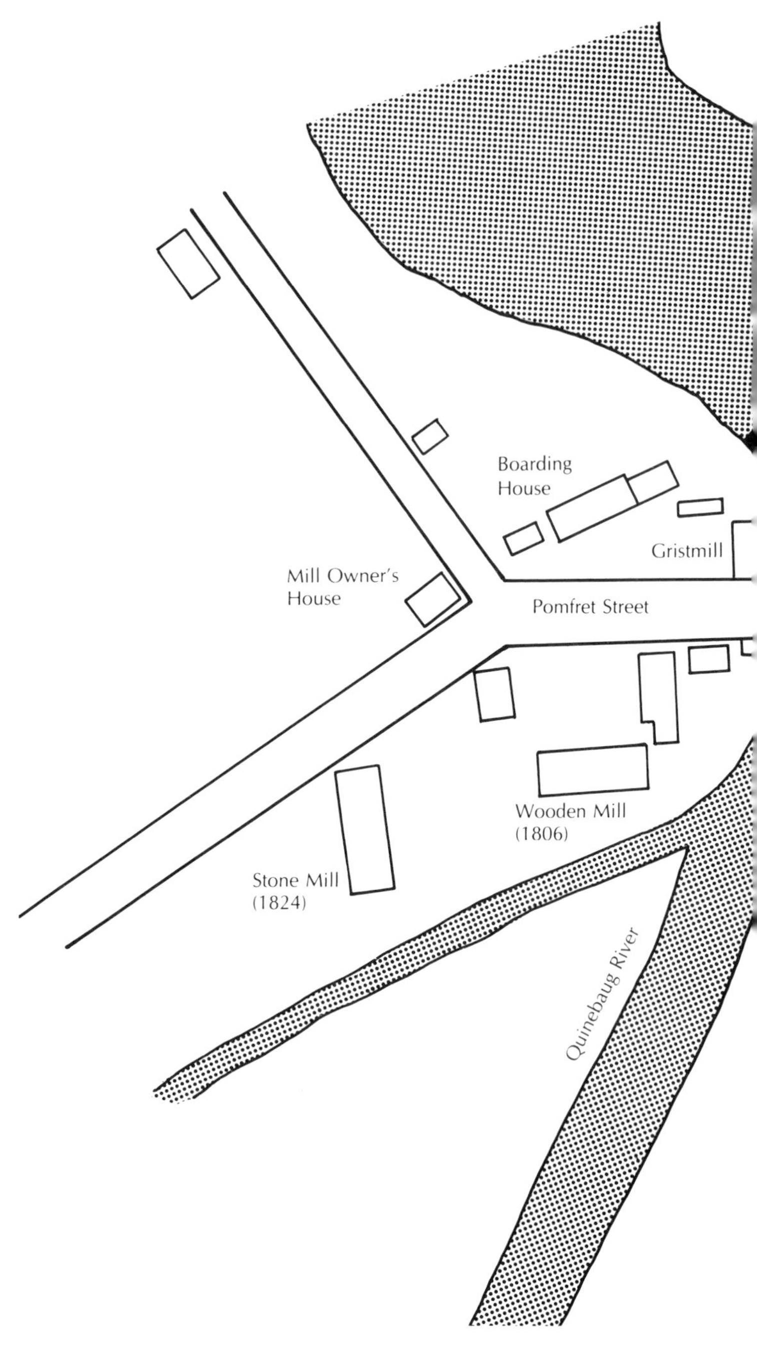
Boarding
House
Gristmill
Mill Owner's
House
Pomfret Street
Wooden Mill
(1806)
Stone Mill
(1824)
Quinebaug River

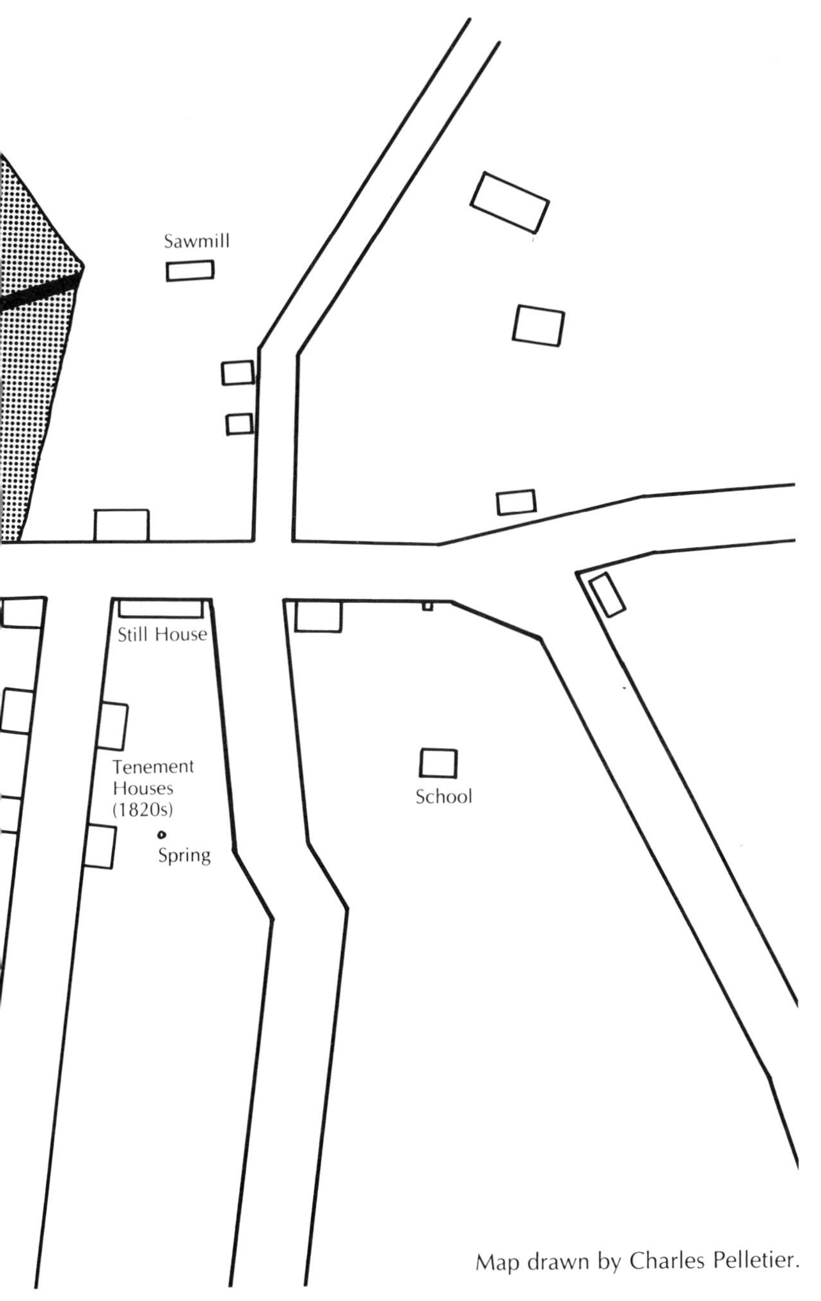

Map drawn by Charles Pelletier.

Globe Village painting, ca. 1822, attributed to Francis Alexander. By permission of Jacob Edwards Memorial Library, Southbridge, Massachusetts.

Sturbridge Manufacturing Company

Ledger (1811–1812)

"I think highly of your stock in a cotton factory," wrote Linus Leonard to his brother Zenas of Sturbridge, Massachusetts, in 1814. "I have no doubt of its being the best property during the continuance of the war." The War of 1812 had brought a boom to New England cotton manufacturing, and groups of small investors had hurriedly joined together in dozens of communities to build mills and put them into operation. The Sturbridge Manufacturing Company was established just before the war, in late 1811, with Leonard, the local Baptist minister, as its agent. The only member of the group with any previous experience in manufacturing was Nathaniel Ryder, who contracted to build the machinery. Other members furnished building materials and labor on the building to help pay for their shares. Later entries in the ledger from which this excerpt is taken show some shareholders also helping to build a boardinghouse for the employees and operating the wool carding machine that the company ran as a service to local farm families.

Despite the initial difficulties recounted here, the Sturbridge Manufacturing Company managed to earn a profit for a time, declaring a dividend of more than ten percent in early 1814. Neither the initial profits nor the original owners lasted much beyond the war, but the mill was in operation under a succession of proprietors for several decades. (See the advertisement for the sale of the Sturbridge Manufacturing Company reprinted in this collection.) (RP)

The ledger of the Sturbridge Manufacturing Company is in the Leonard family papers, Old Sturbridge Village Research Library, as is the Linus Leonard letter quoted in the headnote.

December 5, 1811. The Sturbridge Manufacturing Company, being regularly formed by signing a mutual compact, or Articles of agreement, proceeded to choose [Rev.] Zenas L. Leonard Agent & Franklin Rider Clerk. The Comy. bought of Moses Fiske the gristmill, with a water privilege, & a certain tract of land (with a view to erect a factory on the same) for $645, as the deed will specify.

Chose Jephthah Plimpton, Nathl Rider, & Lieut. John Plimpton a committee to have the sole care of the gristmill, to put it in repair, & to make returns thereon to the Comy.

The Comy. agreed with Nathl. & Franklin Rider to build 128 spindles of cotton machinery & procure all the attendant preparations for $14 per spindle—Also a double wool carding machine for $600. The whole of said machinery, they warrant to be completely good in every part & equal to any in the commonwealth & also to have the same in readiness to run by the middle of June next. The abovesaid machinery to be turned into the Comy. for shares.

Also Lieut. John Plimpton, Stephen Newell, Zenas L. Leonard, Moses Fiske, Jephthah Plimpton, Comfort Freeman, Ziba Plimpton, & Moses Newell agreed to dig & stone the raceway, wheelpit & canal & to erect a building 30 feet by 45, three stories high for a Factory & also to cause to be built the waterwheel & geering & to have the whole completed by the first of June next. (N.B. If Nathl. & Franklin Rider wish to assist, in some small degree, in digging the raceway & canal, it is agreed they might do it.)

Agreed to dissolve the meeting.

Those Eight persons above-named completed their agreement, in its various parts, in the month of June 1812 & got all things in readiness to receive the machinery & put it in operation. The double wool carding machine was started in the month of July 1812 & the 128 spindles of cotton throssels, with the carding, drawing & roping, were set in motion the latter part of August 1812. Nathl. Rider spun his own cotton for two weeks in order to try the machinery. The company then engaged Nathl. Rider to further progress in trying the machinery by working company stock. He began to spin for the company on the 7th day of Septr.

1812. After a few days of experiment, said Nathl. Rider informed the Company the machinery performed well & he pronounced it to be good. He accordingly wished the Comy. to accept it & make a general settlement. A part of the Comy. having obtained the opinion of a number of gentlemen, who were acquainted with machinery that it was greatly deficient, wished to have him consent to have judges brot on to say whether the machinery was genuine or not, & if not, to point out wherein, that it might be put in ample order, if possible, for spinning good yarn. Said Nathl. stated over & over again that if the Comy. would consent to have a general settlement made, he wo'd warrant the machinery to be equal to any in the Commonwealth, viz. it sho'd spin as good yarn, do as much in a day & last as long, & also the term of proof or trial sho'd be one year. And in case the machinery on experiment sho'd fail to answer this warranty, it sho'd be taken back & all damages to the Comy. sho'd be paid. Or otherwise if the comy. chose, he wo'd see that it was made genuine in every part & remunerate the Comy for all damages which might arise to them on account of deficiency; & furthermore offered to get bondsmen for the performance of their engagement. This offer appearing to be fair the majority at length tho't that it was best to accept it & make a general settlement. Accordingly,

The comy. having met at the house of Lieut. John Plimpton on the 26th day of October 1812, after hearing the above statement & offers from Nathl. Rider again & again, Franklin Rider being present, voted to accept the machinery on the following terms for further trial—Viz. Nathl. Rider warranted the machinery to do as good work, do as much in a day & last as long as any in the Commonwealth: & in case it sho'd fail of this, he positively engaged to take it back & pay the Comy. all damages, or otherwise to make it good according to this warranty, & remunerate the Comy. for all the damage which it might be to them on account of its not being good at first. The Comy. then proceeded to make the following settlement:

Shares		$	C
2	Stephen Newell, credited for materials for the Factory, work & money	500	00
2	Lieut. John Plimpton, credited for materials for the Factory, work & money	500	00
2	Eleazer Rider, credited for machinery delivered by Nathaniel Rider	500	00
2	Zenas L. Leonard, [credited] for materials for the Factory, for work & money	500	00
4	Moses Fiske, [credited] for the gristmill & land & for materials for the Factory, for work & money	1000	00
1	Jephthah Plimpton, [credited] for materials, for the Factory, for work & money	250	00
2	Comfort Freeman, [credited] for materials for the Factory, for work & money	500	00
3	Nathaniel Rider, [credited] for machinery	750	00
1	Ziba Plimpton, [credited] for materials for the Factory, for work & money	250	00
3	Franklin Rider, [credited] for machinery & for work & money	750	00
1	Moses Newell, [credited] for materials for the Factory, for work & money	250	00
23	Shares amount to ——	$5750	00

Octr. 26th 1812

Map (1830)

This map illustrates the location of mill villages and center villages along the Quinebaug River in the Massachusetts towns of Sturbridge (est. 1738) and Southbridge (est. 1816). Previous to extensive industrial development in the early nineteenth century, the human landscape consisted of little more than farms with a few sawmills and gristmills scattered along the river and its tributaries. The factory village of the Sturbridge Manufacturing Company was started in 1811 by a series of local landowners, some of whom were connected by marriage to the Massachusetts towns of Medfield and Medway. The Medway Machine Company may have been the source of the machinery for this mill. Successive villages were built downstream during the next ten years by shifting partnerships of the original owners, their close relatives, and other locals with money to invest. (TZP)

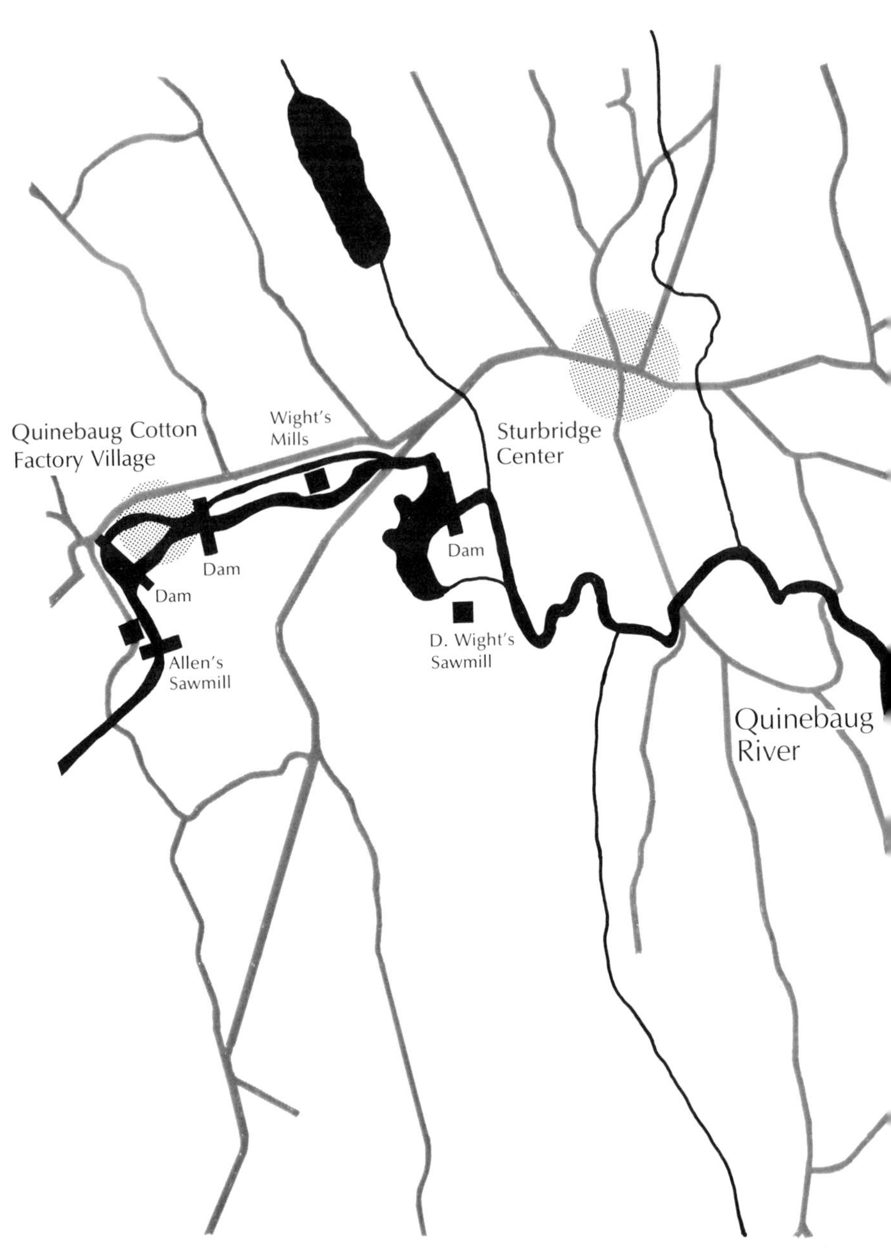
Quinebaug Cotton
Factory Village
Wight's
Mills
Sturbridge
Center
Dam
Dam
Dam
Allen's
Sawmill
D. Wight's
Sawmill
Quinebaug
River

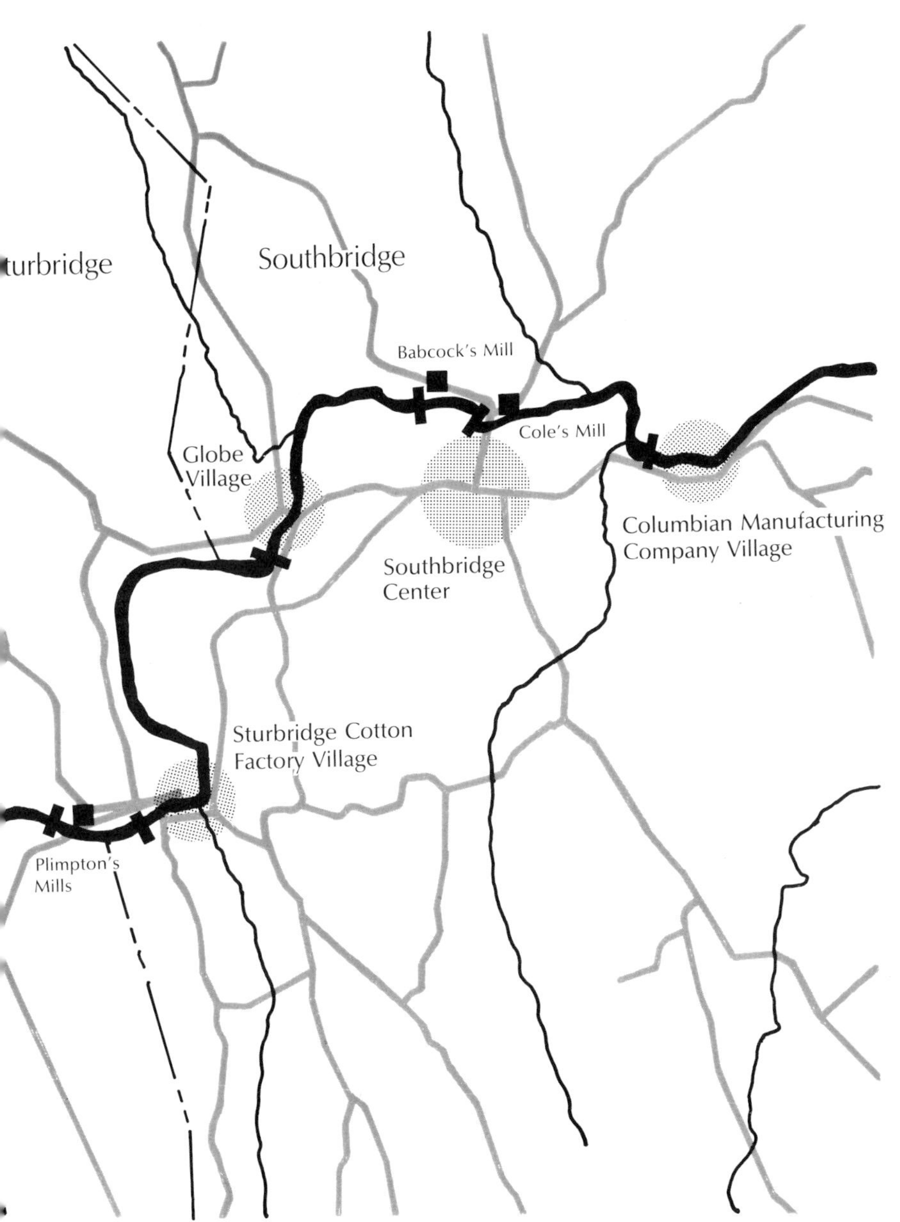

Shaded lines represent roads.

Map drawn by Charles Pelletier.

The Phoenix Mill, Phoenixville, Connecticut, built in 1823, is one of very few early textile mills that survive structurally intact. Old Sturbridge Village photo by J. Alan Brzys.

Land Records (1812–1852)

No private company papers have been found for the small cotton mill that operated at the crossroad today called Phoenixville in the town of Eastford, Connecticut. Nonetheless, a variety of public records, such as deeds, mortgages, attachments, executions, probates, tax records, and newspaper advertisements, are available to help researchers understand how this small village came into being, grew, and declined, all within the first half of the nineteenth century. Similar records can be assembled to write the history of other now obscure communities. The seven documents that follow have been selected to demonstrate the sort of information conveyed in the public record.

In July 1812, Rufus and Pardon Sprague left their home in Johnston, Rhode Island, and traveled to Ashford, Connecticut, where they purchased a water privilege from Bilarky Snow. The sale price and the bounds are included in the deed from Snow to the Sprague brothers, along with the right to raise the dam. The brothers took some local landowners and businessmen into partnership and petitioned the Connecticut legislature for an "Act of Incorporation" in 1815. Included in the growing village at the time was the wooden cotton mill, put into operation in 1814 and later known as the "Red Mill," a sawmill, gristmill, and a few dwelling houses and outbuildings. Brief mention is made of the mill in the 1820 federal census, which also notes that the "present owners purchased this Manufactory of the Original owners & Builders at a sacrifice. . . ." The Sprague brothers had sold out their interest in the company and moved west.

The reorganized company expanded, adding a second stone mill to the older wooden mill in 1823. At about the same time, additional dwelling houses, barns, and perhaps a company store were added to the community. The company continued to operate for the remainder of the decade only to end in bankruptcy in 1830. The inventory of the company recorded in probate records provides the most detailed surviving view of early village cotton mills. Along with a description of the land and buildings, this inventory also lists the contents of the mills and the debtors and creditors of the company.

The mills were reorganized under new ownership as the Phoenixville Manufacturing Company, but they were soon encumbered by a series of mortgages that made it impossible to operate at a profit. The special federal industrial census of 1832 ("McLane's Report") shows that the two mills employed 25 men, 50 women, and 10 children under the age of twelve and supported a community of some 200 people. A newspaper advertisement run in 1835 offered the factory village for sale.

The inventory of the estate of Benjamin Warren indicates that soon after his death in 1852 the Phoenixville mills and dwellings were sold to separate owners and stopped operating as a complete unit. The contrast between the value of the items listed in that inventory and the comparable items in the 1830 inventory makes clear that, by mid-century, the mills and machinery had lost most of their value. The wooden mill collapsed some time in the late 1800s, but the stone mill, although empty, still stands today. (TZP)

Phoenixville Deed

To all people to whom these presents shall come Greeting Know ye that I Bilarky Snow of Ashford in the County of Windham & State of Connecticut for the consideration of Eleven hundred Dollars received to my full satisfaction of Rufus Sprague and Pardon Sprague, both of the State of Rhode Island, Do give grant bargain sell and confirm unto the said Rufus & Pardon and to their heirs and assigns forever one certain tract or parcel of land lying in sd Ashford bounded as follows Beginning at the North west corner of Zwinglas Bullards land on the East side of Natchaug river thence Northerly to a bunch of chesnut staddle standing in the south side of the Connecticut and Rhode Island Turnpike road thence Easterly on the S. side of sd road to a point South of Levi Works line thence north across sd road to sd Works land thence Westerly to a large rock in the pond marked B.S. thence Northerly up sd River on the East bank of it to a point East of land belonging to the heirs of Ephraim Spalding decd. thence west across sd river to sd heirs land thence westerly in the South side of sd heirs land to the East side of where a Committee from the Windham County Court laid out a highway that was not accepted thence Southerly where sd highway was laid in the East side thence till it comes to sd turnpike road thence Southerly across the road last mentioned to the land of Smith Snow thence Easterly to a large hemlock tree and stones on the West bank of sd river thence southerly on the west bank of sd river to a point west of the first bound thence East to the first mentioned bound with the appurtenances thereof particularly the water privilege mentioned in

Land Records, Ashford, Connecticut, volume 16, p. 332.

Ephraim Spauldings Deed to me dated 13 April 1789 recorded in Ashford records 11th Book page 740 also they the Grantees their heirs &c are to have the privilege of raising the dam on the premises as high as I now have a right to do, and I reserve this privilege that the present turnpike road as now traveled is allways to be kept open for publick travil.

Petition for an Act of Incorporation (May 1815)

To the Honorable general Assembly of the State of Connecticut, now in session at Hartford, we your Petitioners of Ashford in Windham County humbly shew that we are Manufacturers of Cloth and Traders in company in Ashford under the name and firm of Sprague Manufc. Co. and wishing to be incorporated humbly crave your Honors to take our request into consideration and grant us a charter for the above purposes under the name of sd. firm, as in duty bound shall ever pay.

Ashford, 20 May 1815

Petitioners
Rufus Sprague
Edward Keyes
John N. Sumner
Benjamin Palmer, Jr.
Mason Palmer, acting agent

Connecticut Archives, State Library, Hartford, Second Series, Vol. 1 (76a).

Federal Census Schedule (1820)

Raw Materials Employed

1. The kind? Upland Cotton wool
2. The quantity annually consumed? 33,600 lbs.
3. The cost of the annual consumption? $6,720

Number of persons employed.

4. Men? No. of men employed, 6.
5. Women? No. of women ditto, 11.
6. Boys and Girls? No. of Boys 4, Girls employed, 16.
   Total Number of Persons Employed 33.

Machinery.

7. Whole quantity and kind of Machinery?

2 Mules	192 spindles each		384
4 Throstles	96 spindles each		384
1 ditto	72	ditto	72
			840 Total

   8 Power Looms—

8. Quantity of Machinery in operation?
   All in operation except 1 Throstle of 72 spindles.

Expenditures.

9. Amount of capital invested?	$27,200.
10. Amount paid annually for wages?	4,500.
11. Amount of Contingent Expenses?	1,475.

U.S. Federal Bureau of the Census: Fourth Census (1820), Manufacturing Schedules. Record Group 29, National Archives.

Production.

12. The nature and names of Articles Manufactured?
    Brown Shirting and Sheeting.

13. Market value of the Articles which are annually manufactured? $14,400

James H. Preston, Agt.

14. General Remarks concerning the Establishment, as to its actual and past condition, the demand for, and sale of, its Manufactures.
    The present owners purchased this Manufactory of the Original owners & Builders at a sacrifice from the original cost—the factory was Built in 1814 and the cost cannot be precisely ascertained—but according to the Best Knowledge we can get the Original Cost was Thirty Four Thousand Dollars.

Cotton	$6720
Wage	4500
Contingent Expenses	1475
	$12695
Production	14400
Expenditures	12695
Gain	$ 1705

Interest on capital $1632 at 6 per cent.

Bankruptcy Inventory and Appraisal (1830)

The annexed is an Inventory and appraisal of the real estate, machinery, goods and effects belonging to the estate of Sprague Manufg. Co., in Ashford, assigned to Thomas Clark in trust for the benefit of all the creditors of the Sprague Manufg. Co., viz—

Real Estate viz—

About 90 acres of land situate in Ashford, Eastford society together with the two Cotton Mills, & Grist Mill, one Saw Mill, eleven dwelling houses, one store, with all the out Buildings, all the water priveleges, main water wheels, upright and main horizontal shafts, at both Factories.

14758.00

Machinery in Stone Mill viz—

18 Looms @ 60—$1080—1 warper $25—1 spooler $25	1130.00
6 Throstle frames of 432 spindles $5	2160.00
2 Mules containing 384 do @ 2.25	864.00
1 Yarn Reel $10—1 Dresser—$150—13 Cards $100	1460.00
1 Picker $150—6 drawing frames @ 180—2 speeders 800	1130.00
1 Regulator $30—5 Stoves and pipes $50	80.00
1 Turning & fluting Engine and wood Lathe	75.00
1 Cutting Engine $3.00—3 Iron Vises $9	12.00
1 Spindle Lathe $10—1 Burr Saw $2	12.00
1 Oil Can $3—1 Indigo kettle $1.50—1 Yarn Sizer $3	7.50
1 Wood wheel clock $8—116 Tin drawing Cans 4/—77.33	85.33
392 Warping spools $5.88—1 Lap Sizer $1	6.88
18 Harness & Reels 5/ $15—1 Dresser in Attic Story 100	115.00

Records of Court of Probate, District of Pomfret, Book 17, pp. 520–522, Pomfret Town Office, Pomfret, Connecticut.

In Red Mill

10 Looms $60—600—8 do—$45—360—1 warper $15	975.00
1 Spooler $20—5 Throstle frames 456 Spindles $2 = 912	932.00
2 Mules 384 spindles $1.50—576 1 Picker $100—1 Reel $10	686.00
13 Cards $50—= 650 3 drawing Heads $30 = 90—	740.00
3 do. do. $25—75—2 speeders $250 = 500	575.00
1 do. Hinds make (broke) 50—1 Dresser $125. 1 Reel $10	185.00
6 Stoves & Pipe $7— = 42—1 Iron vise 3.00 1 Regulator 50—	95.00
1 Cloth press & screw 10—1 Wood wheel clock 8.00	18.00
1 Lap Sizer 0.75—109 Tin drawing cans 58[c] $63.22	63.97
18 Harnesses & Reeds 5/—15 241 Warping Spools 1 1/2¢ 3.61. oil can 2.00	20.61

In Black Smiths Shop

1 Anvil $8—1 pr. Bellows $8—2 vises $6—1 Beek iron $2	24.00
1 pr. Iron shears 1.50—1 Anvil from Mr. Richards 139 lb. 5.50	7.00

Sundry Articles of Mill Furniture

2 Bitt stocks & 1 sett Bitts 6.50—2 hand axes 4/6 1.50—	8.00
2 [Muckle] Wrenches 1/6—, 50—4 chisels, 12[c] 1 drawing knife .50	1.12
7 Hammers 1.17—6 Small Wrenches .75	1.92
2 Saws $1.50 1 Iron Bar 1.17—1 Saw Mill Bar 3/	3.17
1 [Churn] Drill 1—13 old Water pails @ 9—1.62	2.62
24 Oil cups 48[cts] 1 Sizing Tub 6/—20 Lanthorns $15	16.48
4 Reflectors $5—130 Tin Lamps 5[c] 6.50	11.50
2 Grindstones & cranks 4.00—1 pr. clamps 8/- 1 Trying Square 9	5.46
1 Jointer 20[ct] 1 Smooth Plane 30[c] 1 Round Plane 40[c]	.90
1 Iron Bitt Stock 1.25—3 Tapering Reamers 20[c] = 60	1.85
1/2 Sett Turning Tools .50—1 1/2 doz. Drills, 4.50—1 Trowel 17[c]	5.17
1 Jam Plate, set dies, & Taps 5.00—6 Large Yarn Baskets .25—	6.50
5 Drums 3.00—1 old press for spool heads, 80	3.80
3 Old cast iron Wheels 2.30—1 spindle Frame, 30—	2.60
Old Lumber in the Beams at Grist Mill—	5.00
1 Old Bobbin Machine at Grist Mill	5.00
1 pr. Old Black Smiths Bellows do. do.	2.00

About 500 lb Old Iron Rollers &c. do. do. 12.00
Old Gearing Wheels & castings do. do. 6.00
Callender cast and wrot. Irons do. do. 25.00
1 Large Iron Shaft do. do. 10.00
1 Old Drum & Shaft & 1 small shaft do. do. 6.00
Old Iron & castings at the two Mills, except what belongs to Clark & Hibbard 6.00
1 Road Scraper 1.50—52 Iron Scythes in store chamber 10^{c} 5.20— 6.70
3 Linen Wheels in the Garret 6/-3.00 1 Quill Wheel .25— 3.25
Bottle & Oil Vitrol. 50—1 lot paper Bitters .25 1 Straw Hat .25 1.00
1 Box damaged Hats .25—104 Weavers Reeds 20^{c} 20.80 21.05

Cotton Goods—

3987 Yds 4/4 Brown sheeting 9^{c} 358.83
7257 1/2 do. 3/4 do. shirting 6^{c} 435.45
460 lbs. Cotton yarn assorted Nos. 20^{c} 92.00
5 lbs. do. do. Refuse 15^{c} .75
115 lbs. Cotton yarn assorted Nos. loose in store 20^{c} 23.00
About 4 acres Phillip Allen Land $10.50 42.00
1 share in Ct. & Rhode Island Turnpike Stock 25.00
David Keyes Bond for reconveyance of Mumford land 917.50
10 Acres Wood Land $7 per acre in Eastford near Vine Goddells 70.00

Notes &c—

Dua Spauldings Note due on it Feb 11th 1830 11.81
Chauncey G. Gurleys acceptance of Dua Spauldings order dated Sep. 28. 1829 25.30
John B. Adams acceptance of James H. Prestons draft due April 3rd. 1830 154.50
James Howards Note due on it Feb. 11th 1830 3.02
Darius Wilcox Note due on it Feb. 11th 1830 3.36
Darius Hutchius order on Levi Work Esqr. due on it Feb. 11 1830 3.30
Mason Shermans Note due on it Feb. 11th 1830 14.19
Luther Warrens Note due on it Feb. 11th 1830 83.00
Jonathan K. Rindge Note due on it Feb. 11th 1830 .98
John Griggs Note due on it Feb. 11th 1830 9.90
Henry Works Note due on it Feb. 11th 1830 15.50
Benj Bosworth Note due on it Feb. 11th 1830 8.11

Pardon Parkers Note due on it Feb. 11th 1830		7.19
Luther Warrens Note due on it Feb. 11th 1830		3.98
Theophilus Smith Note due on it Feb. 11th 1830	85.67	
Timothy Holts Note due on it Feb. 11th 1830 about	7.08	
David Deans Note about—in hands of J. B. Esqr.	12.00	
Andrew Watkins Note due on it Feb 11th 1830	36.70	
Andrew Watkins Ex^{n} due on it Nov 3, 1827	6.24	
Phillip Allens Note supposed to be about	4.12	
William Griffin in the hands of J. H. Preston about	1.00	
Dan S. Fletcher Note in hands of J. Bulkley due credit about		42.55
Joseph Vails Note due on it about (J. H. P.)	32.00	
Nathan B. Lyon Note due on it about (J.B. Esqr)		1.89
J. & J. Westgate Note due on it about (E. Stoddard)	36.88	
Saml. Collins Exn. due on it	28.90	
Burnham & Howard Note (Ohio) due on this Note—Feb. 11th 1830 uncertain how much due if anything		
James Scarborrough Note.		2.75
Amos Justin do. 11.28—Benj. Holloway, J. B. Esqr—	21.28	
Danl. P. Harris Exn. 57.90—Saml. Douglass Note 32.27	90.17	
Freeman Snow Note 14.48 Dudley H. Snow Exn. 2.67	17.15	
Jacob Willson, Jr. Note 4.25—Weedon Holloway do. 37.37—	41.62	
Amos Weeks Exn. 9.83—Ephm. Bowens Exn. 3.06	12.89	
Benj. F. Eastmans Esqr.	8.35	
John Griggs, Jr. Note		2.46
Hopkins Kline Note	9.09	
Baruch Smiths Exn. about (A. Clark)	46.00	
Benjamin Eastman Note 120—do. do. 50. do. do. 5.53	175.53	
Ebenezer Eastman do. 3.65 do. do. Exn. 35.65	39.30	
Walter Lyons Note 4.07. Arnold Osmers do. 10—	14.07	
Nathan Weeks Note 2.01, Dan Swifts Note 18.34	20.35	
Benj. Batten Exn. 8.87—Thos. Young Exn. (A. Clark) 40—	48.87	
# 323—23 prs 635 1/2 yds. 4/4 sheeting 9^{c}		57.19
		$28812.89

[Note: Liabilities are not subtracted from the total of the assets.]

The above and foregoing is an inventory of the Estate of the Sprague Mg. Co. taken by the subscribers under oath for that purpose.

Ashford Feb. 27, 1830

Weedon Clark
Alva Simmons

The Annexed is an additional Inventory and appraisal of goods belonging to Estate S. M. Co as made by us the 30th day of August A.D. 1830 being under oath for that purpose, viz—

1 Lathe 9 Head Blocks 15.00—1 Patent Ballance 6.00	21.00
6 Iron ash pails 6.00—1 Hammer at Black Smith Shop .25	6.25
2 stoves & pipe at store 16.00—37 Yds. poor Chambra 10^{c} 3.70	19.70
12 do. do. do. 7^{c}—.84—Ira Barrows note dated 10 April 1830 44.80	45.64
3 Cast Iron Wheels at Grist Mill 10.00—Cash on hand 1.94	11.94
140 Chestnut Rails in the Lane 1 1/2	2.10
Cast Iron Crown Wheel 2.00—1308 feet Pine Boards assorted 13.56	15.56
20 Gals oil at both Mills, 64^{c}—12.80—100 lbs starch do. do. 7.00	19.80
1355 lbs. Cotton Yarn, Roping &c.	203.25
80 Yds 3/4 Shirting on Beams Old Mill 6^{c}	4.80
1851 lbs. Yarn roping &c. on Beams, spools, at New Mill	277.67
460 Yds. Shirting at do. do. 8^{c}	36.00
Benj. S. Deans Note at 4 months dated May 5th 1830 half cash & half produce 6.75	6.25
Burnams & Howards Note due upon it August 1st. 1830	317.60
1 Stone drag	.33
	$987.89

Alva Simmons
Weeden Clark.

McLane's Report (1833)

Name of Agent. James H. Preston
Location or Residence. Ashford
Amount of capital. $26,000
No. of mills. 2
When built. 1814, 1823
No. of Spindles. 1700
No. of Looms. 36
No. of yarn. 20 1/2
Yards of Cotton Cloth. 275,000
No. of males. 25
Wages per day. $1.00
No. of females. 50
Wages per day. 35 cts.
Under 12 y'rs of age. 21 cts.
Whole number supported. 200
Pounds cotton worked. 66,000

Louis McLane, *Documents Relative to the Manufactures in the United States,* Vol. I (Washington, D.C., 1833), pp. 986–987. Document 9.-No. 1.-Schedule of Cotton Manufactories in Connecticut, with some details of their annual expenditures and performances.

Advertisement for Sale (1835)

COTTON MANUFACTURING ESTABLISHMENT FOR SALE The proprietors of the Phoenix Manufacturing Company offer for sale their Cotton Establishment, and its appurtenancies, situated in Ashford, Conn. on the stage road from Hartford to Providence, distance about 33 miles from Hartford and about the same from Providence, and about 26 miles from Norwich—it is situated on a never failing stream of water, and the privelege perfectly secure from freshets and affording an ample supply of water at all times. On the premises are 11 dwelling houses, about one half calculated for two families each, 1 large and convenient store, 2 barns, 2 wood houses, waste house, blacksmith shop, and all other necessary out buildings—2 large and commodious Factories, containing 1800 spindles, 36 40 inch looms, and the proportionate amount of machinery, and all now in full operations. Said machinery has mostly been new within the last year, and of the latest improvement, and the remainder thoroughly repaired, which makes it all equal to the best improved machinery—it is a neighborhood where fuel and other necessaries for the operator can be procured at a lower rate than in any other section of the country, which, with its other privileges, will enable the operator to manufacture goods as cheap as any establishment in New England. There is also connected with said establishment a grist and sall mill in good repair; and at about 100 rods a first rate dam, and convenient place for building, and water power sufficient to operate 2000 spindles. It is also a first rate stand for a country store, as the present proprietors are desirous of closing their business, would feel disposed to sell on good terms, and require but a small part paid down.

For further particulars enquire of Mr. Truman Beckwith, Providence, or the agent, John H. Preston, on the premises.

Ashford, April 25, 1835.

Manufacturers' and Farmers' Journal, and Providence and Pawtucket Advertiser, V. XV, no. 57, Providence (July 16, 1835).

Benjamin Warren

Probate Inventory (1852)

Real Estate:

One undivided third part of the Phoenixville Manufg. Co., consisting of about 60 acres of land, 2 cotton factories, 1 store & ten dwelling houses & water privelege $2200.00

(in addition to a house and farm of 140 acres, a 13 acre wood lot, 20 acres of cedar swamp and a 31 acre parcel called the [Plan] Lot, all valued at $3078.00)

Machinery and other property belonging to [Thomas] & Warren:

43 looms	$ 200.00
23 cards	105.00
3 1/2 mules	296.00
13 spinning frames	579.00
2 Dressers, Spoolers & warpers	45.00
1 Lapper	50.00
1 picker & 1 whipper	10.00
2 Taunton speeders	60.00
2 Drawing Frames	150.00
1 Turning engine	10.00
1 Lathe & tools	5.00
1 cutting engine	3.00
	$1518.00
Blacksmith tools &c. for repairing	24.00
	$1542.00

[continued]

Connecticut State Library, Hartford. Estate of Benjamin Warren. Town of Eastford, 1852, no. 4.2. Eastford Probate District.

One undivided third part of the Machinery &c.	514.00
1788 lbs. Mobile Cotton 10¢	178.80
890 lbs. warp yarn on beams	166.20
80 lbs. Roping on spools 14¢	11.20
250 yds. cloth on looms 5¢	12.50
66 lbs. copp filling 16¢	10.56
22 cords stove wood 2.25	49.50
1 pair Iron Bed screws	14.00
29,492 yds. cloth in baling room	147.47
	$590.23

One undivided half of the above stock in process &c. in the Cotton Factory at the time of the decease of Benjamin Warren.

	295.11
Real Estate	5278.00
Personal Estate	888.35
	$6975.46

Map (ca. 1825)

This map of Phoenixville was drawn by combining historical map information with field measurements. The pattern of growth recorded here was typical of many small manufacturing villages in the first quarter of the nineteenth century. With the exception of a gristmill erected a few years previously, no buildings were located in the valley of the Still River at this crossroad when the first cotton mill went up in 1813. It was soon surrounded, however, by an unplanned mixture of service buildings and dwellings. The stone mill was added to the community in 1823, just a few years after the power loom was introduced to American cotton factories. At the same time, an orderly row of tenements was added nearby for the workers. Compare the maps of Phoenixville and Pomfret, Connecticut. (TZP)

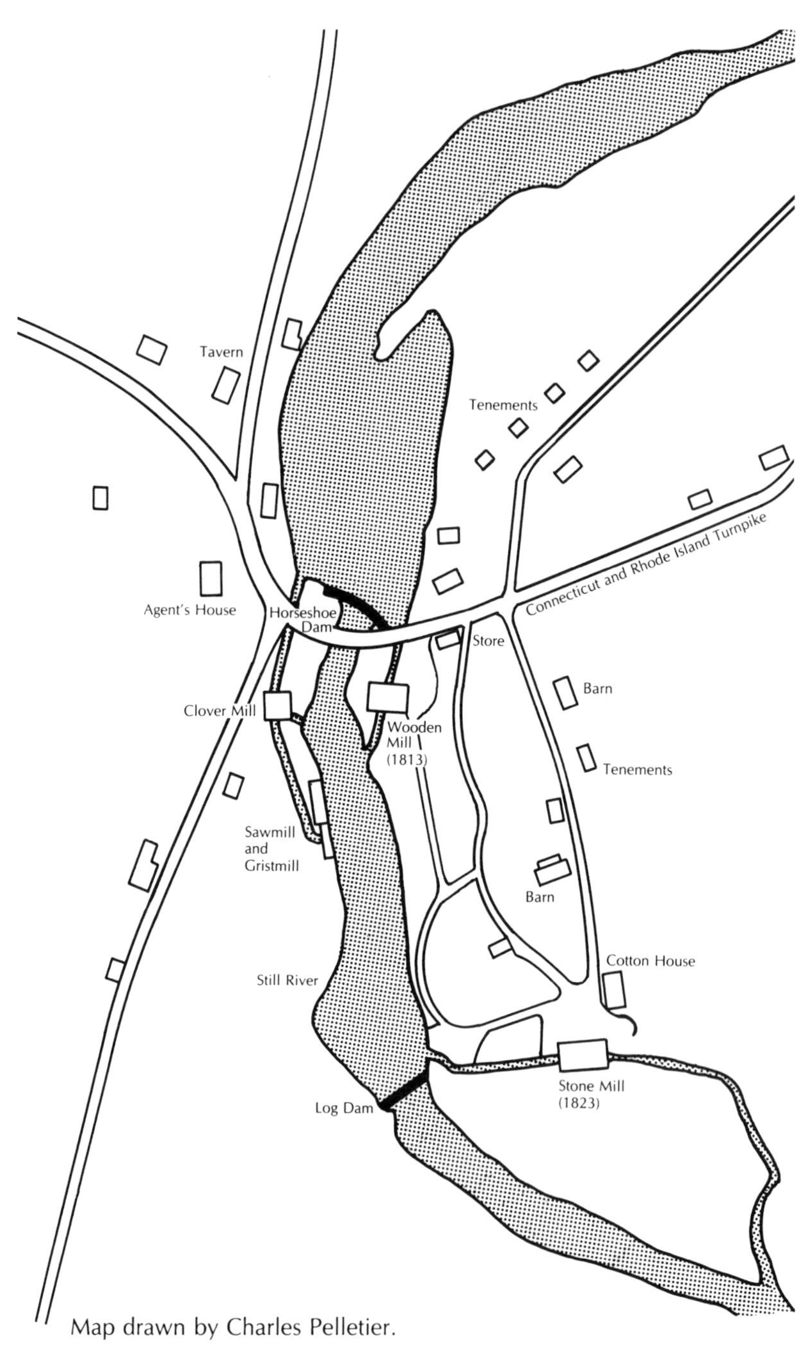

Map drawn by Charles Pelletier.

View of mill tenements and Horseshoe Dam at Phoenixville, Connecticut, ca. 1920. From the collection of Ed Jerzerski.

Ware Manufacturing Company

Records (1824–1825)

The Ware Manufacturing Company was a short-lived firm that developed a cotton mill village on the Ware River in central Massachusetts. This village had been established by an earlier unsuccessful company (see the Ware advertisement reprinted in this collection under "Mill Villages for Sale"). Although the Ware Manufacturing Company failed during the recession of the late 1820s, the community survived around a succession of cotton and woolen weaving and knitting operations, with numerous fluctuations from prosperity to crisis. The Ware Manufacturing Company mill village was remarkable in several ways, most notably because it was backed by a Boston-based corporation rather than a Providence-based partnership. As a result it had access to much greater capital than most village mills. But it had to be administered and supplied from afar. Until the coming of the railroad, long after the Ware Manufacturing Company had collapsed, shipping between Boston and Ware went by way of coastal boat to Hartford, Connecticut, then by canal boat to Springfield, Massachusetts, then overland.

Communication by letter between on-site management in Ware and the directors in Boston was extensive. A letter book, in which each letter to Boston was copied, survives, as do a great many letters from Boston and a variety of other documents relating to the business of the company. Several give very specific details and instructions for the organization of a new mill village community. (MBF)

The papers of the Ware Manufacturing Company are a recent acquisition of the Merrimack Valley Textile Museum, North Andover, Massachusetts.

A New Factory (ca. 1824)

The Boston origin and inspiration for this company is suggested by the fact that it chose to build with brick. Rhode Island mills were first built of wood and then generally of stone, while the Boston textile corporations, starting with the Boston Manufacturing Company in 1813, constructed mills most often of brick. Note that this itemized estimate calls specifically for "Boston glass."

350.335 Bricks— at $3.50 [per 1000] =	1225.82	
Laying do [ditto] ------------------------------------	704.70	
Lime for do --	409.77	
Foundation including wheel pit stoning &c---------	1000.00	
	++++++++++	$3340.29
Roof timbers & framing	672.13	
Covering do: including boards, shingles, nails, trimmings, gutters &c. ------------------------------	907.80	
Floor timbers --	610.47	
Floor boards & laying do -----------------------------	887.62	
Window frames & sashes ------------------------------	226.50	
Glazing 2664 feet of Boston glass 7×9 at 14¢ per hundred feet ----------------------------------	400.00	
setting 6090 panes & priming-------------------------	167.46	
Lathing & stock---------------------------------------	247.50	
plaistering & do --------------------------------------	350.00	
painting --	200.00	
staging & nails --	150.00	
	+++++++++++++	4819.88
Dressing room, including wheel, furnace gearing &c: 75×36 feet--		1800.00
		$9959.17

A Factory House (ca. 1824)

Like the estimate for a new mill, this tabulation of expenses for a domestic structure (written in the same hand) shows the Ware Manufacturing Company calculating, down to the penny, the costs of building a new industrial community.

Expense of a House 28 by 28

1344 E [?] boards @ 100 [i.e. $1.00 per 100 feet]	$13.44
2000 Clapboards @ 100	20.00
10M [thousand] of Shingles @ 258	25.80
1000 feet Roof board @ 80	8.00
13 windows --- @ 350	45.50
1500 feet Floor boards --- @ 130	19.50
14 Doors ------ ------ @ 150	21.00
576 feet of timber @ 4 Cts [per foot]	23.04
468 feet Rafters @ 2 1/2	11.70
330 feet Sleepers @ 2 Cts	6.60
400 feet of joice [joist] @ 100	4.00
800 feet coller beams @ 90 Cts	7.20
182 feet coller beams @ 3 Cts	5.46
150 feet partition plank @ 150	2.25
2 M feet of ceiling boards @ 180	36.00
150 [lb.] Nail ----- 80 Cts	12.00
4144 feet trimming & Laths @ 60	24.86
7M Bricks ------ @ 4 Dollars	28.00
oven ledge & Mantletrees	4.00
8 Hhds [Hogsheads] of Lime	32.00
Mason work	25.00
Diging & stonning Celler &c.	35.00
Carpenter work to finish all [illegible] Chambers	101.69
well	15.00
	526.94

1. Oven ledge is a term meaning the stove lip on a bake oven. Mantletree is not commonly understood to mean "fireplace lintel," but that is what it seems to mean here.

House Lot Gift (ca. 1824)
Boardinghouse Agreement (1825)

These two documents reveal two different strategies for inviting and accommodating a working population in a new village. The house lot gift agreement appears to have been a draft or copy. It is unsigned and appears to have confused one recipient, Preston French, with another named Scott. Note the Second Street houselots on the Ware village map attributed to French and Scott.

House Lot Gift

Whereas the Ware Manufacturing Company, have by their Agents, duly authorised, located a street, on a part of their estate in the town of Ware, in the rear of their present line of dwelling houses; and have laid out a certain number of house lots, containing about half an acre each—which said lots have been offered gratuitously to such persons as incline to occupy and erect houses thereon—and—

Whereas Preston French of said Ware has elected to take one of said lots; and erect a dwelling house thereon, for his own use & occupation; & has made his contracts & collected his materials for the same—

Now we hereby engage and promise the said French in behalf of the said Ware Man. Co. that a free deed of gift of said land shall be executed & delivered, passing all the interests which the said Company have in to the same as soon as the said house shall be erected and fit for occupation by the said Scott—

Boardinghouse Agreement

This Agreement made this fifth day of April A.D. 1825—
—by and between the Ware Manufacturing Company, on the one part and Luther Brown of said Ware on the other part—Witnesseth—
—that the said Brown agrees to undertake the superintendance and direction of the new Boarding house, occupied by him the last

year; and to provide all the necessary help in the family, at his own cost, for the boarding of as many persons as the house will reasonably accommodate, including the washing for all—

—that he will furnish suitable furniture, & bedsteads, beds & cloathing for the accommodation of forty men—

—that he will keep all his accounts with regularity & accuracy and under a weekly account at the Counting room of the names of his boarders & the charges against them—

—that he will use the greatest possible care and accuracy, in preserving the provisions & groceries supplied for the house, and in supplying the boarders—

—that he will preserve the house and furniture from injury and waste—

—that he will cultivate a garden, and raise such vegetables for the use of the house, as can conveniently be done—

—that he will help and tend as many swine as the offals of the home will justify—

—that he will preserve order at all times among the boarders—

—that he will admit the Agents of the Company to a free access at all times—

And the said Ware Manufacturing Co. on their part hereby agree and promise the said Brown—that they will provide suitably at all times for the boarding of such a number of persons as they may send to the house, and also for the supply of the members of said Browns family, a competent number of whom, are to be provided with board free of any charge—

—that they will furnish a cow & provide for her keeping—and—that in consideration of the services and responsibility, recited in the first part of this agreement, to be performed on the part of said Brown, the Company hereby agree to pay him the sum of five hundred and twenty five dollars for one year commencing with the aforesaid fifth day of April—

In witness whereof the parties have hereunto affixed their names.

Thomas A. Dexter for the Ware Mfg. Co.—

In presence of
Josiah Flagg

Articles of Religious Association (ca. 1824)

Just why a copy of such an agreement should appear among the papers of a manufacturing corporation is made evident by subsequent documents in this volume. Corporate influence over the patterns of community and private life was pervasive, but absolute hegemony could not be achieved, as the deletion of the word "every" in this draft suggests.

Whereas upon the mature deliberation of ~~every~~ most of the Inhabitant of the factory village, it has been deemed desirable and expedient, to establish regular religious exercises on the Sabbath at the School house, until some more convenient place is found— and it having been ascertained by a Committee appointed for this purpose, that such an association is strictly legal, and that the taxes assessed upon each individual, for the support of public worship can be withdrawn from the existing parish, and applied to the payment of such preacher or preachers, as may be invited to officiate—

Now We whose names are underwritten, hereby agree to associate for the purpose of forming a religious society in the Ware factory village, to be called the East society, and to organize ourselves accordingly, without delay—

Map (1825)

Comparison of this early map of Ware village and the maps of Pomfret and Phoenixville, Connecticut, shows that the absentee Boston corporate developers of Ware were firmly mathematical and geometric in their organization of the community. In comparison, the new industrial structures in Pomfret and Phoenixville appear to fit into the much more informal and traditional land use patterns of preexisting agrarian towns. (MBF)

Proposed Road to New Braintree
Third Street
Brooks
Kentfield
Graves
Road to Hardwick
Brakeridge
Haynes
Scott
Shaw
Company
Leonard
French
Second Street
Thwing
Company
Washburn
Spooner
Company
Brainard
Magoon
Wetherell
N. Ellis
Dr. Goodrich
Hotel
Road from Northampton to Boston
Front Street
Common
Market House
Ware River

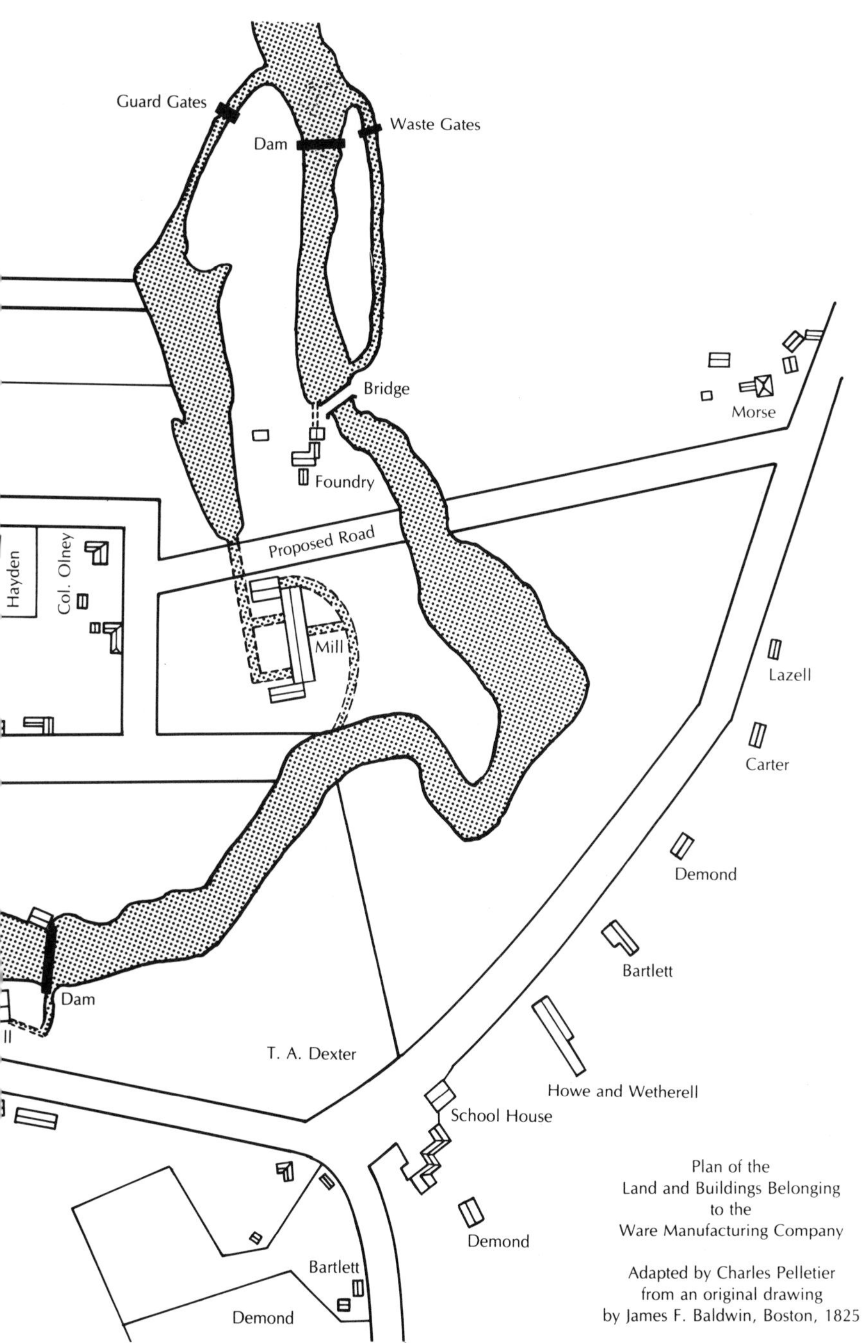

Plan of the
Land and Buildings Belonging
to the
Ware Manufacturing Company

Adapted by Charles Pelletier
from an original drawing
by James F. Baldwin, Boston, 1825

View of textile mill, Ware, Massachusetts, probably 1830s. Mill Village Collection, Research Department, Old Sturbridge Village.

Woolen mills, Ware, Massachusetts, early 1840s. The firm of Gilbert and Stevens was a successor to the Ware Manufacturing Company. Mill Village Collection, Research Department, Old Sturbridge Village.

View of Ware Village, Massachusetts, from Prospect Hill, ca. 1837. Drawn by P. Anderson. Mill Village Collection, Research Department, Old Sturbridge Village. Photo by Donald F. Eaton.

S. V. S. Wilder

Ware Factory Village (1826–1827)

Sampson Vryling Stoddard Wilder (1780–1865) was a Christian businessman and founder of the American Tract Society, whose brief tenure as resident agent of the Ware Manufacturing Company revealed the ways in which traditional cultural forms, especially religious institutions, interacted with the new imperatives of industrial management during the early decades of American manufactures. The eagerness of Wilder and his corporate associates to sponsor a pious manufacturing community is remarkable, though their methods and goals were as rational and manipulative as they were spiritual. They used the power of their money and their organizational logic to impose a church of their choice on a plainly dismayed work force.

The special quality of this entrepreneurial benevolence is suggested by several details. Wilder manipulated the resolution of a disagreement over where to place the new church with a cash inducement rather than proposing, for instance, a democratic vote. The dispute over released time for religious observance was settled by a study of labor productivity, not by an appeal to religious scruples. And the corporation's magnanimity in putting up $3,000 to help build the new church apparently involved a somewhat speculative loan rather than a gift. Wilder did not recall that the stockholders' resolution appropriating the funds concluded: ". . . it being understood that when the building is completed the pews shall be sold at public auction, and

Records From the Life of S. V. S. Wilder (New York, 1865), pp. 196–202, 208–209. The quotation concerning church costs is from a true copy of the stockholders' meeting resolutions, February 21, 1826, Ware Manufacturing Company Papers, Merrimack Valley Textile Museum.

the excess, if any, over the costs and interest, shall be applied for the use and benefit of the Religious Society which may worship therein." The company was apparently fronting the money for the church and hoped to get it back, with interest. Nothing in Wilder's tone or his company's policy indicates cynicism, much less hypocrisy, but theirs was a religious sincerity wholly compatible with good business sense.

Wilder's memoirs, including recollections of his eighteen months in Ware as agent of the Ware Manufacturing Company (1826–1827), were edited anonymously and published after his death. Lewis Tappan, whose reminiscence is quoted in this excerpt, served variously as clerk and treasurer of the Ware Manufacturing Company. (MBF)

It was a strange episode in Mr. Wilder's life, that of his stay in Ware Factory Village, a total change from any surroundings or employment heretofore familiar. Here he found the stir, the gossip, the animation, the regularity of a manufacturing place, and was among a population so much thrown together in their daily employment as to favor the rapid spread of evil or of good. Here, on the retiring of the principal agent of the company, Mr. Wilder was called to combat error in a form new to him, a form seductive and plausible to a superficial mind, as well as tempting to a heart still clinging to sin and to the world. "The fool has said in his heart, There is no God;" may not the same truly be affirmed of those saying, There is no hell?

Into details of business arrangements which occasioned Mr. Wilder's removal for a time to Ware Village, it is not necessary to enter. That it was God in his providence who guided his way, is evident from the result.

He had been induced by friends in Boston to invest largely in a manufacturing company having its works in this village, but with no thought of giving personal attention to the matter, except such as might be required at stated meetings of the stockholders, where his experience among French manufactories gave his voice much weight. Being however soon elected a director in the company, and in September, 1825, its president, it became his duty occasionally to visit and inspect its factories, and afterwards, as will be seen, to make his home for a season in the pretty, square-built cottage belonging to the company, having a piazza all around it, supported on pillars of natural cedar-posts. This cottage was pleasantly located a little out of the village, not far from the rapid little stream carrying the works below, and which was bordered here and there by sheltering groves, destined ere long to become vocal with songs of praise, in place of the Sabbath-breaking revelry which had been wont to disgrace them.

Of his early visits to Ware Village, and of the events connected with his life there, Mr. Wilder writes:

"I soon ascertained that our head-machinist and agent were uncompromising Unitarians, and that they had placed as overseers in all the departments either Unitarians or Universalists; that most of the workmen, consisting of about two hundred in one machinist establishment, were of the same persuasions; that they attended no church or regular preaching, and that on each Sunday some hundred or more young men and women were in the habit of going on board the company's scows and rowing up the large pond of the establishment for a mile or two out of town, and having at a groggery what they called a jollification, thus desecrating the holy Sabbath. I also found that the old inhabitants, who were located from one to two miles around the village, and who were mostly orthodox, stood entirely aloof from associating with the establishment in any religious exercises.

"At a meeting held in Boston of the directors and principal stockholders, a large majority of whom were Unitarians, the question was who among the number would assent to go to Ware to have a supervision over the establishment until a suitable agent could be obtained. Gardner Green, one of the most wealthy men then in Boston, the president of the company, was in the chair. As I had retired from the busy concerns of mercantile life, I was at once fixed upon by those assembled to assume the responsibility of attending personally to the concerns of the establishment. But as I peremptorily declined to comply with their request, Mr. John Tappan and Samuel Hubbard, two of the orthodox directors present, whispered in the ear of the president that, in order to induce me to undertake the overseership, he would do well to propose that $3,000 towards building a church be subscribed, on condition that the good people in the neighborhood would raise $3,000 in addition, and then to invest me with full powers to superintend the erection of said house and of settling a pastor. When I heard the motion seconded, and saw twenty Unitarian hands elevated in behalf of carrying the motion into effect, I did not dare to disregard the obvious call of Providence, and at once gave my assent. On returning home I told Mrs. Wilder that I should leave the next morning for Ware, and hoped that she would cheerfully consent to follow me with our little family as soon as our establishment at Bolton could be placed in a situation to be left for six months, the cottage at Ware being ready furnished waiting our reception.

"On arriving at Ware, it seems that the news had already reached the village of the decisions of the Boston company, and in walking through the various workshops and factories on the afternoon of my arrival, I think I never beheld so many sorrowful and wry faces. On the following evening I assembled all the old inhabitants of the neighborhood, who gave me a most cordial reception, made known to them the object of my mission, and stated to them that I had brought with me $3,000 towards building a meeting-house, on condition that a similar sum should be raised among themselves for its completion.

"In addition to this sum, I was authorized to select a spot on the company's lands for the location of the house. A subscription paper was immediately drawn up, and by heading the paper with $500 on my own account, I had the satisfaction of seeing subscribed that very evening $2,700, and in three days the $300 additional was made up. In ten days I had my plans for the house drawn out, and a contract for building it completed.

"There were however two parties, one of which wished the house to be located on the hill, and the other in the valley nearer the centre of the village. Foreseeing that it would require $500 more to complete the house according to my views, I gave out that whichever party would first subscribe the requisite $500 in addition to their former subscriptions, the house should be located in conformity to their wishes. The hill party having raised this sum, the house was located on the hill.

"In consequence of the lawyer of the village having invited his brother-in-law the Rev. Parsons Cooke to preach for a Sabbath or two, I providentially found this orthodox clergyman at the village on my arrival there.

"The location of the church being decided on, at noon on the following day I dismissed all the hands, both male and female, consisting of some five hundred, and with Mr. Cooke on my right-hand we proceeded, with all the employees and some of the neighboring inhabitants, to the hill; and after addressing the assembled multitude, stating to them our object, and Mr. Cooke's offering an appropriate, impressive prayer for the blessing of God to descend and rest upon the contemplated sacred enterprise, we proceeded to stake out the ground and to consecrate the spot for the worship of Jehovah, Father, Son, and Holy Ghost.

"In less than six months this temple was erected and completed, and I trust, by divine grace, has proved none other than the house of God, the very gate of heaven, to hundreds who now are worshipping in that sacred temple above.

"The next question was to make choice of a suitable pastor. After hearing the Rev. Parsons Cooke for several Sabbaths, the orthodox members of the church, being the majority, with great unanimity fixed on this clergyman, distinguished for soundness of doctrine, superior abilities, and eminent piety, as their first pastor.

"But few months had elapsed before there was a wonderful display of the power and grace of God in that highly favored village, in the conviction and conversion of sinners by the faithful, pungent preaching and parochial visits of the reverend pastor and deacons of the church. A glorious revival of religion was the result; and while hundreds were anxiously inquiring what they must do to be saved, a thrill of sacred joy penetrated the hearts of others, who had found peace, comfort, and consolation in believing on the Lord Jesus Christ. Among these were a good majority of the most inveterate opposers at the commencement of the work, and who afterwards, as far as my knowledge extends, evinced the sincerity of their professions by a corresponding conduct.

"This glorious revival extended into most of the neighboring towns in the years 1826 and 1827, and continued, to a greater or less extent, until the year 1831; at which time the cheering conviction prevailed in those regions that some thousands had been brought by the grace of God to a saving knowledge of the truth as it is in Jesus. Unto God alone be all the honor, all the praise, and all the glory. Thus ended my feeble endeavors to promote the blessed cause of the Lord Jesus Christ in that highly favored village, where I passed eighteen months, instead of six months, as first contemplated."

*

The following valuable testimony to the success of Mr. Wilder's. efforts in Ware is given, among other reminiscences, by Lewis Tappan, Esq.:

"At Ware were a number of large factory-buildings in full operation, under the superintendence of Mr. Wilder, as temporary agent of the corporation. The directors lived in Boston; and although there were in the direction Gardner Green, Esq., Hon. Samuel Hubbard, and John Tappan, Esq., a majority were Unitarians.

"Intelligence had been conveyed to the Board by some of the overseers in the factories who were of the Universalist persuasion, at a time when a revival of religion prevailed at Ware, that the agent sacrificed the interests of the corporation by allowing many of the hands to attend the religious meetings during the regular work-hours. Being about making my annual visit to Ware as treasurer of the corporation, to examine the accounts and look after the general interests of the concern, I was particularly instructed by the directors to inquire into the facts reported, and lay the result before the Board on my return.

"Belonging as I then did to the Unitarian body, I determined to make thorough inquiry, and report the precise facts. Accordingly, after spending several days at Ware, and completing the usual business that called me there, I arranged to have the overseers of the different rooms separately at my room in the hotel, that I might learn from them the true state of things. I went into an examination of the number of hands employed in the different rooms that year and the previous year, of the quantity of cloth and yarn manufactured in each year, and followed up this investigation from evening to evening. On making up my report for the directors, I found to my surprise that the work actually accomplished during that year, when religious meetings were so abundant, considerably exceeded the quantity produced by the same number of hands the year previous. I stated this to the overseers and to the agent; the former being more surprised at the result than I had been, while the agent expressed no surprise at all, it being about what he had supposed.

"On receiving my report at Boston, the directors were well satisfied, mysterious as the result was, that the agent had not been wanting in his duty to the corporation, while he promoted the religious interests of the work-people, who had made up loss of time by greater diligence and faithfulness."

Moses Brown

Deposition (1824)

This deposition by Moses Brown, the wealthy Providence Quaker industrialist, was entered as testimony in complex and acrimonious litigation between Pawtucket mill owners over the fair apportionment of water power on the Blackstone River. Such litigation was common in nineteenth-century New England whenever mill owners built dams.

Brown's testimony gives notice of two separate, though historically overlapping, forms of conflict over waterpower. The first, which led to the excavation of Sargeant's Trench (see testimony) and later to the declaration that the Pawtucket was a "public river," generally pitted backcountry farmers against the owners of blast furnaces and, later, textile mills. Farmers sought to protect a customary entitlement to fish by regulating the construction of dams that obstructed seasonal fish runs. Furnace and textile mill owners, for their part, required large and continuous amounts of water to run their works and were unwilling to open their dams during the spring fish runs. Farmers who owned land upstream of new dams also protested the flooding of their land, flooding that frequently accompanied dam building. Charges by

Moses Brown's deposition (dated January 27, 1824) appears in the records of *Tyler, et al.* v. *Wilkinson, et al.,* 24 Fed. Case 472 (No. 14, 312) (C.C.D.R.I., 1927), also known as the "Sargent's Trench Case"; these records can be found in the Federal Records Center, Waltham, Massachusetts. Legal records arising from the Jenks and Bucklin action are *Kennedy, et al.* v. *Bucklin, et al.,* Court of Common Pleas, Bristol County Massachusetts, Vol. XII, pp. 214–215, located in the Bristol County Superior Court, Taunton, Massachusetts. Research on the issue of water rights led to the preparation of a paper by Gary Kulik, "Artisans, Farmers, and Early Mill Owners and the Uses of Land and Water," delivered at the Smith College Conference on the "New" Labor History of the New England Working Class, March 1979. For further reading, see Morton Horwitz, *The Transformation of American Law, 1780–1860* (Cambridge, Massachusetts: Harvard University Press, 1977).

farmers and large manufacturers, which were brought to the courts and the legislature well into the nineteenth century, reflected, at their root, deep antagonism between two separate economies and two distinct ways of life.

A second form of conflict revolved around the competing claims of waterpower users. Though present throughout the eighteenth century, such antagonism increased in the late eighteenth and early nineteenth centuries, as the power of farmers to defend their rights to fish declined. In 1792, two members of the Jenks family, whose ancestors, as ironworkers, had occupied the lower mill privilege at Pawtucket since the 1670s, and John Bucklin, a grain miller, destroyed the new Slater Mill dam, claiming that it obstructed the flow of water to their gristmill and forge. Invoking the traditional common-law remedy for the removal of public nuisances, the three artisans freely admitted destroying the dam "as Lawfully they might." Though the artisans won their case, the decision turned on technicalities, and the rebuilt dam stood. By the 1820s, after textile mill owners had bought the best water privileges, conflict over waterpower took place mainly between mill owners only—testimony to their growing power. (GK)

I, Moses Brown of Providence in the County of Providence and State of Rhode Island and Providence Plantations in the Eighty sixth year of my Age being affirmed according to Law Testify and say That for many years I have been acquainted with Pawtucket River, the Several Manufactories and Various waterworks on Each side of the same near the Bridge and with many of the people concerned therein. That in 1791 & 92 I was concerned with Oziel Wilkinson and Thomas Arnold in the Purchase of a Water Privi- 1
lege on the said River, on the west side Where the old Cotton Mill and other Mills now stands and including those mill Privileges above the Bridge. That the Dam called the Upper Dam was erecting but not planked above the waters running when on the 31st of 8th month [August] 1792 as was given in Evidence a considerable part of the Dam was cut down, which lead the parties concerned therein and the owners into a unpleasant Lawsuit at 2
Taunton. In the course of the Evidence in Justification it was stated and proved that there had been in the night a plank connected with the frame of said Dam to turn Water into the Trench called Sergeants Trench: (which trench appears to have been made by William Sergeant in the year 1714 by blowing and removing Rocks, for the purpose of Letting Fish get up into the River above the Bridge) which turning they stated was a diversion of the water from its natural course and thereby rendered the Dam a Nuisance which the Court in their charge to the Jury stated in effect that by that transaction the cutting down was Justifiable and the Verdict was accordingly given for the defendants although nothing appeared that the owners of the Dam or any other person for them or by their consent, was concerned in countenancing the placing the plank or turning the water.

John Andrews Esqr aged about 80 years in the course of his Evidence stated that he had when a youth lived at Pawtucket and that it was practised by the Owners of both sides of the river in scarce times of water to pass across the bed of the river and repass to take down and up the materials their dam was built of, which was considered to have an Effect in the Case. It appearing from general confession that there had never been any settlement of the Division of the water or what quantity or proportion of the water of the River, the owners of the works on the said Trench had a

right to, and much controversy among the respective Claimers in a Scarce time of water having occurred I was very desirous of putting an End to that source of mischief in the Neighbourhood and accordingly before the next Trial at Court took much pains to effect by attending four days successively at the Village of Pawtucket in conversing and Labouring with the concerned and
3 as Stephan Jenks Father to the one now in the case and I were intimately acquainted, WE United and procured an agreement before the next trial for the settlement of the proportion or quantity of water to Sergeant Trench owners as per the agreement signed by all the parties under the date of the 6th of October 1796, that owned on the West side of the river and in the course of this business we made such progress towards a final Settlement of the case in Court, that while it was sitting Stephen Jenks Judge Lincoln of Worcester and myself, he having been looked to by both parties, met together by ourselves. We agreed on and he wrote another agreement respecting the Government of the water in the River and the Case in Court was agreed to be settled, each paying their own Costs and we paying to the owners on both sides of the River Five hundred dollars for their Establishing the dam regulations of the water and the Settlement of the case.

John Bucklin an owner on the East side of the mill priviledge there below the Bridge, not fully uniting in the agreement and Settlement, provision was made afterwards for his family he having deceased before the Agreement was signed by all parties on this or West side of the river. this signing was done about four months after, Vis on the 10th of 2d mo [February] 1797 as appears by the Agreement or the Record thereof.

*

I have seen the Bond of William Sergeant to blow up the Rocks to get the Fish up and took off the date which by my minutes was dated September the 4th in the year 1714 before which I believe the waters of the Main River in common or Low times never passed down the Trench I understood they did not succeed in getting up the Fish. Since which viz. in August 1773 as appears of record, An Act of the General Assembly was passed granting liberty to any person to Blow Rocks for that purpose and next year at June Sessions 1774 Stephen Jenks then a member of the Assembly procured a Regulating Act and Stephen Hopkins, Darias Sessions and myself were appointed to restrict or regulate the Blowing up the rocks to prevent injuring the Mill Privileges on both sides the river below the Bridge and I recollect that I went up and got under the Bridge where they were taking the Stones into a Scow that had been blown and directed the Blowers where not to blow, so that the water priviledges might not be injured and the People appeared satisfied.

I am sensible that Pawtucket River is not only an ancient River but was declared by our general Assembly a Public River in their Act of August 1773. . . . 4

*

As to the State of the Dam just above the Bridge as long ago as I recollect, I answer that there was no regular Dam of Timber or Plank across the River between the Bridge and Upper Dam but there was stone, Gravel, Straw and some wood placed up and down the River on each side from Rock to Rock forming above so as to turn the water into the flumes of the mills through the Abutments of the Bridge on each side, which dam was moveable by each owner crossing over from time to time and letting in or out more water as they thought their occasions required. After Blowing up the Rocks as mentioned in 1773 and the fixing the Passage ways on both sides by restricting the moving of the Rocks in 1774 before mentioned, the owners of the Mills that takes the water through the Bridge Abutments, I conclude, built a more permanent Dam with logs across the river above the Bridge to accomodate their respective works. this dam was fastened to the Rocks as much as it now is.

Afterwards soon after the building the Upper Dam, There was about Two feet added on the Dam last mentioned with a view as was believed to Cause back water to flow on the Wheels above which giving much uneasiness to the Village as well as to the owners of the Upper Dam and Works thereto belonging. Application was made to the general Assembly for its removal and a Committee was appointed to view the premises and decide upon it, they finding that it neither served those who erected it, or any others, but was a dangerous Addition lessening the Passage for the water to [pass] under the Bridge might in time of a great Flood turn the water across the Road to the Damage of the Houses, Shops and Water Works west of the Main River including those on the Trench. It was ordered to be removed and was done immediately. Judge Daniel Owen, Rufus Hopkins and If I mistake not Judge Bales, all well acquainted with Water Works were on this Committee.

1. This was the Old Slater Mill. Its dam was referred to as the upper dam to distinguish it from the lower dam at Pawtucket Falls. Brown's partners were Oziel Wilkinson, the blacksmith whose son, David, invented the slide-rest lathe, and Thomas Arnold, a fellow Quaker who built a flour mill in Pawtucket in the 1790s. (For information on the Wilkinsons, see earlier documents in this volume.)

2. This was *Kennedy* v. *Bucklin* mentioned in the headnote.

3. Stephan Jenks was a noted artisan in Pawtucket; he fabricated muskets during the Revolution. His son, also named Stephan, was one of three who destroyed the Slater Mill dam.

4. See Rhode Island Colony Records, Vol. 9, pp. 61, 118, and Rhode Island Petitions, Vol. 15, p. 105, Rhode Island State Archives, State House, Providence. Backcountry farmers had succeeded in having the river declared a public river and in legalizing the removal of rocks that impeded the progress of migratory fish. In response to this, Stephan Jenks's Regulating Act was intended to protect the interests of artisans who used the waterpower at Pawtucket Falls.

5. Rhode Island Colony Records, Vol. 14, pp. 274–275; Rhode Island Petitions, Vol. 27, pp. 83, 116. See also J. R. Bartlett, ed., *Records of the State of Rhode Island,* Vol. X (Providence, 1865), p. 508. Artisans at the lower dam had raised the level of their dam to obstruct operations of the Old Slater Mill. These were the same artisans who had initially destroyed the upper dam.

Joseph K. Angell

Law of Watercourses (1833)

The first edition of Joseph K. Angell's *A Treatise on the Common Law, in Relation to Watercourses* was printed in 1824. The second edition, which is reprinted in part here, was published in 1833. In the preface to the later edition, Angell states that more decisions probably were rendered on the subject of water rights between 1824 and 1833 than "all of an antecedent date put together." That the years spanned by these editions represent a period of dramatic growth in textile manufacturing and factory villages can be verified by comparing the 1820 Census of Manufacturing with McLane's special industrial census of 1833. As factories were almost universally powered by water at this time, it was natural that the growth of manufacturing be accompanied by an increase in the litigation of cases pertaining to the law of watercourses.

The rights to water were of great importance to early industrialists. Not only was water required to turn mill wheels, but it was also needed for a variety of manufacturing processes. In the Sargeant's Trench case, factory owner was pitted against factory owner in the struggle for waterpower. Damming rivers for waterwheels caused a variety of problems for landowners along rivers, of whom millowners were just a few. In the case of the *Wolcott Woollen Manufacturing Company* versus *Upham,* the cotton mills of Southbridge, Massachusetts, came into conflict with a local farmer when they flooded his mowing lot by releasing water from a reservoir upstream. (TZP)

Joseph K. Angell, *A Treatise on the Common Law in Relation to Watercourses* (Boston, 1833), pp. 133–137.

Under the statutes of flowage in Massachusetts, it has been determined, that a mill owner is authorised to create a reservoir of water for the use of his mill, by erecting a dam remote from that at which the mill is situated, and the owner of land lying between the two dams, which is overflowed by the water from the reservoir, must apply for damages in the mode provided by the statutes. This decision was made in *Wolcott Manufacturing Company* v. *Upham*,[1] in 1827; and, as it is the first and only case upon the subject of the right to make reservoirs, it will no doubt be satisfactory to the reader to present the case just as it has been reported.

This was an action of trespass brought by the Wolcott Woollen Manufacturing Company, and certain individuals doing business under the firm of the Columbian Manufacturing Company, against the defendant, for breaking and entering, on the 10th of August, 1825, into their close in Sturbridge, situated on both sides of the outlet of Watkin's pond, and shutting down the plaintiff's gate in their dam, thereby preventing the water of the pond from running to the plaintiffs' grist mill and woollen and cotton mills, situated on the Quinabaug river, in Southbridge; by reason of which the plaintiffs had suffered damage for want of sufficient water to drive their mills.

The defendant pleads, that long before and on the 10th of August, 1825, he was possessed of an ancient mowing lot, situated on both sides of a brook, flowing from Watkins' pond to Quinabaug river, on which lot he, and those heretofore possessed of the same, have been accustomed to cut and make into hay large quantities of grass, which from year to year grew thereon, free from any flowing of the same, and that the grass

1 Wolcott Woollen Man. Co. v. Upham, 5 Pick. Rep. 292; and see Butz v. Ihrie, 1 Rawle's Rep. 218.

being then ripe, &c. he cut a part of it, &c. and which was then lying on the lot, and not sufficiently made into hay to be carried off, and was cutting the remainder, &c. and the plaintiffs having recently erected a dam across the brook above the lot possessed by the defendant, and thereby raised a large pond of water, and having on the 10th of August raised the gate of their dam, and thereby let down so much water as to overflow the mowing lot, so that the defendant could not make the grass into hay, &c. for the purpose of stopping the flow of waters on and over the mowing lot, and of enabling himself to make the grass into hay, &c. he on the 10th of August entered upon the premises of the plaintiffs and shut down the gate, &c.

The plaintiffs, admitting the facts stated in the plea, reply, that on the 10th of August, and long before, they were the owners and occupants, in severalty, of two mills situated on Quinabaug river, about three miles below the ancient mowing lot of the defendant; that for the purpose of raising a suitable head of water for driving their mills, they, on the 31st day of December, 1822, purchased jointly, in fee simple, and have ever since been jointly seised and possessed of a tract of land in Sturbridge, about a quarter of a mile above the mowing lot; that on the same 31st of December they erected upon this tract of land the dam and gate mentioned in their writ, by means of which a suitable head of water was raised for driving their mills; that for the purpose of using the water, they from time to time, and as often as occasion required, opened the gate, so that the water flowed down in sufficient quantities to drive their mills; that on the 10th of August, having occasion for the water reserved and collected by the dam and gate, for the purpose mentioned, and the same being a suitable and proper occasion for the use of the water for such purpose, they opened the gate, whereby the water flowed in the brook and over the ancient mowing field to the plaintiffs' mills in such and no greater quantity than they had occasion to use in driving their mills to the best advantage; and that on the 10th of August, while the water was so flowing, for the purpose mentioned, the defendant entered, &c. and shut the gate, whereby the plaintiffs were prevented from using their head of water to the best advantage. To this replication there was a general demurrer.

Per curiam. The facts stated in the plea are a sufficient bar to the action at common law, and the question is whether the replication brings the plaintiffs' case within the statute of **1795**, *c.* **74**, or any of its supplements, for the support and regulation of mills. The encouragement of mills has always been a favorite object with the legislature, and though the reasons for it may have ceased, the favor of the legislature continues. By ***St.*** **1824**, *c.* **153**, the provisions of the mill acts are extended expressly to damages caused by flowing lands below any mill dam. The phraseology of this statute, which may be considered as a legislative exposition of the former ones, is very general. "Whenever any person shall complain, &c. that he sustains damages in his lands by their being flowed, whether said lands shall be situated above or below any mill dam," &c. We think that if the facts stated in the replication do not bring the case within the former statutes, it certainly is embraced by these words.

But it is objected that the operation of the statutes, under this construction, will be inconvenient; that it will be impossible for the jury to assess suitable damages. It is true, it would be difficult in regard to future damages, but the same difficulty occurs in other cases under these statutes; especially where the flowing is below the mill dam. The amount of damages will depend on the use to be made of the water, and on the times when it is to be drawn off; but it may be estimated from the damages which shall have been actually sustained. It is said also, that the jury cannot determine how far it will be necessary to flow the lands, and during what portions of the year. There may be a difficulty as to the time, but there is none in regard to the height of flowing. The jury might however find, that the mill owner shall not flow the lands in the season proper for making hay. But the inconveniences insisted on cannot be very great, as the assessment of damages, &c. will be subject to alteration upon the application of the party thinking himself aggrieved. We do not perceive that they are greater than in many other cases of this nature, and, as before observed, the present case comes clearly within the meaning of the statute of 1824. It is very common that two or more ponds are required for a mill, though they are not often so remote from each other as in this instance.

Town Clock (1828)

In almost every mill village, the only record of public time was the mill owner's factory bell. Advertisements for workers in Pawtucket often carried references to a work day, not defined by the total hours worked but by the factory bell. Such a system led to abuse, as mill owners recorded time to suit their needs. The erection of a public clock, therefore, would implicitly counter the owner's exclusive power to define public time.

The following editorial raises the issue explicitly. It gives notice of a clock to be built by public subscription and located in the belfry of the village's new Congregational church. Designed to reduce confusion, to serve as a "regulator," and always to give *"the time of day,"* the public clock would place limits on the power of mill owners. Its construction is further evidence of the suspicion and hostility that mill owners faced in Pawtucket. (GK)

Pawtucket Chronicle, October 18, 1828. The "Report of the Committee on Education," reprinted in this collection, contains evidence that the conflict over factory time in Pawtucket continued.

TOWN CLOCK

We are highly gratified to learn that our citizens are about to purchase a time-piece; the expense of which will be $500, and that subscriptions are now solicited of the inhabitants to defray the expense, which, so far as the paper has been shewn, has been promptly and liberally subscribed to. A deficiency of about one third the amount, we are told, now exists, which we have no doubt will be readily and cheerfully made up. A time-piece which can be depended upon as a regulator, located in so central and public a situation as the new Congregational Church, will be of great utility in this village. All are aware of the vexatious confusion occasioned by the difference of time in the ringing of the factory bells at the time, and which can only be remedied by erecting a clock that will aways give "*the time of day.*"

IV

Owners and Agents

Most of the documents in this section indicate what the experience of village manufactures looked like from the management point of view. They reveal the attitudes, stratagems, and arguments of the men who owned and superintended the mills. A number of them also speak in the first person of the private experience of such men, who spent their lives struggling with an uncertain economy, recalcitrant machines, independent-minded workers, unpredictable weather, and sometimes conniving business associates. The one sharply negative statement in the group, the excerpt from Thomas Man's "Picture of a Factory Village," is probably also a voice from the world of management. Its author appears to have been an alienated member of the family that established the mill village of Manville, Rhode Island, where the poet was born.

Autobiographical works of early industrial owners and managers are rare. Those printed here are the most important available from the entire history of the New England mill village. George S. White's apology for "The Moral Influence of Manufacturing Establishments" is, in contrast, an example of a very common kind of document. Such pro-industrial arguments appeared widely in the middle-class and commercial press. The excerpt from Zachariah Allen at the beginning of this volume is another example. Volume 1 of this series reprints a variety of such arguments in support of "the factory system," along with the most pertinent and influential attacks on that system during the decades when it flourished in New England villages. (MBF)

William Fisher

Memoirs (1878)

A number of early nineteenth-century textile manufacturers worked their way up through the ranks in one way or another. Smith Wilkinson, for instance, though a member of a prominent family of early New England manufacturers and a brother-in-law of Samuel Slater, began his career as a child working in Slater's first mill. William Fisher (1788–1878), an acquaintance and detractor of Smith Wilkinson, was another manufacturer who started humbly. He went to sea at seventeen in 1805 and intended to make a life of it. Two years later the Embargo of 1807 put him back on shore, and an offer to work in a recently built cotton mill launched him permanently in a new direction. He began in supervisory jobs in cotton mills in southeastern Massachusetts and later operated a mill in Killingly, Connecticut, under contract to the owners. After working in the industry for more than twenty years, accumulating experience, capital, and contacts, he entered into a temporary partnership and built his own mill and village (Fisherville, now North Grosvenordale, Connecticut), later becoming the sole owner. He remained in business another quarter of a century, but like many of his contemporaries, he eventually failed.

Fisher's memoirs are one of the few accounts in which an early manufacturer outlined the course of his career. In the years following those covered in this extract, he and his wife were detained in Georgia during the Civil War. They were finally rescued by a Union gunboat and eventually returned to live in Fisherville, where he wrote his memoirs and died in 1878 at the age of ninety. (RP)

These portions of William Fisher's memoirs were first printed in the *Flyer* of Slater Mill Historic Site, volume I, no. 6(September 1970) through volume II, no. 3(March 1971). They have been reedited here by Michael Brewster Folsom, with permission. Spelling, capitalization, and punctuation are as in the original, but paragraphing has been introduced.

About this time [1807] the Dr. [Nathaniel Miller of Franklin, Massachusetts] invited me to call and see his little cotton factory, where he had been preparing to manufacture thread. The machinery at this time was standing idle, in consequence of disputed rights of ownership, which soon after was contested by arbitration and became the property of Dr. Miller. The Dr. asked me how I should like to superintend the concern? My answer was, if I were acquainted with the business I should like it. He said there were about 500 spindles, and the necessary preparation for running them. The operators were all somewhat experienced, and I could learn to manage them. At any rate I might make the trial and, if successful, we would afterwards agree on wages. I accordingly commenced, and soon got a little insite in the operation and management of the machinery, and in fitting the goods for the market. The wages allowed me was not quite satisfactory. I however saw no prospect of going to sea again, & concluded I might as well abandon the idea, and improve myself in the art of manufacturing cotton.

I continued with Dr. Miller to the end of the year, which was about the 20th of March, 1809. I attended church in Medway the sabbath before leaving Franklin, and there took the Measles, which confined me about a week at my Br. Ellis's in Attleboro, after which I contracted with the Attleborough Manufacturing Co. to superintend their carding department for one year, at one dollar per day. Board was one dollar and fifty cts. per week. Wages were small but my location was pleasant and prospects promising. I had a growing attachment for the young school teacher before alluded to, and convenient opportunity to visit her.

I remained in this situation about three years, with increased wages from time to time, in the course of which, I married and became a father. There were a number of owners in this manufg. Co. among whom were Maj. Ebenezer Tyler agt. Danl Babcock Elias Ingraham & Edward Richardson of Attlebo. Nehemiah Dodge, Abner Daggett & Maj. P. Grinnell of Providence, all of whom except Daggett and Ingraham were Free Masons, which circumstances gave me a favorable opinion of the Fraternity and eventuated in my becoming a member. I rec^{d} three degrees in Bristol Lodge, Mass., and four degrees in St. John's Chapt. in

Providence, R.I. (The different offices that were conferred on me from time to time I will not record, but I will say I always had a favorable opinion in Free Masonry).

About this time a Misunderstanding between Maj. Tyler & m[y]self in regard to lighting the factory occurred, which caused a separation for a season. Very little loss of time, however, and no diminuation of wages occurred by the change. Not many hours after I had resigned my command in the Mill Lemuel May Esq. called and offered me a situation in the Attleboro Falls Cotton Mill which I accepted, and in a few days found myself and little family domiciled in that village. My expense in moving were pd. and the same wages allowed that I had been having. My business was to superintend the whole mill, which I found some more laborious than the oversight of one room. I however saw that nothing would be lost by a thorough knowledge of all the different branches of Manufacturing, and therefore tried to content myself and go ahead.

In the course of my stay here, my good wife presented me with a second son, a fine red haired cherub. I now began to feel sick, and to look ahead for the manner of training my little offspring. I had even before the first one was born, made up my mind that should it be a son, to give him a liberal education, and prepare him for the Gospel Ministry. But all this was hid in the Sequel.

I could see many ways to use money and wished to increase the means of accumulation. I visited a place of amusement, where racing of horses and men had been notified, and saw much gambling in process. Persons of my acquaintance seemed very fortunate in taking up twice as much money as they laid on the dice board, and thus I was tempted to try my luck as the saying is.

At first I was successful, and thought I had got into the right track, but the scene was soon changed, and I found myself a looser. I was then invited to try a game of cards, which proved no more lucrative. The day was past and it had now become nearly daylight and I had lost more than thirty dollars, which indicated very plainly that I had done wrong, and more than the loss of money, I was troubled in my conscience, for having so basely treated my dear wife. I was so smitten that I crooked my elbow, and promised my God, and my dear wife, never to do so again, which promise

I have never broken. On my return I found my darling wife in tears, and nearly sick from loss of sleep. I made a due confession of my guilt, and a promise of future regard for her happiness, which I have uniformly kept. Her forgiveness was religiously rendered and is still fresh in my memory.

I continued on my contract about ten months when I received a letter from one of the owners of the Attleboro. Mill, inviting me to return to my former employment, and offering me an advance of wages. My reply was that I was under engagement till the first of April, and could not consistantly leave before. But another letter soon came, asking me to see what damage would be required should I leave my present employers immediately? Accordingly I put the question to my employer which was negative. I then told him I should leave on the first of April at all events when my engagement was fulfilled. He then said if I would procure a satisfactory person to take my place, he would release me.

It so happened that a man of experience with a large family appeared and I was released without further ceremony. It was now the first of March 1813. My wages was increased fifty cts. per day and I found employment by the Attleboro. Co. for three years more. I found a welcome return by the owners and other friends. The machinery had been badly managed in my absence and some of the cards nearly spoiled, so that I had plenty of business in making repairs and restoring order. But everything seemed to go on satisfactorily.

Our new agent, Capt. Joshua Rathurn, was a very pleasant man, and exerted himself to the best of his ability, but he found the management of a cotton mill, very different from the management of a ship, which had been his employment for many years. But during the War he discontinued his maritime business.

During my residence here I was blessed with another son and a daughter. I now felt that something besides day wages might be necessary to support my growing family. I accordingly invested five hundred dollars in a small cotton mill that was then being built, and after it was prepared for operation the war was ceased & there was such an influx of foreign manufactures, that nearly all the cotton mills in the country were stopped and I saw no way to support my family with necessaries for at least six mo. and go to the Grand Banks a cod fishing. This idea was thought to be rather wild, but I started for Plymouth in the month of March on foot, the ground being frozen and very rough, a distance of forty miles, which I intended to reach before I slept, but late in the P.M. I stopt at an Inn for refreshment. On rising to pursue my journey I found myself too lame to proceed and put up for the night. The next morning I rose early and found the ground covered with snow several inches deep, and very hard travelling, but I reached P. in time for breakfast.

After resting an hour or two, I visited the wharves in pursuit of a good Banker for cod fishing. I soon had a fine schooner pointed out and the owners' whereabouts. I was not long in making a contract and to take charge of the vessel in a few days. Having accomplished my business I had now a bad job before me. Snow deep, and very lame, to get home to Attleborough. There was no direct conveyance, but a stage was going to Boston, and I could get there in time to take the stage for Attleborough, which I did, and when I reached home to my great surprise I was informed the Company had concluded to start up the Mill, and were depending on me to opperate it. This was pleasant news, but how to avail myself of the birth I knew not. I was however acquainted with a Cape Cod Fisherman who could be well recommended, and I soon made an arrangement with him to go to Plymouth and take my place, which was satisfactory on both sides, and a profitable season to the owner and crew.

I was now all right again, and apprehended no further trouble. It was however not many Months before the Mill was stopped again, and no one could tell for what length of time. I thought it might be a long time and some other business must be resorted to. I tried fishing along shore from New Port to main without much success, and I hired a horse and lumber box, and procured a quantity of cotton thread with a few other articles which gave me a tolerable prospect of success in my new calling.

The sleighing was good, and I commenced a Northern Tour, through Worcester, over Rutland Mountain on one of the coldest Fridays that ever existed in this latitude. I however, after reaching the top of the hill, which I walked to favor my horse, and keep myself from freezing, hastened to get my horse under cover, and myself in a comfortable resting place, where I could wait for a favorable change in the weather. The next morning though very cold, was a very little softened, and I put forward towards Kene N. Hampshire where on my arrival, I found a number of stores, and made little trades, but not enough to make me proud of my calling.

From Kene I journeyed to Charleston, where I found a merchant that had plenty of good butter & cheese, and was ready to barter for my goods, which when I saw an opportunity to make a small profit I was not long in closing a bargain. I exchanged the most of my goods and found I had added much to the weight of my loading. But I was willing to walk up hills, and favor my horse when necessary.

I got along very well the first day, but at night the weather again changed, a warm rain set in, and the travelling was wretched. I walked the most of the way in snow and water, but reached Attleborough before I slept. I retired to bed with a firm resolution never to be honored with the occupation or title of a pedlar, if I could obtain any other possible means of livelyhood.

I soon after visited our little new factory, and conversed with the principal owners about putting it into opperation. They were all willing enough if it could be done without loss, or at least for a small gain. This I think was about the first of March 1817[.] Abiather Richardson had then the care of the Mill, and proposed to have me accompany him to Marble Head where he had business, and on our return to Boston, we would make enquiry about the state of the Market, the price of Cotton, &^{c}.

Accordingly we did so, and found encouragement. We therefore purchased a few bales of cotton and started the Mill, which was trusted to my management, and in a few days I had sufficient quantity of yarn ready to try the market. I went to Boston with it and disposed of it so favorably that we called a Company Meeting and voted to continue Manufacturing under my agency, on a salary as agreed.

I continued to make yarn for a season, when it was thought best to purchase looms, and make cloth which we found more profitable. At the close of my third year, I made up my mind to change my residence, and to try to find a more profitable business, as my family were multiplying.

At this time I had my fifth child, a daughter three weeks old. It was not long before I heard of a place in R. Island, where an experienced operator was wanted, and I hastened towards the place, but stopping in Providence I fell in with David Wilkinson who learned my business, and in a very impressive manner said, hold on, I am engaged for a few minutes and will see you again. In due time he returned and said the Killingly Manuf^{g} Co. had a very fine mill standing idle, and he as one of the company would be very glad to have me go & look at it, and he would go with me, but first he would introduce me to Mr. Elisha Howe their Agt. who was then in town.

This having been done, and it being late in the day, I agreed to meet him at the factory the next day, and see what was to be done, and made up my mind what I must have for Manufacturing cloth per yd. On my way to Providence, I studied the matter faithfully, and fixed the price per. yd. that I must have for spinning and weaving, but when I saw the Agt. he thought I was extravagent in my demands, and could not comply with them at all, but he found I knew something about the cost of Manufacturing, and began to reason on the subject which brought him to near my terms, that I concluded to close a bargain, which was to be kept secret between the parties.

I then returned home and found an experienced weaver who thought I had made a good bargain and wished to join me in equal copartnership, which I consented to, and articles of agreement were entered into for one year, under the firm of Wm Fisher &^{c}.

We, after I had settled my business with the Farmer's Co. started for Killingly, where we arrived on the first day of April, 1820 about 10 oclock P.M. On our way we found that the secret had been divulged, and discouraging salutations were ready for us before we reached the factory. Even Smith Wilkinson one of the owners, who had been long in the business, said the agent had done wrong in letting the mill to us, for we could never manufacture at that price, and would fail, and would only make the Co. trouble. He had long experience & he certainly could not do it so cheap.

The mill had been operated about seven years, and had run the owners in debt about $14,000. The machinery was much out of order, and required much repairs before we could operate it. We however soon started up the spinning, and found it worked well, but the old fashioned cotton picker, situated in the garret was a miserable consern, and required as much power to operate it as all the other machinery. Besides it had a very bad effect upon the speed of the mill.

We soon petitioned for a new picker but Wilkinson said he never would pay a cent toward one, it was as good as his. We therefore concluded to by one and put it in opperation at our own expense, and we soon felt remunerated by it. It did not cost half as much to pick the cotton, made less waste, and reduced the strain upon the water wheel and all the shafting. By this time we had 24 looms in opperation, and began to turn out the cloth, and we felt well satisfied that our labor was not in vain. The first year soon rolled away, during which time our worthy Agt. Mr. Howe lost his life by falling through a scuttle in the upper story of a cotton store in Providence. He was succeeded by Col. John Andrews who renewed our contract for a second year.

Much satisfaction was manifested from the quantity and quality of the goods manufactured, and no dimunition of price was asked. The second year soon run round, and the third was commenced upon the same terms. Before the close of the second year, Mr. Wilkinson called at Dr. Grosvenors one of the owners, as he was on his way home from Providence late at night, to inform him that the Killingly Co. were paying no interest money.

The third year soon passed away, and the fourth was equally as prosperous, and the Co. made semi annual dividends. This year brot me another son, and some of the Co. thought we were making too much money and made some attempts to make us work cheaper, but did not succeed. I thought the increase of numbers in the family would not justify a dimunition of the means of support. We however continued through the fifth year without any change.

But the idea of our making a great deal so excited the envious disposition of S. Wilkinson who in the first place said we could make nothing, that he encouraged Augustus Howe his brother in law, when it was about time to make a new contract, to supercede us, by offering to manufacture at a less price. And what was very strange a blind advertisement had been published in the Prov[d] papers that a Factory of____Looms____spindles about 30 miles from Prov[d] could be let for a year from the first day of April ensuing. Not a word had been said to us that a change was desired, much more contemplated.

It however from some accidental cause reached us that there was mischeif brewing, which should be immediately attended to. It was now P.M. and snowing tremendously. I with a smart horse and sleigh started for P. at the half way house in Smithfield the Landlord asked me if I was aware that we were to be superceed on the first of April. I answered no.

He said A. Howe had gone to Prov[d] for that intent, and said he intended to propose a dimunition of half a cent a yd. I thanked my friend, and started on. I soon met Howe on his return but did not stop to talk. I reached Prov[d] too late to see the Agt. that night, but next morning I met the Agt. and some of the owners. Among others Wilkinson, with whom I had some unpleasant words, and intimated that I felt their treatment to be very ungentlemanly, especially in the blind advertisement. If they didn[t] want to let their mill to us, all they had to do was to say so, and we would take ourselves out of the way at the proper time.

I was now about ready to leave to Killingly but Wilkinson wished me to call at the counting room where the Company were in session. I did so, and was informed by the Agt. that they on opening the cealed proposals had found an offer half a cent per. yd. cheaper than they were paying us, but they would let us run the mill for another year for one quarter of a cent deduction which was agreed to.

I then returned and laid the matter before my partner who thought a quarter of a cent per yd. would make rather small profits for us, to which I agreed, and offered to give up the business to him, but he said no, I am the one to go. You obtained this situation and generously took me into company with you, and I have done well. Thus we agreed to separate. I continued the opperation of the Mill and Mr. Sparrow purchased a farm and moved on to it.

At the commencement of our 5th and last year we made a contract with Judge Eben^r Young to opperate his Chestnut Hill factory for one year by the yd. We had but just got it nicely under way, when we found we had expended more than $5,000 of our own money, and we had occasion to go to a furnace in Prov^d for castings. I was asked how the Judge was progressing? and after answering I was told that they should put their claim against him in suit immediately.

I enquired the am^t which they said was about $300—I told him I thought he was good for that amt. Will you garante it? I sd. yes & started for home, and We held a council with the Judge, and told him that we had put his Mill in nice order, and we were making money, provided he would secure us in our own pay, we would relinquish our contract. We showed him our expense acct. and also profit, and convinced him that we were making money, and that he could do the same by assuming our contract, which we would give him, if he would secure us for what he owed us, which he agreed to do, and gave us satisfactory security for our claims and we abandoned all claims on the Chestnut Hill Factory, so that nothing stood in the way of his [Sparrow's] farming and my manufacturing.

I now put Shoulder to the wheel, and soon found that I was making more money than I had ever done before. There were more Spindles in the Mill than were necessary to supply the looms with yarn, and the company consented to put in 8 more looms, which was a great enhancement to my profits.

I went through the year and made money for the Company as well as myself. The 7th year was commenced and continued at the same price. In the course of this year I purchased at auction, some real estate in the village of Thompson, which seemed to give the Killingly Manuf^g Co. an idea that I was making too much money, and something must be done to cut me short, another plan must be resorted to, and under bidders must be found which seemed a very easy matter.

I went to Prov^d at the proper time for making contracts, and found that I must work cheaper or lose my place. The agent, however, told me plainly that they would give me a quarter of a cent per yd. more than any new applicant. The price was named, and I was told I could have the contract continued from one to five years, but I thought three years would be long enough, and accordingly papers were drawn and executed for that length of time.

I was fortunate in the length of time as there was a great falling off in the market price of such goods the last year. Soon after the commencement of this contract I was invited to call up to Thompson and look at a water power that was for sale, which might be purchased cheap. I accordingly called on Judge Nichols who with Darius Dwight accompanied me to look at the premises. We found a gentle fall of about 5 feet on the French River with a dam commenced, and a farm attached of 160 acres which we could purchase with the dam completed for $3,300—

We found by examination that by taking the water out of the pond and by conveying it by trench 100 rods below would give us 10 ft. fall, or head and fall, and give us an elegant building lot for factory and houses, and adjacent to stages and other means of conveyance. We took the matter home with us, and after due consideration, conceeded to purchase.

As soon as convenient the purchase was made, and deeds executed and recorded. Judge Nichols was appointed Agt. and the work of digging trench and erecting a store was immediately commenced. In a short time we had a nice two story brick store ready for use.

At first we thought of building small, and not make large debts. We divided the property into $1000 shares. J Nichols took ten, Self ten, and Dwight five, but after consideration, it was thought best to build large enough to improve all the power, and if we found it necessary could take in other partners. We agreed however to keep a majority of the stock in our hands, lest we should get into trouble by being out-voted.

But in a very short time application by two brothers by the name of Fenner from Providence was made for as much of the stock as we would sell. We agreed to let them have 20 shares and that would leave us a majority of five shares, or votes. But it was not long before the Fenners began to show a disposition to take undue advantages, and were anxious to take the lead. It was not long before we had notice that Dwight had sold his interest to them, and they had gone to Sturbridge and ordered a different sized wheel without consulting us (Nichols & self). They went on to Providence and sold the factory to their creditors and failed.

[Fisher continues with a lengthy account of the stratagems and compromises necessary to extricate the Thompsonville mill property from debt—including a conference with his creditors "on the line between Con. & R.I., where we could talk secure from arrest." Once the creditors were satisfied and the factory successfully started, Fisher entered into protracted negotiations to buy out Judge Nichols, and then to take into partnership a Colonel John Andrews of Providence.—MBF]

My family had now become very expensive. One son in College, and several other children away from home at school, I found that my Salary was not sufficient to support us. I thought I had better sell out and go to the West. I accordingly proposed to sell my partner who seemed willing to buy, and asked me to set a price. I offered my interest in the concern for $30,000 which he said he would give, and I went to Providence to give him a deed and take my pay.

But, after talking the matter over, the Col. said the price was low enough, and he was willing to pay it, but said he I think at your time of life, it would be better for you to buy my interest, and take your boys into company with you, and you can soon own the whole property. You can give me a Mortgage on the Estate and take your own time to redeem it. I thought the matter over and advised with friends who encouraged me to buy which I did and soon after took my sons Wm. & Ellis in the firm of Wm Fisher & Sons. From this time we had but very little encouragement for manufacturing. We had poor markets and many bad debts. And worse than all, my sons who were my partners were in bad health.

In May 1841 I was chosen a representative to the General Assembly and 1842 again and while at New Haven my second wife died. I rec^{d} the notice by a special messenger, and returned home to attend the funeral, after which I finished the session at N.H. [New Haven] and attended a special session in the fall. At the close of this Session I first saw the lady who afterwards became my third wife. My Son W^{m} on whom I depended much for aid in the transaction of my business, was now evidently declining in health, and I was obliged to resign him to that Great Being who gave him to me. He left a wife but no children. from this time my business seemed to be waining. My son Ellis was but little acquainted with Manf^{g} and I was obliged to let the factory out to different contractors who did not do me justice.

My oldest son after graduating at Amherst studied and commenced the practice of Medicine and Surgery in Prov^{d}. His health has never been perfect since he was 18 years old, but from many changes in life he "still lives". In the course of the winter 1843 I married the lady before mentioned and found in her a kind and pleasant companion. Ellis my third son married his first and second wives, and lived with them until they died in Thompson. His health became bad, and he went South and married a third wife in Georgia, where he died of yellow fever. On the banks of the Big Satilla he closed his earthly career Oct 5th 1854. My oldest son by my 2^{d} wife Geo. A. married a daughter of Dr. Bengass in Norwich and was after very unhappy, and unsuccessful in business. Francis, my youngest son, went to live with my son-in-law, Penniman, in N. York but did not succeed well. He finally died of consumption, Decm., 1862, and was buried in Fisherville, Cont. Augustus my oldest son Married the only daughter of Judge Nichols and settled in Providence where he commenced the practice of his profession as above stated.

I continued the manf^{g} business in F.V. [Fisherville] until May 1854 when I was obliged to assign what I had left for the benefit of my creditors. In 1855 I left Fisherville and went to Brooklyn, N.Y. where I and my wife embarked for Satilla, Georgia. After remaining here several months, Penniman sold his Mills, and we all returned to N. York, where we resided Several months, and then moved back to Satilla, Penniman having repurchased said Mills.

[Fisher's memoirs conclude with a long narrative of hard luck, involving numerous journeys back and forth between Connecticut and Georgia, complicated disastrously by the Civil War, which trapped the family in the South.—MBF]

N. B. Gordon

Diary (1829–1830)

In contrast to William Fisher, who rented and eventually owned a cotton mill, N. B. Gordon was a salaried manager of several small factories in New Hampshire and Massachusetts. His diary, which he kept from at least 1828 to 1834, is a rare source of information about the inner workings of an early textile mill. During the years covered in this selection he was the agent of the Union Cotton and Woolen Manufactory in the southeastern Massachusetts town of Mansfield. In his continual efforts to get his mill running smoothly—often without success—Gordon was closely involved in the problems of water and raw material shortages, broken machinery, and absent workers. His was not a life into which industrial employment brought monotonous regularity. We have no direct evidence of how his employees felt about their work, but some of them seem to have maintained more independence than Gordon liked. (RP)

The portions of Gordon's diary quoted here are in the Gordon Papers, Baker Library, Harvard University. Another segment of the diary is owned by the Winterthur Museum, Winterthur, Delaware.

[1829]

[January 3] Extreme cold day.—One weaver sick.—4 Looms stoped

[January 5] One weaver sick 4 Looms stoped.—Water wheel froze up this Morning.—Took untill 8 o clock to start it

[January 6] One weaver still sick.—4 Looms stoped.—Water failed some about 4 o clock P.M. & stoped the old spinning & set Prudence weaving.

[January 7] 2 weavers sick.—Old spinning stoped.—Prudence weaving—by so doing pond held out

[January 8] Last night & yesterday weather warm, which gave plenty of water this day for all hands One weaver sick.—4 Looms stoped.

[January 13] Six weavers—2 Looms stoped.—One weaver sick Sally & Mary Ann Leonard out 3/4 of the day by permission Mr Gilbert loaded for Boston

[January 14] Looms all in operation Turned the corner Loom took 3/4 day.—little rain.

[January 16] Fine warm day.—Almira Lovell absent this day & also to be tomorrow, to bury her grandmother.—P.M. went to Mr Carvers to get Harness made

[January 17] Spinning cleaned.—Nany & Betsy White absent 3/4 of the day, Loviza & Alpheus Lovell all day to a funeral

[January 18] Lords Day.—Mr Gilbert returned from Boston at 2 o clock this morning

[January 19] One Breaker tender absent 1/2 day

[January 23] N° 2 weaver lame hand[,] N° 5 sick & absent 3/4 of the day

[January 26] All hands present.—Sent a load of goods to Boston

[January 30] Went to Sharon by request of Mr. Turner, to see 4th reservoir, absent 5 hours All hands to day.—Mr Gilbert arrived from Boston.

[January 31] All hands.—Fixed & filled the safety Hogshead in the Spin^{g} & weaver room.

[February 2] All hands—little short roping

[February 3] All Hands—Roping plenty.—Fine snow last night—6 in

[February 9] N° 3 weaver absent untill night One spinner absent A.M. Some short of filling.

[February 10] Some short of filling.—Emptyed & filled the Picker safety Hogshead.—Water smelt Bad.—All hands.

[February 11] All hands.—Filling plenty sent Picker to Pawtucket to be repaired.

[February 13] All Hands.—Short of filling & short of roping.—Roping short on account of Beater being out of repair.—One old spinning frame stoped 1/2 day

[February 15] Lords Day.—1/2 of it spent in the factory covering & repairing sising rollers.

[February 17] N° 4 weaver absent 1/4 day Short of fill^{g} & roping the cause of the roping being short is mostly owing to cold weather.—One old sp frame stoped all da

[February 18] N° 5 weaver absent 1/4 day Short of filling, Mule spinner Lame.—Roping come up with the spinning A.M.

[February 21] Snow storm continued with increased violence, Blocking up the roads & even the river Got in some of the hands and hoisted about 9 o clock, but had not water to get speed with only 1/2 the card room 12 Looms, Dresser & mules Pond at night down 15 or 16 in

[February 23] Back water & small pond 2 spinning frames & Picker stoped.—Dresser stoped at 4 o clock for want of L. Beams

[February 24] Made out very well this day on a/c of water.—N° 4—5 & 6 weavers out in the evening on a/c of the weather.

[February 25] Dresser stoped this day 3 hours for want of yarn Beams.

[February 26] Violent rain storm last night.—No stirring out without wading half leg deep, 5 of the hands got in on sled about 9 o clock.—No. 4 weaver sick—Large Cog wheel to the gate broke this P.M. Repaired it in part

[February 27] N° 4 weaver sick.—Cold—Finished repairing cog wheel.—Spent a good share of the time for 10 da past in the machine shop; making gudgeons to yarn beams & Iron pins for cards in all, worth as Pawtucket folk charge $20.

[March 5] All hands.—Some short of filling.

[March 9] All hands.—Filling short.—Loom spring Broke.

[March 11] Roller hook to Loom broke N° 5 weaver absent.—Finished working the G cotton

[March 12] All hands.—Great rain which caused an extra freshet Back water P.M. had to stop 1 old spg frame & Picker

[March 13] All hands.—Back water Mill run very well without runing Picker. Ladders erected at each wing of the factory.—16 cords dry wood last untill to day

[March 16] All hands.—Dresser stoped 3 hours for want of yarn beam

[March 23] N° 4 weaver absent 1/2 day N° 5 & 6—1/4 da each.—a job repairing dresser.—Extreme cold.—Lovells children did not get in untill 1/2 past 7 A.M. on account of water in the road

[March 24] Mr Thayers party last night broke up 3 o clock morning hands in consequence come in late & one,

N° 6 weaver not untill noon.—H. Kingman commenced repairing the old looms

[March 25] One of the Speeders out of order, part of yesterday & part of today.—roping short.—One old spinning frame stoped all day—All hands

[March 26] All hands.—Short of roping part owing to speeders, but most to the girl who tends has give out word that she will not hurt herself.—the old sp. frame started middle P.M.

[March 27] Geering to the old speeder give out & went to Taunton & got new, roping short of course, Prudence out after breakfast, Patty 1/2 the day.

[March 30] All hands.—Mule men arrived P.M.—Isaac Flagg finished work to night

[March 31] All hands.—Sally Leonard took the Dresser @ 15/ Mary A. Leonard @ 12/—Schuyler Skinner left the mules in a dirty state. Isaac Flagg left for home Evening Morse cleaned the old mule

[April 1] Taped the 65 Gal. Oil cask.—Morse took the mules.—All hands.—Mr Gerry here

[April 4] Short of Roping.—one old spinning frame stoped all day, Prudence wove part of the day.—The new spinning stoped an hour or two for want of sp. Bobbins.—Warper stoped till middle P.M.

[April 13] Jane E. Bailey sick 3/4 of the day Mr Gilbert Loaded for Boston Evening speeded 4 cards Worked better than 17 hours

[April 14] Mr Whitman here.—18 Looms started.

[April 15] One new sp Frame stoped all day for want of roping—Mr Gerry arrived at night

[April 18] N° 6 Looms stoped, no weaver N° 5 weaver quit.

[April 19] Went to Norton after weavers.

[April 21] N° 5 & 7 Looms stoped—no weavers.—New spinning badly tended.

[April 23] One weaver Short.—Kingman altering creel to mules.

[April 24] One weaver short.—Kingman building bobin shelf to mules

[April 25] Kingman finished mules.—repairing bobbins & help raise water wheel.—One weaver short. sp cleaned

[April 27] Stand to the drawing broke went to Easton to get a new one, finishing at about middle P.M.—All the looms in operation

[April 30] Repairing Loom 1/2 the day

[May 1] Card Room short of bobbins

[May 2] Card Room short of bobbins at night

[May 4] Old spinning stoped, spinner absent without leave. One of the looms turned.

[May 5] Mary A. Leonard began on the old spinning. Maria Williams & Almira Drake three looms each Kingman 1 1/4 da turning Loom Patty Thomas tended spooler & wa͡rper.—Training Card Room & Mules stoped 1/2 the day

[May 8] Dresser & Warper, yesterday & to day stoped equal to one day for want of Beams.

[May 11] Two of the new sp frames stoped P.M. Simeon Thayer went out sick.—1/3 of the day repairing pulley to mule

[May 20] All hands,—Out of Cotton in the Card room 1/4 day. Stoped about an hour to repair couplings.

[May 27] Election day run 1/2 of it one weaver absent all day.

[June 1] [. . .] Factory stoped a short time to repair couplings.

[June 8] Since last Wednesday the mules have been very troublesome, by the bands running off as was supposed.—The real cause has this P.M. been found to be done by the back Piecer Lewis Kingman, throwing them off & at last cuting the spindle & one of the main bands.—The weavers from the above cause have not had fill[g]. for more than 1/2 the day For Boldness & Cunning the above tricks surpass all description

[June 9] All hands. Mules run well

[June 13] Filling Short.—Mule spinners quit work at 3 o clock & went to Dedham.

[June 15] Sally Williams not in untill 6 o clock (Morning) No fill[g] weavers come in after Breakfast Extra repairs on Loom worth $1.

[June 16] All hands.—Went to Norton after Harnesses.—Water down 10 inches.

[June 17] Pond filled but 5 in last night Shut down at 4 o clock Lost 1/4 of a day, Pond down 13 inches went up to reservoir.

[June 18] Pond filled 6 in last night Run handsomely all day Pond settled but 3 1/2 inches down 10 1/2 inches.

[June 19] Pond filled 3 1/2 inches last night Down 12 1/2 in. at noon & deemed it advisable to stop.—P.M. filled 3 in. Altering geering on tumbling Shaft to new mule & enlarging a pulley—This day at noon closes 2d year with 105753 yds short of 1st year 831 yds

[June 20] Pond filled since yesterday noon 6 in.—Run all day Pond at night, down 12 1/2 in. [. . .] Factory stoped 3/4 of a day this week for want of water, being the 1st

[June 24] Pond filled but 2 1/2 inches last night.—Did not hoist this day Repairing Mule drums, Harnesses, Looms & Temples.

[June 25] Pond full to within 3 in this morning.—Tolman Reservoir gate hoisted 1 hole yesterday All hands, water down at night 15 inches.—Went Doct Pecks Foxborough with Mrs Gordon.

[June 26] Pond filled 5 in. Factory stoped Wrote Mr Gerry.—Repairing & Dressing harnesses, also Looms & Temples.

[June 27] Pond full to within 2 inches At night down 9 inches All hands to day. [. . .] Factory stoped 2 days this week for want of water.

[June 29] Pond lacked 1 1/2 in. being full,—some showers yesterday. Fine showers to day.—All hands.—Pond down 4 1/2 inches

[June 30] Water plenty. All hands Fine showers.

[July 3] Water plenty.—All hands Breakers out of Cotton after 8 o clock. out 1 day

[July 4] Water plenty. Morning Cotton arrived.—All hands A.M.—P.M. Factory stoped to celebrate da

[July 6] Fine rain since Saturday An overflow of water.—All hands.—H. Kingman new hung the Bell.

[July 9] All hands.—Boston girl left.—water down 8 in

[July 10] All hands.—Water at night down 13 in.

[July 11] N° 5 weaver absent, Mule Spinners absent. Water failed at 12 o clock, shut down. Mr Gerry arrived

[July 12] Mr Gerry left.

[July 13] Loviza Lovell out, mother sick A.M. Went with Thomas to help survey Dawson reservoir P.M. in the Factory.

[July 14] A.M. went again as above. water failed at Noon, Loviza Lovell out. P.M. Went to Norton to Pratts & Bates Factories.

[July 15] Factory stoped.—Jobing in the Factory part of the day Winslow's put in dam this day

[July 16] Run the factory this day.—One Breaker tender & N° 7 weaver out.

[July 17]	Factory stoped water short Went with Mrs G. to Norton & then to town. Repairing harnesses & making out pay Roll
[July 18]	Run the factory.—1 Breaker tender & N° 7 weaver out. [. . .] Factory stoped 2 1/2 days this week for want of water.
[July 20]	Run the factory.—1 Breaker tender out all day.—1 Drawing tender out 1/2 the day.—N° 7 weaver out 1/2 the day & N° 3 all day.—Pond out & dismissed the hands.—Notified Mr Stone that the pond would be kept to the old mark 2 days (viz) 21st & 22d
[July 21]	Went to Rhode Island visiting factories
[July 22]	Went to several Factories in Warwick
[July 23]	Returned to Providence & took a trip to Newport & back, in the steam Boat Rush Light.
[July 24]	Visited several factories on Pawtucket river to Slatersville.
[July 25]	Returned home through Wrentham A.M.
[July 26]	Lords Day, Mr Gerry left. Factory stoped the past week since Monday. Pond drawn 2 days for Mr Stone to dig Ore.—by my absence lost 1 day of water.
[July 27]	Pond full. Hoisted. A M 8 Looms stoped P.M. 5 stoped old spinning frames, stoped 1/4 the day.
[July 28]	Worked untill noon 7 Looms stoped.
[July 29]	To[ok] down 6 main drums to repair.
[July 30]	Went to Easton for cuplings for drums.
[July 31]	Repairing Drums
[August 1]	Repairing Drums No waste weighed this week Factory stoped 4 1/4 days could have worked 1 1/2 day if the drums had been rep
[August 3]	Finished & put up the 6 drums

[August 4]	Pond full,—Hoisted.—Fine rain this day—N° 3 & 6 Looms stoped being 5 Looms
[August 8]	Four Looms stoped.—Pond out at 4 oclock & shut down Fine shower at night.—The past 4 weeks. Factory stoped 10 1/4 days—1 1/2 day by my absence to R Island, & 1 1/2 day while repairing Drums, the 10 1/4 da for want of water. Short of weavers.
[August 10]	Full Pond. 4 Looms stoped
[August 13]	Pond Low,—Stoped.—Spinning cleaned.—Wrote Mr Gerry Shower to day
[August 14]	Pond full.—Hoisted.—4 Looms stoped.
[August 17]	Water run to waste.—A.M. 4 Looms stoped.—P.M. 2 stoped Pond 1 1/4 in more than full in the morning but down at 6 o clock.
[August 18]	Pond for an hour 1 1/4 in to high 6 weavers to 18 Looms
[August 19]	Pond 1 in cover for 30 or 40 m in the morn.—18 Looms—6 weavers.—Made a Calender roller to the Drawing, which took 4 hours.—Wrote Mr G
[August 20]	Pond 1 in over.—All well 2 Iron jobs on Looms.
[August 22]	N° 5 weaver quit at noon. N° 6 weaver sick.
[August 24]	5 weavers only,—Almira Lovell out sick
[August 27]	5 weavers.—1 spinner short.
[August 29]	5 weavers. 1 spinner short—Pond out at noon & shut down.—Bailing goods
[August 31]	Mr Gerry arrived yesterday Factory stoped on a/c of the hands wages being cut down.—Turned 3 looms Mr Gerry left after dinner
[September 1]	Factory run, card Room help all in.—But one spinner.—5 weavers. Turned & started all the looms but 3.

[September 2] Two spinners short.—5 weavers.—Finished turning the Looms

[September 4] Looms all in operation 3 & 4 spinning frames in operation, had to wait for the spooler and warper. Mule spinner out 3/4 of the day, his piecers lounging about all this time.—Card room stoped P.M. Sam fixing to throw off spinning bands. Lovell and Wilbur wheeling in wood.

[September 7] 3 girls & William King spinning. Kept only part of it in operation.—Went with Mr Thomas to see Esq Boyden respecting reservoir.

[September 17] Repairing Drums & plastering

[September 18] Put up drums, cleaned stove funnel & sundry jobs

[September 19] Went to Sharon to attend a referrance on complaint of William Tolman for flowage Factory stoped 4 days this week 2 days water too low

[September 21] Started Factory, old spg. stoped.—New Card Man New Mule Spinners and 1 new weaver.

[October 6] Mules spinners absent 2 training Recd. of Mr Patten $9.24 on a/c of Reservoir. [. . .]

[October 9] Training, carding room stoped 1/4 of a day.

[October 12] All hands,—Prepared one Doz pair Pickers, being all on hand

[October 14] Water Low, factory stoped. Repairing Rollers & spooler

[October 16] Muster Day.—Water low. Factory stoped. Spinning & new mule cleaned.—covered the spooler cylender

[October 31] One Spinner short.—Speeder girl absent—Fine rain this day, with very strong wind from N. E.

[November 2] New spinner in to day, a raw hand, one frame stoped

[November 6] N° 2 weaver absent.—Quarterly meeting, shutdown at candlelighting, time to be made up by over working hereafter

[November 9] One Breaker tender short 1/2 the day—A.M. Went to Quanticot after hands P.M. cleaned the clock.

[November 23] 65 Gal oil lasted from April 1st to this day. 7m 22 days Taped a cask of 28 Gal.——Warm, wind fresh at south with smart showers & some thunder A.M. [. . .] Short of fill^{g} on account of Mr Cobb's absence to move his family to Mrs Leonard Weavers worked but about 2/3 of the day.—Prudence tending one of the new frames, wages in addition 4/— 1 of the frames stoped part of the day.

[November 24] All hands.—Prudence tending 4 sp. frames.—This day repaired & dressed 7 old harnesses.—Abigail C Fuller left weaving

[November 27] N° 4 Weaver not in untill noon. No 5 sick & no weaver to N° 6 Looms.—Need 9 qts varnish.—Killed the hog Wt. 409 lbs.

[November 30] Full compliment hands.—Geering to the last head of the old Drawing gave out at noon

[December 1] All hands.—Went to Taunton to get new geering for Drawing Card room help kept the full supply of roping with only half the drawing

[December 7] Nancy Williams started her frames about 9 o clock

[December 25] All hands. Warm day, no fire in the spinning room

[December 31] Williams boy absent 1/2 day Absent to an auction 3 or 4 hours.

[1830]

[January 11] N° 2 weaver absent, Mother sick.—N° 4 absent 1/4 day.—The carding room help could not get in untill 9 o clock on account of the water flowing the road caused by anchor frost.

[January 14] N° 2 weaver absent.—28 Gal Oil lasted from Nov. 23d to this day 1m 22 day's, or 46 working days, nearly 3 qts per day.

[January 19] Nancy F. Thayer left weaving in the morning.—Nancy Ellis too her looms in the evening.

[January 20] All hands.—Maria Newcomb come in after breakfast to learn to weave.—Mr Gilbert retd from Boston.—Taped a barrel of Sp. Oil (32 gal). Used 96 gal of sp. oil since the 1st day of April 1829.

[January 22] All hands.—H. Kingman left weaving at night.

[January 23] Maria Newcomb took N° 6 Looms.—Prudence Thomas absent 1/4 day.

[January 28] N° 2 weaver came in after breakfast.—Bell rope in the geering of one of the old spinning frames & injured one of the wheels which took 4 hours to repair.

[January 29] N° 1 weaver absent 1/2 day on account of the visit of a brother.—A new wood spring made to one of the breaker combs.—Repaired a pair of temples.—One of the loom springs gave out this evening which stands as the first job for tomorrow.

[February 1] Snow last night, rain this morning.—Gates in the factory well sealed with ice.—took untill 1/2 past 8 to start the mill, notwithstanding every precaution was taken to guard against frost, Saturday night.

[February 4] N° 4 weaver come in after dinner.—Warper girl absent 3/4 of the day by permission.—Prudence called out 1/4 of the day.—Treadle gave out & repaired. Good sleighing, which makes the hands uneasy to go out.—Poor lot of weavers at this time & bid fair to be worse.—Short of roping 1 frame stoped.

[February 8] Cold & violent snow storm from the North—Brought in the Lovell girls with horse & sleigh after Breakfast, otherwise all hands.—No trouble in starting this morning. At night fair weather, sent the girls home on a sled by L. Thomas Snow too deep for them.

[February 10] All hands.—Warm day wind at southwest. Went of Attleborough after shuttles & to Wrentham to get reeds repaired.

[February 13] N° 5 weaver absent 1/2 the day Two blacksmith jobs on the looms.

[February 14] Cold, fire in the factory all day.—Lords Day.

[February 15] N° 4 weaver absent 1/4 of the day N° 5 half the day.—Raw hand began on N° 3 Looms.

[February 22] Nancy Williams not in untill after breakfast.—Fine day.

[February 24] Prudence out 1/4 day by permission.—Not cold.

[March 1] All hands except myself absent a few hours to town meeting.

[March 3] N° 2 weaver & spinner absent 1/2 the day to their sister's wedding.—Two stirrups & a take up Loom laver made took 1/2 day

[March 9] All hands.—Mr Gerry left.—Repairing harnesses.

[March 10] All hands.—Esq. Pratts Factory burnt about 11 o clock last night; in it about 31500 yds cloth Wrote Mr Gerry (via) Easton. Went up & see the ruins of the above factory.—25 Pickers lasted from Dec. 16, 1829—12 weeks—2 doz pair now hand.

[March 11] All hands.—Sundry jobs repairing looms.

[March 20] All hands. Washed the windows of the spinning room.

[March 26] Violent snow storm which prevented the hands coming in, in season.—Lovell & Wilbur girls, brought in on a sled; out 1/4 of a day.—Williams girl (Nancy) & N° 2 weaver absent until 11 o clock N° 4 weaver absent all day.

[March 27] Nº 4 weaver absent. her looms supplied 3/4 of the day—No 5 weaver or Nancy Ellis quit about 11 A.M. M. F. Thayer took the looms at noon.

[April 1] Almira Lovell out 1/2 the day sick.—Fanny Wilbur supplied her place.—Planted Peas

[April 5] All hands except myself absent 4 1/2 hours to town meeting Nancy Williams out P.M. S. Thayer Jr supplied

[April 8] State fast.—Dull day Mr Gerry left at noon.—Spent 7 hours putting collars on to new mule.

[April 13] All hands.—Cold, dull Dark & rainy weather.—Working hours cut short by it.

[April 14] All hands.—Record's girl came in to learn to spin.

[April 19] Prudence left Sat. 2 new spinners undertook, One leared [?] at 9 o clock—Most of the frames (or Prudence work) looked untill noon like a sheep with 9/10 of her wool off & the other in tatters P.M. Tucker girl come in, took 3 sides of the old frames at 12/ fair spinner, & the frames look natural, but she is not Prudence.

[April 20] All hands.—Went to Sharon in the P.M. & stoped the leak in the Tolman Dam.

[April 26] Nancy Williams went out lame at 4 o clock P.M.

[April 27] All hands, by Simeon Thayer taking Nancy Williams place

[April 28] Emeline Tucker & Mary Howard tending all the spinning

[May 3] Machinery all in operation.—Little short of fill[g], on account of one mule spinner absent to town meeting.

[May 6] Mary Freeder sick her Looms stoped A.M.—P.M. tended by Abiah Leonard.—Some short of fill[g], roping run bad cause not know, unless it is the cotton or weather, perhaps both; cotton coarse staple.—Simeon Thayer finished learning spinning, By employing him as extra has cleaned the frames, with stoping only one frame two days.—It has usually stoped the whole one day for cleaning.

[May 12] All hands.—Plenty of irons in the fire, some will burn.—Pratts factory raised.—Almira Lovell quit for good—sick.

[May 13] All hands.—Alpheus Lovell began tending drawing. Daniel Wilbur the Picker & Fanny Wilbur the breakers,—Pinched for fill^{g}.

[May 17] Rainy day—N° 4 weaver absent Looms still.—Abigail C Fuller began tending Spooler & Warper.

[May 18] Great rain finished this morning.—No. 4 weaver out Looms still.—Beater roller stand broke, went to furnace to get a new one, but could find none, one to be cast tomorrow

[May 19] Put in a wood stand and started Picker, run a short time, then the under feed roller broke, went to Taunton & had it mended & to Furnace for stands Daniel Wilbur failed as Picker Boy—hired George Thomas to whip the cotton

[May 20] A.M. Repaired Picker & went up in town & obtained a Picker boy, George Thomas picking cotton.—P.M. made a new treadle & sundry jobs.—No. 3 weaver out sick, looms stoped.—Fine shower

[May 21] Picker Boy sick of the mill, paid him 8¢ & sent him home.—James Wilbur took his place No 3 Looms stoped weaver out.

[May 24] N° 6 weaver out sick—No 7 a raw hand.—Paid off the hands.—Lucy started for Exeter Emeline Tucker gave notice to quit in 2 weeks

[May 26] Election day, Factory stoped—Oatis & myself jobing 1/2 the day.—I hereby enter my dissent to this day, being one spent in a useless & worse than useless manner.—I could not peaceably work the mill as all hands seemed determined to have the whole day

[May 27] N° 6 weaver come in after Breakfast.

[May 31] N° 2 Looms started up by the Perry girl, for the past 4 weeks N° 9—Alfreda Perry come to spin.

[June 1] All hands.—Augustus White run his head against the Dresser & broke it [dresser].—took 2 or 3 hours to ment it.

[June 2] All hands, such as they are.—Loom spring to mend.

[June 3] Warper-tender sick E. Tucker took her place, part of the spinning poorly tended.—Mary Hayward on the old frames.

[June 5] Tended Warper myself P. Fuller tended spooler 3/4 of the day.

[June 7] Looms full, also spinning, tended warper myself, but not able to keep up with spinning Quiggle out sick.

[June 8] Full supply of hands, but a poor lot of them.—Straw & brimstone the rage at these times.

[June 9] All hands, webs work bad.

[June 10] All hands.—Mrs Gordon Went to Easton after the Littlefield girl to tend warper.—Rain in the forenoon.

[June 11] The warper girl a coarse concern, obliged to tend it for her most of the day; but few good hands in the mill

[June 13] N° 5 weaver quit at 1/2 past 10 o clock.—Mr Gerry arrived in the evening with P. Temples

[June 14] Mary White or N° 5 weaver come in at 9 o clock.—Mr Austin put on 6 pair patent temples.

[June 21] The old maid H. Perry come in this morning & bid us good by.—3 Looms stoped.—Fine rain in the A.M.

[June 22] 3 Looms stoped being 1 weaver short

[June 23] One weaver short.—Mr Gilbert went over this town & to Bridgewater after one

[June 24] Mr Gilbert did not go to Bridgewater untill to day.—Went all the four points of it, but found no weaver.—Mr Whitman called, he is 4 Wea^{r} short.

[June 25] One weaver short.—hands scarce.

[June 26] Mr Gerry left. Paid the hands.—One weaver short

[June 29] One weaver short, & one absent 1/4 day.—Tending Spooler part of the day.—Water failing fast.

[June 30] One weaver short. Tended Spooler & Warper.

[July 1] One weaver short.—Shower.—Water down 11 in.

[July 5] Betsey White took the vacant looms.—N° 4 weaver absent 1/2 day.—Alpheus Lovell quit yesterday.—One hand short in carding Room.

[July 6] All hands.—Short of fill^{g}. Pond down 14 or 15 inches.

[July 7] All hands.—New spinning & Mules, cleaned.—short of fill^{g} Stream failed for the first time this season at noon, & shut down for 1 1/2 day.

[July 9] All hands.—More rain last night, pond full to overflow this morning.

[July 12] Otis Harris left Saturday.—Mr Belcher came in & took charge of the carding room about 9 A.M.—short of filling

[July 13] All hands, excepting the Mule spinners Boy went out sick.—Very short of fill^{g}.—N° 1 weaver out 1/4 of the day on account of fill^{g}—Rain yesterday & to day.

[July 14] N° 1 weaver out 3/4 of the day & all the weavers stoped 1/4 of the day for want of fill^g.—Quiggle carried A. White home & hired O. Harris

[July 15] Warper girl run off last night.—N° 1 weaver out waiting for filling.—Tended warper all the time I could get.—Mr Gilbert loaded for Boston.

[July 19] Old spinning stoped, being one spinner short.—N° 4 weaver out without notice.—2 new hands one at spooler & 1 at warper.—Almira Lovell went out sick.

[July 21] Spooler girl cleared last night.—One O. sp frame stoped.—Fill^g plenty.—Hot, the most so of any day yet.

[July 24] Old sp. frames stoped.—N° 6 weaver out all day.—Water failed at 4 o clock & shut down.—Loss 1/4 day—Lost on a/c of water 1 1/4 day this week

[July 26] Showers to day—N° 3 & 4 weavers come in about 10. N° 2 & 5 out all day.—Old frames stoped.—3 Pane boys tending 1 frame & hard work at that.

[July 27] The two White girls come in at noon, made excuse that it was wet.—Old frames stoped.—3 boys at one frame.—dull day.

[July 28] Looms well supplied. Started one of the old frames, boys made out to tend one side with one new frame.—dull weather which affects the sp. as well as all other machinery

[July 29] Fill^g. & Weavers.—Boys with my help now & then tended one new & one old frame, like all raw hands Rainy day wind east.

[July 30] Fill^g., & weavers Boys gain some, tended the same as yesterday with less help.—Storm continues.

[August 5] All hands & fill^g.—John has worked badly to day & I fear will be poor help.

[August 9] N° 4 weaver out untill 9 o clock, nothing however uncommon with her.—All other hands in, in season.

[August 12] Hoisted at 9 o clock All hands excepting N° 3 weaver & J. Quillin.—Put Mary Ann Abbot to weaving.—Mr Austin began to repair the looms.

[August 14] Unwell, but able to be in the factory & labour some, Mr Austin fixing pattern for iron to picker treadle & repairing looms.—Belcher, making bags & bagging waste, & prepairing card cloathing.—At night went to the furnace. [. . .] Water failed last night. Stoped 2 3/4 days this week

[August 18] Mary Ann tending N° 2 Looms after 6 when she got into the mill.—Dolly at spooler. Mr Austin in my stead being at the Cobb house, which I could not avoid, had fair luck, for so heavy a building this day making the 3d of my time

[August 19] Pond failed at 9 o clock & shut down for 1 3/4 days.—Mr Austin repairing Looms.—Fixed the house on wheels, the job looks rather discouraging. Mr Gerry arrived in the eve^g, & put new Life into all hands.

[August 20] Factory stoped.—Teams tried to move the house but failed in strength & wheels.—at noon left it.—Went into the factory & worked 1/4 of a day.—Mr Gerry here all day.

[August 24] All hands.—At noon Mr Belcher began to repair speeder bobbins.—Quiggle in the card room. Mr Stone this day waved all claim to having the pond drawn for him to dig ore the Season.

[August 25] All hands,—Mr Austin repairing Looms.—Mr Belcher repairing bobbins.—Threatened to storm all day. Water low, P.M. Part stoped & speed slow.

[August 26] Pond fill^d., but little last night, not up to the old mark by one inch, dismissed the hands, untill Saturday if no rain come, but if rain only untill to-morrow.—In the course of the A.M. a violent rain storm commenced from the North North east & continued through the day. Examined & fixed water wheel Mr Austin repairing looms

[August 27] Water in abundance some stormy in the morning 3 weavers to begin with.—N° 5 come in about 10.—Drake girls at noon.—White girls & Sally Williams about 9.—Mr Gerry arrived in the evening

[August 30] Fine showers this morning, on account of which hands come in late, but all in between 7 and 8 o clock

[September 2] All hands.—5 webs hung to day.—Mr Gerry here & house moved part way Belcher repairing top cards.

[September 6] Elmina Drake, William Snow & Sally Williams not in in season, otherwise all hands.—Belcher finished Cards,—worked on them in all 3 days.

[September 7] All hands.—Fine rain to day.—Belcher repairing bobbins.—More breakages this afternoon than 3 hands could repair.—Dresser & one speeder now stoped for the same.

[September 16] Bell rung 1/4 before 5.—All hands—shut down 25m past 6

[September 17] All hands.—Bell rung at 5 o clock.—Hung two webs—Made 2 1/2 Gal varnish.—Repaired 2 harnesses & dressed them—Shut down 25m past 6.

[September 18] All hands.—Bell rung at 5. Shut down 5m past 6.—2 or 3 hours job repairing dresser binder besides numerous others.

[September 20] Two Drake girls not in untill near 9 o clock, other hands all in.—Mr Gerry left at 5 A.M.—Bell rung at 5.—Began to work evenings. Shut down at 8 o clock.—my time taken up paying hands repairing Lamps and measuring house lot.—Sara Grant sp arrived in the evening.—Gave William Snow notice to quit in two weeks.

[October 1] Bell rung at 5 o clock All hands.—Sara Porter arrived at night & went to Mr. Thomas to board

[October 2] Bell rung at 5 o clock—All hands.—Fanny Drake left & Sara Porter took her place—Mr Austin repairing Looms & put one card grinding.—Took a/c of stock.

[October 5] Bell rung 1/4 past 5.—All hands.—Bolt which holds one of the picker rods gave out about 2 o clock, will take untill tomorrow to get it started, being a bad thing to make, also destitute of a Blacksmith, coal, and suitable Iron

[October 8] Bell rung 20m past 5.—The upper dam took up last night by some one, of course pond quite low.—All hands but at noon dismissed the spinners (except Sara Grant) & Warper, on account of water Set S. Grant to wiping spinning frames.—Hannah Crossman weaving on one loom.—Took out stove Funnels and cleaned them.—Some cold to day.

[October 9] Dark morning Bell rung at 1/2 past 5.—All hands.—Set 12 or 14 squares glass.—Fine rain to day. Adelice Maria Sanington arrived at 11 o clock A.M.—Gilbert returned

[October 13] Bell rung 25m past 5.—2 hands off to muster, one mule Spinner & Picker boy, hands however to keep all going.

[October 18] Bell rung 1/2 past 5.—Mary A Abbot & Sarah Grant quit the mill.—Adelia M. Swinington went to N° 2 looms.—Sara Porter 3 Looms.—Hannah Crossman went to the spooler.—One spinning frame stoped 3/4 day being short of roping

[October 26] Bell rung 1/2 past 5.—Elmina Drake left after breakfast & Hannah Eaton took her looms James

Wilbur quit in the morning Roping come up & spinning all started.

[October 29] Bell rung 35m past 5 All hands, but poor weavers

[November 10] Boys rung the bell 10m after 5—All hands—sp frame stoped unlucky day—Mule band broke—Web burnt by James Wilbur All the old looms stoped in the eve[g].

[November 12] Bell rung by the boys about 1/2 past 5.—All hands.—Fill[g] plenty—Burnt web repaired & started, lost by the accident the weaving of about a cut.—Mule Spinners began work at 1/2 past 3.—7th dull day

[November 15] Bell rung at 5 o clock. Currier girl better, took her to my house to board, not able to work.—come before dinner Betsey White come in.

[November 16] Bell rung at 20m before 6—Cynthia Currier went into the mill after breakfast helping H. Eaton & S. Porter.—11th dull day—Looms arrived from Taunton.

[November 17] Bell rung 1/4 past 6. Mr Hixon took the Lap Waste sent him 11 bags Wt estimated 40 lb bags to be left at J. Gould's store in Sharon.—Mr Austin began to set up the Taunton Looms.—All hands No rain, but dull kind of weather

[November 22] Bell rung befor 6—All hands.—Nancy Snow began work on the spooler.—James & Daniel Wilbur & Alpheus & Loviza Lovell gave notice to leave in 2 w

[November 24] Bell rung 5m past 6 Mariah Newcomb left about 11 o clock P.M. Hannah Crossman took her Looms.—Old sp frame started at noon.—Warper stoped 1/2 the day, the same moved Mr Gerry arrived in the eve[g]. Mr Thomas stated in my presence & in the presence of Mrs Gordon, that he would give $5 per year to have 56 spindles stoped, being in answer to Mr Gerry. What the damage was to said Thomas.

[November 26] Bell Rung 45m past 6—Hannah White went out sick at 11 o clock, Lucy took her place otherwise all hands.—Dull day. Bethiahr Clefton arrived in the evening & went to Mr Kings to board, to supper.

[November 29] Bell rung 10m past 6—All hands.—A.M. Swinington left & Abiah. Leonard took her looms.—Davis Albion gave notice to quit in 2 weeks.—William King & Alpheus Lovells, wages 12/ per week. Loviza Lovell's 6/ Do.—Weavers to have a premium of 30¢ for 15 cuts & 35¢ for 18 cuts per week.

[December 1] Bell rung 20m before 6—All hands.—Shut down at 5 o clock, making the day short 2 1/2 hours, to be made up here after by working extra hours.—Short of roping. 1 sp. frame stoped

[December 2] Thanksgiving.—Mr Austin & H. Kingman put up a drum on the main shaft to carry new looms.

[December 3] Bell rung 20m past 6.—All hands, except Bethahr sick with the mumps.—Warm day, wind South, no sun.—Shut down at 9 o clock Short of roping 1 frame stoped William Snow gave notice

[December 4] Bell rung 1/2 past 6.—All hands except as yesterday. Short of roping 1 frame stoped Abigail Glidden & Sara Porter gave notice.

[December 7] Bell rung at 6.—Mr Gerry left in the morning—All hands.—Went up to see Mrs White & Mrs King to take boarding house, but no success.

[December 10] Bell rung at 10 past 6 All hands.—Mr Belcher moved into the Cobb house.

[December 16] Bell rung 10m past 6 One of the old sp frames cleaned to day.—Mr. Fuller moved into the Cobb house & Belcher moved into Fullers house

[December 20] Bell rung 20 past 6. Went to Rehoboth after a Spinner Almeda Goff; could not get her, Ret[d]. by Taunton, for screws.—Henry Kingman Jr. come in to spin for 2 days.—Snow yesterday.—Abigail Fuller Maria Newcomb & Eliza McAlester come in to weave H. Eaton went to the spooler Betsey White warping—Nancy Williams gave notice.—Mr Fuller took Eliza McAlester Maria Newcomb, Hannah Eaton, Nancy Williams, Clarisa White & Bethiah Clefton to board & [ ] Baker arrived in the evening

[December 21] Bell rung 25 past 6—Abigail Glidden & Sarah Porter left Mr Thomas' for Boston. Fine snow storm this morn[g]. Prudence Thomas come in to sp @ 3$ per week, to tend 3 frames.—Baker took Hannah Crossmans Looms & she to the warper & Betsey White back to the mules.—The Speeders halves repaired this day by Kingman.—New Sp cleaned.

[December 25] Bell rung 5m past 6.—Warm morning.—Rain & at night snow about gone. All hands. Warper stoped for lack of beams 3/4 of the day, being all full.—But very little fair weather for 2 months past, which has made it dark in the factory, thereby causing my labour a quarter harder than otherwise would be if fair.

Samuel Ogden

Thoughts, What Probable Effect the Peace with Great Britain Will Have on the Cotton Manufactures of This Country (1815)

Samuel Ogden was one of the many skilled British immigrants who played an important role in the American textile industry. He was active during the first two decades of the nineteenth century building textile machinery in Rhode Island. David Wilkinson recalled in his reminiscences (reprinted in this collection) that Ogden was a man of "great experience and good abilities," even though Ogden had advised him to abandon his attempt to build a metal lathe: "You can *ner* do it, for we have tried it out at *ome* [home: England], and given it up; and don't you think we should have been doing it at *ome,* if it could have been done?" Ogden displayed the same British chauvinism when he composed this pamphlet offering advice to American manufacturers in 1815.

The following excerpts from Ogden's pamphlet are the first published observations on the problems of labor management in the American textile industry. Ogden intended to convince American cotton manufacturers that the "best dependence for the permanent security of a trade, is good management, and to depend on aught else but that, is uncertain." For Ogden, the key to good management was reliance on British practice—a clearly defined set of work regulations, leaving the employee with no uncertainty as to his or her responsibilities. The centerpiece of effective discipline was the piece rate

Samuel Ogden, *Thoughts, What Probable Effect the Peace with Great Britain Will Have on the Cotton Manufactures of This Country* (Providence, 1815), pp. 1, 6–7, 19–20, 26–33. On Ogden, see William Bagnall, *The Textile Industries of the United States,* volume I (Cambridge, Massachusetts, 1893), pp. 407, 536, 537, 538, 546; and *Transactions of the Rhode Island Society for the Encouragement of Domestic Industry in the Year 1861* (Providence, 1862), pp. 76, 108; and *in the Year 1864,* pp. 79–80. For Andrew Ure's discussion of Arkwright, see *The Philosophy of Manufactures* (London, 1835), pp. 15–16.

combined with a premium or bounty given for speedy, effective work—measures that would be codified in the scientific management tracts of the early twentieth century.

The importance with which the British treated management policies was reemphasized twenty years after Ogden's pamphlet by Andrew Ure. Ure, the foremost English promoter of the factory system, considered the invention of industrial machinery secondary to the establishment of factory discipline. In *The Philosophy of Manufactures,* he observed what he considered the true accomplishment of Richard Arkwright:

> The main difficulty did not . . . lie so much in the invention of a proper self-acting mechanism for drawing out and twisting cotton into a continuous thread, as in . . . training human beings to renounce their desultory habits of work, and to identify themselves with the unvarying regularity of the complex automaton.

Ogden's extensive description of British factory rules and his invocation to American mill owners to adopt them strongly indicate that such elaborate work rules were rare in the United States. Historians know too little to fully explain why, but it is likely that the strong equalitarian heritage of this country and the popular suspicion of textile mills made owners hesitate to establish rigid work rules that would have almost certainly increased labor scarcity. (GK)

THOUGHTS,

WHAT PROBABLE EFFECT THE

PEACE WITH GREAT-BRITAIN

WILL HAVE ON THE

Cotton Manufactures of this Country:

INTERSPERSED WITH REMARKS ON OUR BAD MANAGEMENT IN THE BUSINESS; AND THE WAY TO IMPROVEMENT. SO AS TO MEET IMPORTED GOODS IN CHEAPNESS, AT OUR HOME MARKET, POINTED OUT.

BY SAMUEL OGDEN.

PROVIDENCE:
PRINTED FOR THE AUTHOR,
BY GODDARD & MANN
1815.

Objections to the erection of large factories are very commonly made, on the ground that the persons employed therein, are subjected to tyrannical rule, and that the children are deprived of the privileges of education. To satisfy such objections, I will make the following remarks.

By the rules of society, a plan is held out to mankind, the long practice of which has displayed a plain view, that industry is necessary, for our comfortable conveniencies, and our general happiness.

I do not wish to be understood, as holding out an idea, that there should not be any distinction or variety of rank in society; for such an idea, forced into practice, would cut asunder the sinews of social order. But as society has a variety of links, to make the whole great chain; to each link, or class, separately considered, there is an allotted line of pursuit—nor is there a person, existing among a social community, who does his duty, if he does nothing to the advantage of that community.—But hold—my mind is crowded with a cluster of ideas, that draw me from my purpose, and I must pause to get right.—It is not possible by compulsory measures, to induce persons of obstinate and untoward tempers, to continue in the right line of social duty, and persons of a better temper, will not always voluntarily enter the regular path of industry, though their necessities may require it.

To such persons, I ascribe the cause of objections against factories, more than to real causes; for such dispositions are generally apt to stretch beyond the bounds of truth, and they are generally more defective in their conduct, than the managers of factories are tyrannical, and if they have families, they more commonly neglect to have their children taught to read, than their children are deprived of the opportunity to learn.

It must be admissible with every unprejudiced person, that, that which has a tendency to better the condition of a section of society, ought to be supported by favourable report (so far as is admissible with truth) from all persons.

Industry is the only means by which we can obtain a surplus supply of what we need, and we can thereby ease our minds of that perplexing fear, the dread of starvation; for plenty knows not want, nor feels its pinching gripe. Where then is the reasonable man that will say, it is better to be idle and want, than to work moderately, and have plenty!

A cotton factory is a school for the improvement of ingenuity and industry; and the improvements in machinery, combined with industrious exertion, have had a tendency to raise the conditions of the employers, and employed, in this country, and in Great-Britain they have been raised to an unparalleled degree. It must therefore follow, that factories are beneficial to mankind, and ought to be encouraged and continued, while on the scale of society, the balance is favourable to general advantage.

*

In the conversation that I have had with different persons, I have stated to them the following facts:

That, in Great-Britain, a stated price is given for almost every description of work done in manufacturing cotton; and that stimulates the work people to exertion; and when work is wanted, it is very common to give a bounty to every person, that earns above a stated sum per week.

That during working time, there is a continual strife which can do the most work; and every exertion is used, to excel.

That a combination of human invention and industry is brought into action to despatch work; and that to all its depending business due attention is paid, to forward it to good advantage.

That no unnecessary waste is made, and that all stock is worked to the best suitable purpose.

Objections have very commonly been made, by those with whom I have conversed, against an attempt to bring our business into the same way, because, said they, we think it impossible. And I have been repeatedly told that our work people will not submit to any rule that does not comport with their inclinations, or to which they have not been accustomed. That they will not bend to be such slaves as the working people in Great-Britain are, nor work for so low wages as are given in that country.

To such objections I have repeatedly given the following reply:

The leading inducements to incline mankind to work are necessity and interest; and without labour, nothing can be acquired by those who have no other means to support them; and all surplus labour is done from motives of self-interest. Thus viewed, we find necessity the forcible cause, and self-interest the inclinable cause. Both effectually bring us to industry, and when we want employment, we must take it as we can get it, if it does not exactly comport with our inclinations, or that to which we have been accustomed.

To the remark, that our working people will not bend to be such slaves as the British, nor work for so low wages, the following comparison may serve as a reply.

But let me observe, that I will not give it as a supposition, because it is drawn from facts in my own experience.

Place two men to work in one room and at one sort of work, and employ one by the day and the other by the piece, and you will find that the day man is a slave to time more than at his work, and that the piece man is actively industrious at his work, and *takes no note of time*, unless it be to think it short, when work he had allotted out to do is not completed. Take the weekly amount of wages paid to each one, and the quantity of work each one has done, and you will find that the day work comes the highest, and that the piece-man by working four days in a week, will earn more than the day man's wages amount to for six days.

*

Necessity binds the generality of our cotton manufacturing firms, to erect their spinning establishments where they can apply water-power to turn the machinery; and that often happens to be on a spot, near to which there is not a dwelling house; and necessity also binds them to build houses for the work people. A store is also commonly found wanting, which is likewise built, and stocked with the necessary articles to supply those families employed in the factory, with what they may want for their support. All this is admissibly necessary for a new establishment, but to continue it is hurtful to the permanent prosperity of a manufactory, and turns into the spirit of despotism; because those around the spot are directly dependent on the firm that employs them, for every morsel they receive, and the population within the boundary is limited to the will of the establishment. By such restraints, difficulties must often arise from a scarcity of hands, in not having a surplus stock to supply the necessary wants, when a deficiency happens. To remedy such difficulties I would propose, that every possible encouragement should be held out, to induce people of any trade or calling, to settle near the factory; and the best rules that I know of, would be, to pay all the work people in a circulating medium; to pay them weekly, and not have a factory store. That would naturally draw tradesmen to the spot, where so much money would be certain every week, who would furnish a necessary supply of every article that people's wants require. Such unlimited rules would make an opening for a populous town, and every person in it would feel free. It is by freedom and a spirited and liberal encouragement, held out to work people; and noticing and rewarding deserving merit, that the extensive manufacturing inland towns in Great-Britain, have increased to such a numerous population; and that has also given them, the pre-eminent advantage which they at present hold, over the rest of nations, in manufacturing those articles in which they excel all others, both in elegance and cheapness.

For the particular advantage of our cotton manufactures, I will more fully describe the modes, practised by the British cotton manufacturers, to stimulate the work people to exertion, and to keep all hands employed to good account.

The spinning department is the first, in the order of

rotation, and is also the primary cause of cheapness. So considered, if I indulge in a particular detail, bordering on tediousness, that shall be my apology.

The most valuable acquisition to a spinning concern is a good manager, who, by practical experience, when a machine makes bad work, is capable to set it right, and can discover a defect or neglect in any department. With the requisite qualifications, there ought to be combined an industriously close application to business; and nothing short of practice, makes a person fitly qualified. The British manufacturers most generally raise their managers from the ranks of the work people.

Every bag or bale of cotton, is, by British spinners, kept separate throughout the working, and is prepared and spun on the machinery calculated for the sort. When a bag of cotton is spun, the weight of yarn made from it proves what it has lost in working; and if there happens to be too great a deficiency, it is known in what line of machinery the waste has been made, for all the machines are numbered, and the numbers are taken into the account. The yarn made on every spinning machine is also kept separate till it is inspected, and by that means, without inquiring it is known when any bad is spun, where to charge the defect; and the reels also follow, each their own spinning machines.

By such rules, and by strict adherence to them in the department of inspection, all the work people are made mindful to do their work well.

Having taken a cursory view of the rules in a British spinning factory, I will now proceed to detail the modes of practice in a well-conducted mill. It is necessary to observe, that all are not under one rule of practical order; but those that do not manage well inevitably fall into ruin.

The preparation departments for water and mule-spinning are practically conducted in one way, and no notice of difference is necessary to be taken in any process but the spinning.

When cotton is taken from the bag to be worked, it is opened by a devil, a batting or a blowing machine;

and is then taken to be picked. The picking is almost generally done by women, in the roof story of the factory, and is weighed out to them, a dozen pounds at a time, which they make into fleeces in the form of a basoned hat. When it is taken in, the inspector takes every fleece (one by one) and holds them to the light, so that he can see through them, and if one fleece is defective, the whole is returned, to be looked over, and picked better where it is not clean; and that is called licking the calf. During the time of looking it over again, a mortifying bleating is uttered from the pickers, in various parts of the room, and that, together with the trouble, brings them into the way of doing their work well. Every parcel of cotton delivered for carding, is taken into a small room adjoining the card-room, to be weighed ready for feeding; and to the person who weighs it, a ticket is sent, containing the number of the machine on which it is to be carded. To every machine there is a sufficiency of boxes with square partitions in, to put the cotton into when weighed, to keep up a ready supply; all of which are numbered to their respective machines. The quantity of cotton to be spread between two marks on the feeding cloth, when weighed, is put into one of the partitions; and when the box is filled, it is put aside, to be handed to the feeder when wanted. Great precaution is taken, to spread the cotton even on the feeding cloth, for on that, regularity in carding very much depends.—In short, every department in the card-room is diligently attended, and if any one neglects, or is slack in the performance of the necessary duty, the manager immediately discovers it: for there is scarcely a minute during working time, that his attention is not drawn to watch all parts of the work, but when detained to regulate a defect. The rule of a good manager is to be always walking to and fro in the room, or else standing in a place where he can see the greatest part of the hands that are at work.

The work in the carding room, is, in many mills, done by the day; but it is known what quantity ought to be discharged separately from every machine; for the steam-engine has a given speed, and to prove if it deviates, a clock is annexed, the time of which, for the right speed of the engine, is calculated to common time. That clock is called the mill clock, and is commonly fixed in the porter's room, and by the side of it a common time clock is placed. The speed of every motion given, from that first cause (the steam-engine) is taken into account, and by that means, and the rules for detecting bad work, it is known when there is a defect in the quantity or quality, in what particular department due attention has been wanting; and that binds those who work in a card-room (though by the day) to their duty.

The carding and drawing is sometimes taken by the manager of the room, at a stated price per hundred pounds, and in that case he engages and pays all hands employed in his department—but, whether the work is done by the day or piece, he is charged with the whole blame by the general manager for all defects.

The next department in the regular order of working, is stretching, and that is commonly done by the piece. Every stretching machine has its own carding and drawing to follow, and that would be an obligation to bind the workman to his duty, was there no other cause; but working by the piece, causes an exertion that is more free, than that which is produced by force. To each stretcher a regulator is fixed, by which it is known how many *hanks** are put on a set of cops; and by weighing the cops, and dividing the number of *hanks* by the weight, the length contained in each pound is found. To have a regulator to a stretcher is very advantageous in fixing a spinning machine, to make yarn of the fineness wanted.

Water-frame and throstle spinning are commonly done by the day; but all hands are bound to their duty, by the same means, and in the same manner, as before described of the carding room—nor can a frame fall short, in either quantity or quality, and the defect not be discovered, which the following statement will fully prove:

* A hank of cotton yarn is what we call a skein.

Of the speed of the spindles, the regular twist to be given to every number of yarn, and the quantity of every number one spindle will spin in one day; by such given speed, and such twist, the manager has a regular table.

When yarn is too slack twisted, the table proves it by the quantity spun on the frame, which is more than could be done, and give it the proper twist; and when the quantity falls short, it is a proof that the frame has stood, or else, that the yarn is over-twisted.

In mule-spinning the work is all done by the piece, and in that department, an almost unparalleled degree of voluntary industry, and active exertion, is displayed by the hands, when at work. One man works two mules, and during working time, there is a continual strife throughout a room, to gain a stretch† at him who spins on the next pair.

The present state of cotton spinning in that country, proves the mule to have reached that pitch of improvement, bordering the most close on perfection; and should the British ever be rivalled in it, the country that effects it has much to do. By extraordinary exertion, and other means employed to facilitate work, though much is done for the price paid, a man will, in good times, earn three or four pounds sterling in one week.

Would my limits, or the views before me make it admissible, I could prove, that the means used by *master spinners*, were to induce, and not to compel the work people, to such an extraordinary degree of exertion; nor do I think it possible to raise a business to so high a pitch of perfection, by any means but open and liberal inducements. It has been a very common practice with the British manufacturers, to bestow gifts, as a reward for meritorious industry, or doing a particular piece of work well; and the following sketch will prove, that those who work at the cotton business are free.

All persons who work in a spinning factory, whether by the day or piece, are paid every Saturday, and when they are paid there is no certainty of having a single

† A stretch is one draw out of the mule carriage.

hand to work, on the following Monday—and hundreds go to work at another place every Monday.

Reeling is generally done by women, who work by the piece, and every reel is numbered with the number of the machines that it follows; and the reeler is under the necessity to exert herself at work to keep up with the spinning, and supply the frames with empty bobbins. And if any reeling is improperly done, the faults are brought home to the right place, the same as in any other department.

There are but very few British manufacturers who assort the reeled yarn on a balance; nor is it necessary where a skilful manager is employed, and the hands are mindful to do their work well.

In Great-Britain the work people in every branch or business are paid when the work is done, and no time is lost in getting wages when earned, for the money is always ready; and I am certain from experience, that manufactures cannot rise to so high a pitch of perfection, as they are now at in that country, in any country where a different rule is practised.

Where much time is lost to get wages when they are earned, the workman must receive a high price for his labour, or otherwise he cannot subsist upon it; for *time is money, says poor Richard.*

It may be thought too immediate a repetition of similar ideas, to state particularly the general rule of practice in paying work people in Great-Britain. But I think it necessary, in order (if possible) to convince our manufacturers, that it would have a weighty tendency to bring our goods lower into the market, by enabling our hands to improve all working time, which would eventually better their circumstances, admitting that they must do more work for the money than the amount received, in such pay as we at present make them. The usual time to pay the work people, is four o'clock in the afternoon of every Saturday.

In many of the large factories, the wages are carried by a boy to every person at work, before four o'clock, and when the time comes to quit, the hands have nothing to do but walk home, or go immediately to the market. Those factories and work shops, that do not commence paying before four o'clock, will pay all the hands in fifteen minutes, which is done in the following manner—every person's money is ready counted, laid on a table, and when they quit work all hands that have money to draw, repair to the counting-room.

Here I must observe, that a man draws for all of his family that work by the day, and that a mule-spinner employs, and pays his piecers himself.

Only one person is admitted into the counting-room at a time, and the first words used to call them in, are, *come forward*—and then, *another—another*, &c. till all are paid; which is done as quick as a person can go in, take up his money, and walk out again.

By such rules, and by possessing such means as ready money, all working people have a chance to purchase the necessary articles on that day, for the support of themselves or their respective families during the following week. The time is allowed to all persons that work by the day; and they receive their full wages, as though they had worked the day through.

The British cotton cloth manufacturers have established rules, to which they closely adhere, and for the advantage of our business I will briefly state them.

All warps are put out to the weavers, ready warped; and with each warp a ticket is given, specifying the length, and the number of the *reed* it is to be wove in.

One, two, or three days in a week, according to the extent of business a manufacturer does, are appointed to take in cloth from, and give out warps and *weft* to the weavers. All weavers know the days, and do not, therefore, carry in cloth on any other day; and during taking in time, the manufacturer, and often one or two of his warpers are kept busily employed at measuring and inspecting cloth, paying the cash for the weaving, and making the necessary entries in the books.

By the above rules, a very extensive manufacturer of cloth has time to do other parts of his business in, without having one proceeding interrupted by the interference of another.

But I do not think that we can, at the present time, improve in our mode of putting out weaving, of such a quality as we can make to the best advantage, so much to make it an object worth notice, as in the spinning. I will therefore recommend to our manufacturers (as of greater value) to turn their attention more towards improvement and close economy in spinning than weaving.

The foregoing account of the rules and modes of practice in manufacturing cotton in Great-Britain, gives proof of a well-regulated systematical order in every department of the business.

Walton Felch

The Manufacturer's Pocket-Piece (1816)

This poetical curiosity in serviceable heroic couplets is of interest for several reasons. It is a very early example of the intrusion of industrial technology into the poet's storehouse of metaphor. By using the subtitle "The Cotton Mill Moralized" Felch meant not to make the mill a moral place but to find figurative moral meaning in it, drawing from each process and piece of machinery some more or less appropriate precept. One of Felch's precepts is that technical ignorance is the cause of disastrous mismanagement in the textile business, and he is at pains to use his verse as vehicle for a primer in engineering and factory management as well as for general virtue. The contrast between the engineer's precision and the poet's sublime is nicely exemplified in this work. In addition to intentionally playful and unintentionally amusing literary qualities, Felch's verse expresses a particularly conservative manufacturing point of view. He shows the industrial proprietor trying to mix a patriotic argument for American freedom with a tightly constrained law-and-order proposition.

Nothing is known of Felch's life, though he was obviously familiar with the management of a textile manufacturing company. Medway, where Felch published this poem, was a textile town in southeastern Massachusetts specializing in carpet manufacture. (MBF)

Walton Felch, *The Manufacturer's Pocket-Piece: or the Cotton Mill Moralized. A Poem, with Illustrative Notes* (Medway, Massachusetts, 1816). Line numbers have been added, but the footnotes are Felch's.

THE

MANUFACTURER'S

POCKET-PIECE;

OR THE

COTTON-MILL MORALIZED.

A POEM,

WITH ILLUSTRATIVE NOTES.

BY WALTON FELCH.

---

*—————Never on these arts presume
To deal reproach, or haughty words assume;
But, 'mid the scenes where silent preachers dwell,
Your own neglected DUTY ponder well.

PUBLISHED
FOR SAMUEL ALLEN,
And sold by him and the Author, Medway.—*Price 25 cts. single—2 D. per doz.—and 12 D per hundred.*
1816.

OCCASION.

SO recent and complicated are the principles of manufacturing Cotton Yarn, on the Arkwright plan, that it were vain, for a manager to build his pretensions to excellence in the art, upon any course of instruction, or period of practice, unless combining the closest theory, with the most assiduous experimental application. The prevalent ignorance of this fact, is a prolific source of erroneous management.

The various constructions of machinery, designed to effect the same operations, each presenting peculiar disadvantages, generally needless, and often produced by avoiding imaginary evils, are standing tokens of the deficience of an accurately theoretic system, among our Builders.

Add to these considerations, a total want of fidelity, on the part of a great proportion of those employed in Spinning Mills, and you have so many of the domestic inconveniences, under which our manufactories languish.

The assertion is undoubtedly correct, that, through a contemptible lust of authority, or for want of an accurate conception of the human mind, and the genius of a free government, a domineering policy has frequently been adopted; but it is believed that the preposterous assumption of those who, in the destruction of all legitimate regulation, would bar every avenue to profit or character, is, at least, equally unjustifiable. And it may also be allowed, that in every line of business, avaricious speculation is often carried beyond honorable or profitable bounds, and that, in this, many are hired under expectations which they never realize; yet why should this accusation be urged by those who practice a more ruinous course of speculation, the tendency of which is, to impoverish those from whom they draw their support, without the least benefit to themselves?

The fact ought to be distinctly known—that the currency of unfavourable reports, has generally arisen from the credulity of those who have heeded the ranting assertions of an unprincipled, or unreasonable portion of the work-people; whose narrow views incapacitate them to judge their usage, and whose restless dispositions and insatiate prodigality would render the best treatment, or compensation fruitless. A principal evil is, perhaps, the employment of too many of this description, to the ultimate exclusion of their betters.

If the writer's feelings are either way mostly interested, it is in favour of the labourers; he therefore, more freely censures those who become their own persecutors.

To mitigate these evils, the business should be placed under the direction of persons of unquestionable honour and benevolence, and the rules of duty and requital should be thoroughly methodized, previously promulgated, and faithfully executed.

Nothing can be conceived more conducive to health, than the regularity of time and the erect and liberal positions of body, afforded by this manufacturing attendance; nor can a large family, without the possession of property, employ their hours so profitably in any other profession; if, therefore, they are not satisfied, some improvement is either wanting in their situation, or in their opinion of it.

The general opposition of the laboring families, to the interest of their employers, may in certain cases, be palliated by lack of confidence in their friendship and integrity (though never justified.) For a want of friendship to all with whom we have concern, is a want of philanthropy, and a want of philanthropy is the absence of justice, and the destitution of all moral excellence. If we would see people excel, in any profession, merit must be honorably and humanely distinguished.

However hard of digestion these wholesome truths may seem, to the mental functions of some, it may not be erroneous to presurmise, that the acquisition of profitable help, on the one hand, and the general credit of their families, on the other, will, eventually, be in direct proportion to the reputation of these establishments. And it is the summit of folly, to seek for a substantial reputation, in a free and moral community, but by means reasonable, benevolent and virtuous.

If the writer had noticed no failings, he had as well said nothing; and if he had cast the blame all on one side, no one, having twenty years experimental knowledge of the human character, would regard his declamations.

It is, finally, a piteous circumstance, that any should conduct the wheels of business under such repulsive movements. And the writer hazards an opinion, that half the number of hands with which our manufactories have been usually thronged, placed under the direction of a skilful manager.and all acting in hearty concert with the proprietors, would execute an equal quantity of work, and save twice their wages in the improved quality, and the diminution of the common waste of stock and machinery.

The lines to which this is prelusive, were originally designed for a supplement to a prosaic attempt to recommend an improved management of Carding and Spinning, and to investigate such defects and remedies, in the preceding particulars, as occurred to the author, and seemed well supported by argument.

That object being yet delayed, this admonitory trifle is presented in its present form, by request of a friend; accompanied by a few illustrative notes, and this apology for imperfections, in addition to those which the obvious difficulties of the subject involve—that it was mostly composed, in the course of a few days, amid the pressure of cares and the clamor of business, and has passed to the press almost without revision, or arrangement, the writer not having the present benefit of such retirement, as he conceives indispensable to metrical excellence.

Whatever betide it, he has two consoling thoughts,—that the Manufacturer's Pocket-Piece was *spun out* with the hopeful desire of doing some good, and no harm,—and that his ambition of poetic fame is so small, that he could hardly experience very excruciating pangs, if condemned to suffer the healing wounds of a literary *picking machine.*

In protraction it needs only to be stated, that, in reducing this work to a size near the proposal, it has been necessary to omit or retrench many important articles, which may hereafter expand an improved copy.

Medway, July 1816.

[Felch opens his poem with praise for founders of the cotton textile industry, such as Richard Arkwright and Samuel Slater, then proceeds in 400 lines to touch upon each textile technique in his search of moral significance.]

And thou, Rhode-Island! to aspersion known,
Where pious Freedom rear'd her infant throne:
Where founder Williams, from oppression fled,
And pilgrim-bands to humble honors led;
Thou to thy Country gavest the *fibrous skill*,
Their vesting Ark when warring tempests swell.
Where marching floods dispense enlivening powers,
And tolling metal marks the laboring hours,

(*b*) Samuel Ogden, author of "Thoughts on what probable effects the peace with Great Britain will have upon the Cotton Manufactures of this Country," was, probably the first workman among us, who dispensed instruction gratis.

(*c*) If bad habits prevail at some of our Spinning Mills, they were not manufactured there, but brought from other places. Although the young, in these employments, lose a portion of the diffidence of retired life; yet their situation is, probably, in most respects, far preferable to the condition of the poorer class in the maritime towns. They acquire less servility, more skill and industry, and at least, equal dexterity in their deportment and social dealings. I will further add, that the youth in well governed Factories, have many moral duties enforced, and gain an unusual ascendancy over their tempers—except such as are encouraged by their parents and relatives, in disobedient and disorderly conduct.

It is but 26 years, since Samuel Slater, Esq. erected the first Cotton Spinning Mill in America; yet many instances might be adduced, of promising young gentlemen, and ladies, who have been reared in like establishments.

These remarks will suggest the propriety of employing well appointed Managers.

Care's busy notes, and art's responsive sound,
In cheerful concert through the vale resound.
The breaker-lads, (*d*) with voices shrill and clear,
In treble accents greet the musing ear;
The buzzing spindles fancy's tenor trace,
And angry pickers (*e*) thunder on the bass.
My heart, enchanted, to this music beats,
My willing hand the joyful duty meets!
  Within the fabric, as I gladly trace,
The swift revolving wonders of the place,
Hydraulic movements call the watchman's care,
Harmonious discord fills the laboring air,
The wheeling Canns a shining host display,
And clattering engines intervolve my way.
When youthful hands the snowy flakes have spread,
And form'd in neatest style, the wooly bed,
Cylindric fields of smoothly pointed wires,
Revolving, seize and stretch the curling spires.
Through thorny ranks the woolly serpents glide,
Dispersing dusty showers on every side.
The critic's search their eager flight denies,
Till where the labouring doffer (*s*) takes the prize,
The dancing jaws of iron thence release
The rolling porcupine's enlarging fleece.
The silken squadrons, in successive chace,
The snowy tufts that each apartment grace,
And spotless rovings, should forever prove
The purest types of innocence and love!
  The master's voice directs the movements all,
And drives disorder from the busy hall,

(*d*) The Breaker is the Card through which the cotton first passes; the second is called a Finisher.

(*e*) A Picker is a machine by which the cotton is picked. Those on the common construction have a tremendous sound.

(*s*) The doffer is the smaller, or finishing cylinder of a machine card.

And why should Freedom's favoured children learn
Their infant race at wholesome laws to spurn?
Why, where no tyrant hand has power to bind,
With rebel-passions feed the tender mind—
Too prone, alas! from order's paths to flee,
And rush, impetuous, on to misery?
Ah, why should those the helm of business steer,
Who blindly follow vice's mad career?
Why he who wisdom's banner feigns to wave,
To passion bow—a doubly wretched slave!
Some moral problems have solution here,
Where pressing Rollers, (*f*) governed by the gear,
On steady axes, in submission rolled
To devious man neglected truths unfold.
Like him they through their whole extension, join,
When formed upon a just, unerring line;
Else moralizing threads uneven draw,
And broken order spurns the artist's law.

*

The *drawing heads* admonitory move,
And narrow, selfish *singleness* reprove;

(*f*) The draught of cotton by Rollers is, perhaps, the most important improvement in manufacturing. It is performed as follows.—The drawing filaments, or ropings, are received between two cylindrical rolls, or rollers; the under of iron, fluted, the upper covered with leather, and somewhat pliable; thereby closing snugly through their extent, and taking sure hold of the passing spires, or fibres of cotton. In front of these are placed another similar pair at such a distance that the spires will not quite reach from one junction to the other. These have more frequent revolutions, and are generally larger; whence the passing substance is extended in length from twice to fifteen times, according to the construction of the machinery, or the change-wheels affixed by the manager.

The top-rolls are confined to those under by weights; (sometimes by springs.) Neither of them can be made too straight, cylindrical and smooth, run too true and steady, or be kept too clean and dry. The filaments of cotton are drawn through 5 or 6 setts of rollers, before they are twisted into yarn; and at 3 or 4 of them in succession, 2 or 3 pass collaterally. By which process if well carded, the fibres are drawn straight, and disposed in quantities nearly equal.

They intimate Fredonia's (*k*) favorite plan,
Political equality of man;
They bid us rule our lives with even sway,
And neither weak, nor haughty points display,
While rolling years. to level frame. reduce
Our tempered minds—thus only fit for use.
Methinks they call for virtue's cleansing aid,
Where joys can dwell, or fearful woes invade;
To wasteful idleness their works declare
The pressing need of vigilance and care.
These busy samples may we all embrace,
And mend our lives at every broken place.
The fleecy cord. descending to the Cann. (*l*)
Invokes inspiring ACTION, friend of man;
Exhaustless source of competence and wealth,
Life's very essence, and the sire of health,
By moving harmony they rightly fill,
And brisker motion packs it better still.
Thus we by steady action, only, find
A healthful body, or a tranquil mind;

(*k*) A poetic term for the United States.

(*l*) The four principal operations to be investigated by the manufacturer of cotton-yarn, except such as are common to some other mechanical art, are:

1. The digestion of cards
2. The draught of rollers,
3. The drag of bobbins.
4. The reception of running canns, (or the tin canisters which receive and twist the roping, after it has passed the rollers.)

Bad Canns throw the roping into disorderly curls; yet few can accurately explain the secret causes. I shall not attempt particulars here, but observe, however. that regular roping will be found extending with the periphery of its receptacle, while that which is tangled falls in the centre. The former is probably the natural disposition, when the motion is rapid. and the internal parts of the cann and the whole passage of the roping. are smooth. clean. dry and properly shaped. The roping makes an angle as it enters the cann-top, as commonly constructed; and again, in passing to the body of the cann. The form should be such as to present no obstruction to systematic motion. in either instance.

To calculate the twists to every foot of roping, to be wound by hand. I have multiplied by 3 1-2. the square root of the No. of yarn to be spun from it with a ten drt. (or when the spinning frame increases the length tenfold.)

And idle men, on nature's bounty fed,
In point of use, will sink beneath the dead!
When rest exceeds what nervous fabrics crave,
This life's frail engine verges to the grave.

*

In single warp involving powers we trace,
Which seem to claim another's firm embrace :
But doubled cords in four-fold safety dwell
By winding oppositions balanced well.(*p*)
And those in twining love's coherent state,
With stranger force repel the hand of fate.
But love has bounds, or, like the crowded thread,
A tangled fortune rises in its stead.
   To pack the straightened fibres, side by side
In equal columns, for a thread supplied,

(*p*) Every one knows, that, when any quantity of twist is forced into a string, a certain portion will naturally recede to the cord, after doubling ; and, the two opposing twists being balanced, it will remain straight.

To form this Balance-twist, the number of turns put into the doubled, should be to those left in the single thread, as the diameters of the doubled to the single, inversely ; or, directly, as 1 (square root of one thread) to the square root of the number of threads united. And the first twist, being equal to the two final,

The curling locks in glossy cords to bind,
By needful windings (*u*)—else but feebly join'd,
The workman aims; for this the frames are reared,
And laboring music through the mansion heard.
Thus skilful parents draw by slow degrees,
The infant mind, with diligence and ease,
In even lines of justice, truth and love,
And with a serpent's wisdom shield the dove;
Lest jarring vice should rend the thread in twain
And ruined innocence lament in vain!

*

should be to the last, as square root of threads doubled, plus, 1, is to 1. Hence the following Table, which may be enlarged at pleasure.

Balance.	First Yarn twist.		
Thd. twt.	2 Doub.	3 Doub.	4 Doub.
1	2.41421	2.732	3.
10	24.14	27.32	30
11	26.55	30.05	33
12	28.97	32.78	36
13	31.38	35.51	39
14	33.80	38.25	42
15	36.21	40.98	45
16	38.63	43.71	48
17	41.04	46.44	51
18	43.45	49.17	54
19	45.87	51.90	57
20	48.28	54.64	60

(*u*) It is well known that the Rule of Three gives to great a proportion of twist to fine yarn; nor do I know that any rule for different numbers has hitherto been formed. I have lately ascertained, that the quantity of twist required, is in proportion to the diameters of the threads, consequently to the square root of the numbers of skeins, or hanks to a pound.

To find the right number of twists for every inch of warp-yarn, I multiply the root by 5; giving 5 twists to No. 1, 10 to No. 4, 15 to No. 9, 20 to No. 16, 25 to No. 25, 30 to No. 36 &c. This throws the external fibres across the longitude of the thread at an angle of about 34 degs. No. 1 being not far from ,04 inch in diameter. The average angle would be about 25 degs.

Although this rule is mathematically accurate, yet I would observe, that, since a large thread possesses more contractile force than a less one, I commonly deduct a very small allowance from the rule—twist of coarse yarn.

Thomas Man

Picture of a Factory Village (1833)

Little is known of Thomas Man's early life. In 1833, the Pawtucket *Chronicle* identified him as "Thomas Man of Manville," which was a cotton mill village four miles south of Woonsocket Falls on the Blackstone River. The Man family came to be the principal owners of the mill established there in 1811, and Thomas Man was probably related to these local mill owners.

In 1833, Man published the long poem, *Picture of a Factory Village,* portions of which are reprinted here. Two years later he published *A Picture of Woonsocket or the Truth in its Nudity* (Providence, 1835), a caustic prose attack on a specific mill village. On the title page of the latter, Man described himself as a "Professor of eloquence, moral philosophy, and the languages," and the volume included his own translations of some French, Spanish, and Italian romances. In addition to his assault on the factory system, he took an active part in the temperance movement, prosecuting liquor dealers for the city of Providence in 1848. He was also interested in phrenology.

From at least 1824 until his death on December 15, 1880, Man resided in Providence, where he worked as both a phrenologist and a language teacher. After he died the Rhode Island *Press* memorialized him as a "venerable and eccentric teacher of languages."

Man's *Picture of a Factory Village* is as remarkable for its polemic

Some of Man's publications can be found at the John Hay Library, Brown University; others are located at the Rhode Island Historical Society. Sources for his biography include Thomas Steere, *History of the Town of Smithfield* (Providence, 1881); the Providence *City Directory,* 1824, 1832, 1836–1837, 1847–1848, 1854–1855, 1864, 1874, 1880; the Providence *Daily Journal,* December 17, 1880; and the Rhode Island *Press,* December 25, 1880. In this text of Man's poem, line numbers and notes have been added.

vigor as for its poetic incompetence. Fortunately, Man did not take himself too seriously, as he acknowledged:

My drooping muse, with flagged wing,
Croaks hoarse as raven, tries to sing;
Brays like a jackass, in the spring;
On which, once rode Judea's king.

As a social critic, however, he could hardly have been more serious: "Come! help me, oh! thou vilest muse, / Into my verse thy gall infuse." The first several hundred lines of his irregular verse are the sharpest condemnation of factory life to come out of the mill village experience and are as damning as any of the jeremiads that moralists and labor organizers were to aim at the far more notorious factory cities. After this opening attack, most of which is reprinted here, Man rambles on for almost 3,000 more lines, assassinating various characters in the village—apparently real persons, faintly disguised—but hardly mentioning manufacturing. Altogether it is some of the most opaque, incomprehensible, incoherent verse ever penned. Its one redeeming aspect is this opening flight against manufacturing. Note that Man's criticism is not ethical but political. He sees dangerous and oppressive the overwork imposed on people by "aristocrats" and "tyrants." The significance of Man's critique may or may not be enhanced by his claim in a prefatory note that the publication of the poem was underwritten by more than a thousand subscribers, "among whom are the Ex-Governors, James Fenner and Lemuel H. Arnold; Hon. Elisha R. Potter; the mayor and some of the Aldermen; the Attorney General . . .," and more. Whether such gentlemen knew what they were paying for is an open question. (GK/MBF)

PICTURE

OF A

FACTORY VILLAGE.

"No Aristocracy of Wealth has a Right to control Public or Private Opinion."

"To catch the manners living as they rise."

"To hold the Mirror up to Nature."

"Nothing extenuate, or set down aught in malice."

The Muse, dejected, disconsolate, and broken-hearted, sitting on the confines of a Factory Village, taking a perspective view of the *Beauties of Nature*, almost lost in the distance—and the magnificent buildings—casting ever and anon, her last lingering looks, like the sun sinking below the Western horizon, on the *Habitations* of *Innocence*, *Improvement* and *Leisure*—bidding a *reluctant*, but *eternal Farewell*—to *Manufactories*.

For Liberty our fathers' fought,
Which with their blood, they dearly bought,
The Fact'ry system sets at naught.
A slave at morn, a slave at eve,
It doth my inmost feelings grieve;
The blood runs chilly from my heart,
To see fair Liberty depart;
And leave the wretches in their chains,
To feed a vampyre from their veins.
Great Britain's curse is now our own;
Enough to damn a King and Throne.

Hark! dont you hear the Factory Bell?
Of wit and learning 'tis the knell!

God, in mercy, break our fetters,
And let us look upon our betters.
And rend the heavy iron bands,
Which long have pinioned both the hands;
Black with old time's corroding rust,
With which we 've been so long accurs't!
The clanking chains around the waste,
Bound like the lion, rav'nous beast,
And ever praying for relief,
Sunk in despair severest grief.
Now let us quickly cease as slaves,
Or grim death's fell dart point our graves,
Ferried o'er the black Stygian waves.
Hark! don't you hear the fact'ry bell?
Of wit and learning 'tis the knell.
It rings them out, it tolls them in,
Where girls they weave, and men they spin.
Look! See them rising from their beds,
With eyes half open in their heads!
And in their hurry, in their flirts,
One half have left behind their shirts.
Now see them for the mill a-clawing,
Their breakfast in their mouths a-chewing.
Hark! don't you hear the Picker hum?
It would a deaf and dumb man stun!
Sounds like the wailing of the damn'd,
Who in the lowest Hell are cramm'd.
Now, see! their lordling strut along,
Follow'd by his bleating throng.
Out in all the foulest weather,
Trav'lling on their worn out leather;
Thro' snow, thro' dirt, thro' mud, thro' mire,
O'er the ankle, sometimes higher.
With draggled coats about their feet,
Their hoods, their cloaks, all filled with sleet.

1 Now see the asses drive the mules,
Of Aristocracy the tools.
Go in the Card-Room! See the girls!
With cotton sticking to their curls.
See the ladies' handsome Weave-Room,
'Tis capacious—has much sea room;
2 In the centre stands a pulpit;
The girls at noon, how they throng it;
No sound of mercy echoes from it;
But clinking hammers, fixing tools,
To keep at work the wretched souls.
Hark! hear the looms, shuttles rattle,
The girls they stand like frightened cattle:
Like cows beneath the forest oak,
Riven by lightning's blasting stroke.
The thunder loud rolling o'er head,
One struck, a ghastly corse lies dead.
The rest with horror, awe-struck stand,
With terror fastened to the ground!
Direful lowing! awful bellow!
Like ocean's waves, gloomy, hollow;
Whose dying murmurs reach the shore,
In silence hush'd, now cease to roar.
See mighty Vulcan raise his arm,
With blackened brow, like thunder storm,
Which e'en would Jove himself alarm.
Look! see beneath his powerful stroke,
The heaviest iron hangers smoke.
Could forge more spindles in an hour,
Than steam-engine of ten-horse power.
Who, when he strikes the frighten'd steel,
The block and anvil how they reel!
The blows resound throughout the cave,
Like roaring of the mountain wave.
Now see old Vulcan's smutty men!

Ascending from the lowest den;
Ragged, greasy, sturdy devils,
Like tenants of western hovels;
But should they stop, drink some water,
3 Vulcan sets them off a quarter.
4 The Pickerman to by-stander,
5 Looks a very Salamander.
The long loose cotton hangs like hair,
Is sticking to him every where;
Flutters like leaves upon the trees,
Rustles at every passing breeze;
6 Moves like Beggars' fly-trap patches,
Ev'ry thing in contact catches.
He wears a hat looks like the Pope,
His face and hands have ne'er seen soap.
The present one, tyrant Nero,
A scourge on earth, far from hero;
There's nothing strange in his features,
Looks like other Fact'ry creatures.
Has labored lately on a farm,
In agriculture found no charm.
Like Adam would not till the earth,
Prefers romantic Picker birth.
This is the strangest thing I 've seen,
They work their children 'fore they wean.
In the morning, one hour they creep
About the mill, e'en half asleep.
Hark! hear the breakfast bell's loud call,
Like sheep begin to scatter all.
Look here! look there! look all around!
They cover far and near the ground.
Talk of one thing, then another;
How do you do? How 's your mother?
How 's my beau, John, your brother?
Have you laid in your beef and pork?

In factories 'tis grinding work—
Don't you think this is slavery ?
And the basest kind of knavery,
To shut us up in such a place—
Living by such hard sweat of face.
Mary, and is not this despair?
Curse on the Fact'ries every where—
Besides, it causes consumption—
Oh ! that we might have redemption—
O God! we 're ignorant indeed,
Numerous as Abraham's seed,
Nor have we time to learn to read.
Many of us can't write nor spell;
A Fact'ry is a Gothic hell.
E'en a head clerk can't read the news,
Or insolently they 'll abuse.
What shall we do? where shall we go ?
We have no time to knit or sew.
7 Fact'ry owners are Algerines,
And bind the mind with Vandal chains.
My soul is wrung with fell torture,
Dark the present, and the future.
Our life 's in danger, exposed to constant harm,
The wheels tear the hand, pickers take off an arm.
A handsome girl is caught in a cursed drum,
Dash'd from things of sense, into the world to come.
Who would spend their time in such a horrid place?
Worse than Bastile—Inquisition of our race.
Parent of Heaven! take our breath,
Redeem us from this living death,
We 've not time to court and marry,
Which makes me feel very sorry.
The law of nature is to wed,
But, sure I cannot buy a bed.
I cannot muster enough cash,

To buy a dish of suckatash.
After Bell time, at our house,
8 **Call, eat a dish of good lobscouse.**
Good night, we soon shall meet again,
'Tis fair, and hope it will not rain.

[In the next forty lines Man describes the clientele of a factory store, mostly women purchasing harmless trifles. He then triggers the following salvo in italics.—MBF]

A Factory store is a sponging place;
9 ***It is the eel-pot of our sorry race;***
We toil all day, fatigu'd, with might and main,
And the next day, repeat it o'er again.
At the year's end—this you, indeed! well know—
We've not a single paltry cent to show;
You grasping owners, put it in your pocket,
And we, poor vile wretches, can't unlock it.

*

1. A pun is intended in "asses drive the mules." Mules were mule spinning machines. See glossary.

2. The term "pulpit" is obscure.

3. The phrase "sets them off a quarter" means that their pay was docked (reduced) by twenty-five cents.

4. A pickerman was one who operated a picking machine. See glossary.

5. A salamander was a mythological lizard that could endure great heat.

6. The phrase "Beggars' fly-trap patches" is obscure.

7. "Algerines," or Algerians, were reputed to be great despots.

8. Lobscouse was a sailor's stew.

9. An eel-pot is a trap.

George S. White

The Moral Influence of Manufacturing Establishments (1836)

This excerpt from G. S. White's *Memoir of Samuel Slater* deals directly with the popular debate over the factory system. It should be read not simply as a defense of manufacturing or as a reflection of the attitudes of manufacturers and their supporters, though it is both. Rather, it should be read as a combative document designed to counter the widespread suspicion of factories and factory owners. As such, the document reflects a stage in the growing self-assertiveness of America's industrial captalists. Hostility to the new factory system found its most articulate expression in the writings of Thomas Jefferson, John Taylor of Caroline, and other intellectuals whose arguments derived in part from agrarian and Physiocratic thought. Far more pervasive, however, were popular beliefs that equated the textile factory with the poorhouse, the pretentions and antidemocratic dangers of English aristocracy, and a loss of freedom and republican virtue. According to such beliefs, the mills, dams, and factory villages were a threat to a predominantly agricultural social order delicately balanced on the edges of industrial capitalism. The mill owners and their supporters emphasized the benefits of domestic manufacture, the economies of labor-saving innovation, and the industrial virtues of diligence, time-discipline, and sobriety. In another passage of the *Memoir,* White argued that textile mills were sponsors of "Moral and intellectual improvement":

> Hundreds of families . . . originally from places where the general poverty precluded schools and public worship, brought up illiterate and without religious instruction, and disorderly and vicious in consequence of their lack of regular employment, have been transplanted

George S. White, *Memoir of Samuel Slater* (Philadelphia, 1836), pp. 113–135.

to these new creations of skill and enterprise; and by the ameliorating effects of study, industry, and instruction, have been reclaimed, civilized, Christianized.

This passage and the text that follows help to clarify the contemporary terms of the debate over the "moral" effect of capitalist manufacture, a debate that often took place above the heads of the mill workers, who were neither paragons of proletarian virtue nor vicious and depraved. Arguments over morals had their appeal to the urban middle class and the church-going yeomanry. Workers themselves had more immediate and pressing disagreements with the factory owners and their own ways of articulating them. The early debates over manufactures, in which White's *Memoir* is a key text, are documented extensively in volume 1 of this series.

To provide a documentary basis for his argument in behalf of industrial interests, White circulated a set of twenty-one questions "to several heads of manufacturing establishments." At the conclusion of his argument, he appended these questions along with a number of answers. Among White's respondents was Smith Wilkinson of the Pomfret Manufacturing Company; other documents concerning Wilkinson and the Pomfret Company appear in this volume. (GK)

CHAPTER IV.

MORAL INFLUENCE OF MANUFACTURING ESTABLISHMENTS.

"There is no artist, or man of industry, who mixeth judgment with his practice, but findeth in the travail of his labour, better and nearer courses to make perfect the beauty of his work, than were at first presented to the eye of his knowledge."

We have already seen that manufacturing establishments exert a powerful and permanent influence in their immediate neighbourhoods, and time, if not already, will teach the lesson, that they will stamp indelible traits upon our moral and national character. Evidences abound, wherever man exists, that his character is modified by localities, by a diversity of pursuits, by a facility of acquiring a living, by the quality and fashion of the living itself, by a restrained or free exercise of his rational powers, and by restraint on the enjoyment of liberty. Different climates and different countries produce indelible peculiarities. In the same climate and in the same country similar changes appear, from the effects of immoral habits, and from what may be termed artificial or mechanical causes. The effects of immoral habits are well known to all observers of human nature.

Those pursuing different occupations are aware that these exert an influence upon character, producing moral, no less than physical, varieties. For example, butchers become hard-hearted and cruel, and in England are excluded from the jury-box ; those who are confined to a particular routine against their will, peevish and discontented ; those who are always ordered or driven, and expect to be so, exercise little control or discernment for themselves.

Manufacturing establishments become a blessing or a curse according to the facilities which they create for acquiring a living, to the necessary articles which they provide, and the general character which they produce. To set up and encourage the manufacturing of such articles, the use and demand of which produces no immoral tendency, is one of the best and most moral uses which can be made of capital. The moral manufacturer, without the power or disposition to overreach, is in reality a benefactor. The acquisition of wealth in this way, is the most laudable. In point of benevolence and real worth of character, it claims a decided advantage over the cent per cent. process of accumulation.

Some have not the requisite ability to carry on manufacturing establishments; capital, then, with great propriety is loaned to those who have. The moral influence of a community is not promoted by creating or submitting to a manufacturing, or any other aristocracy, solely in the pursuit of interest, in which selfishness is wont to predominate.

The manufacturing interest, in a flourishing state, naturally creates power and wealth. The value of labour and the value of money are then at his disposal; but, in this free country, there is a sufficient counteracting influence to keep up the price of labour and to equalise the prices of their commodities with the value of the products of the earth. Without such a resisting power, a few would abound in wealth and influence, while the multitude would be in poverty and reduced to servitude. But there always exists a counteracting influence in the rival establishments, and the general spirit of enterprise. On the supposition that the manufacturing interest was strictly benevolent and moral, dispensing its favours according to merit and precisely as they are needed, the community might not be losers by such a state of things. This must be always the case where a people are left free to use and purchase according to their free choice. With the common experience of mankind, it could not be expected so. Only a few look beyond their own interest; when that is provided for, the employed who have assisted in the provision, are left to shift for themselves. Benevolence is not so general among mankind as to expect it uniformly. But in the progress of manufactures among us, every department becomes interested in its prosperity, the operatives receive a greater emolument for their services than in any other part of the world, whilst capital receives but a small interest, compared with other branches of industry. With such a power established merely by selfishness, morality is promoted so far and no further, than interest; but the promotion of morals becomes their interest. And if religion appears something in name or in sectarianism, more than in reality, still its promotion is for the interest of the whole community. It is said, on the presumption that the capitalists are aiming at their personal wealth, the facility for acquiring a fair compensation becomes less and less at every pressure. A rise of wages is then adapted to convenience or pleasure. But it must be remembered, that the pressure bears as heavy on the employer as the employed, and renders him liable to lose all the earnings of many years of labour, and the savings of much self denial, and render him poor and dependent. There are two sides to this question, and the operatives in good times ought to lay up for time of need. Then they would not be obliged to bring their labour into market the best way they can, to obtain their daily bread. To take advantage of such a position, is one of the greatest immoralities. The liability of its consequences are as bad in creating discord and producing civil commotions. But the owners of factories are not known to stop their mills till obliged by dire necessity : they generally run them till they become bankrupt. The real power belongs to the labouring class ; no one ought to expect to employ this without paying for it, and no one does expect it. It is power when rightly used, and most often ceases to be so when abused. Those who are so thoughtless, negligent, or squandering, as to trust wholly to the present occasion for a bare subsistence, can hardly be thought powerful compared with what they would be did not necessity compel them to take what they can get for the present occasion. It is a mistaken notion to suppose the manufacturing interest promoted by creating poverty, or, in the end, by heavy reduction of wages. The articles manufactured very soon sink in like proportion, and the profits are swallowed up in the payment of the operative. Besides these consequences, the ability to purchase does not exist, a consideration which more or less affects the value of every article brought into market.

Our day has witnessed the surprising effects of the ingenuity of man, in calling into existence and putting in operation labour-saving machinery. If it would be, in reality, promoting human existence and human happiness in our present character and condition, that our food should come to us ready made, our habitations ready built, our conveyances already in motion, and our understandings already improved—the nearer we approach such a state of things the better.

But if not—if the desires and pursuits of objects be no less blessings than their possessions—if human nature be bettered, and the grand object of existence benefited by employment—there must be a point beyond which to obtain food and clothing and other things, without application, would be objectionable. To be moral and desirable, labour-saving machinery must bring along with it some particular benefit to the community, as well as to individuals.

This may be such as more than compensates for the many losses which are sustained in some countries, in consequence of the improvement. When it was proposed to introduce printing into the Prussian dominions, the king objected by saying, it would throw forty thousand amanuenses out of employment. After printing went into operation, to ameliorate the condition of those who were thrown out of employment, the Prussian government made a law that the initial letters should be omitted by the printers, in order that they might be executed by the amanuensis at a high compensation. That they performed these letters with great ingenuity, and in a manner difficult to be imitated, may be seen from a copy of a bible now in possession of the antiquarian society at Worcester, Mass. It must have been a calamity for so many to be thrown from their pursuits, and be deprived of the means of getting a livelihood. The benefit resulting from the introduction compensated for this loss, more than ten-fold. This is one, among many instances of human invention, which wonderfully adds to the dignity and happiness of mankind.

The first introduction of Hargreaves' and of Arkwright's machinery into England, was not only met with objections, but with popular vengeance. It threatened a speedy destruction to every jenny and water-frame in England, and so in appearance carried in its motions frightful evils. The anticipated evils actually happened; hand spinning met with a speedy overthrow, and those who had earned a few pence per day in following it, were compelled to resort to other employments, and perhaps to be employed in manufacturing on the new plan which they had laboured to oppose.

Similar feelings and similar consequences have happened and are still happening in America. Manufacturing, instead of going on quietly and single-handed in private families, with immense labour, grows into large establishments, which employ and bring into association, masses of population.

This position is moral or immoral according as it furnishes proper stimulants for industry and for exertion, and for improving and directing the mental powers and principles. With little or no inducements or expectation of emerging from a state of ignorance, with no schools, no moral or religious instruction, the liability is great for an introduction of all the evils which the opposers of manufacturing establishments have often predicted.

It is well known that vice grows worse by contact with its kind. If it can be proved that manufacturing establishments tend to accumulate, consolidate, and perpetuate, vicious propensities, and their consequences, on the community, this will serve as no inconsiderable drawback upon the apparent prosperity which is indicated in their immediate vicinity. If found so, the condition must be charged directly to the establishments or to their consequences and abuses. It is evidently an abuse to collect a mass of vicious population, and keep them in a state of ignorance and irreligion. When this is done, the whole community have a right to complain. If it can be shown that such things are frequently done—it is contended that they are not necessary consequences of manufacturing establishments. The owners of such establishments have it in their power to change the current of vice from its filthy and offensive channel, and make peace, order, and comfort among those they employ.

The dependence between the employed and employers should be mutual. But by employing vicious, improvident, and indigent characters, the dependence falls mostly on one side—yet it is a benefit to the community that such a class should find employment and support. Though in some countries, oppression ensues, poverty and vice show their dismal and disorderly features, and then the honest, upright, and intelligent, are driven from the establishment, and perhaps from the employment; better things can be spoken of this country, where the honest, upright, and intelligent, have always a preference. Such are leaving the old world, they are disappearing, and many of them are in the west, engaged in other employments. Pursuing such a policy, by and by, only the dregs are left, and then without looking for the causes, it appears that factories have been the immediate cause of all the mischief. On a candid enquiry, it is seen to be the abuse, and therefore not chargeable to a proper use.

Slater, the founder of the cotton manufacture in America, abundantly demonstrated, that under right management, they had no immoral tendency. On the contrary, he made it appear, that they might be serviceable to the most moral purposes. Following the plan instituted by Arkwright & Strutt in England, taking the oversight of the instruction and morals of those he employed, and instituting and keeping up sabbath schools, he successfully combated the natural tendency of accumulating vice, ignorance and poverty. Such remedies not only prevented their occurrence, but had a tendency to remove them, when they actually existed.

Industry, directed by honest and intelligent views in moral pursuits, and honourably rewarded, holds a very high rank among moral causes. To maintain good order and sound government, it is more efficient than the sword or bayonet. At the anniversary dinner of the public schools in Boston, the following toast was given by Edward Everett—"Education—A better safeguard for liberty, than a standing army. If we retrench the wages of the schoolmaster, we must raise the wages of the recruiting sergeant." So far as manufacturing establishments have promoted industry, and furnished means for an honest livelihood, thus far they have exerted a salutary influence on the character of those who have been employed. Multitudes of women and children have been kept out of vice, simply by being employed, and instead of being destitute, provided with an abundance for a comfortable subsistence.

Those who are furnished with an opportunity, and are trained up to lay by in store—moderate and regular returning means, to be used at some future day—are invariably superior in point of character to those who have not. It is not so when means flow excessive and irregular. Many a youth has been ruined by beginning with large wages, and having in prospect plenty of money.

It is believed that there may be found more young men and women, who have laid up a few hundred dollars, or even a few thousands, by being employed in manufacturing establishments, than among those who have followed other employments.

On the score of employment, manufacturing establishments have done much to support the best interests of society. It appears also, at the present time, that they have done so by their improvements. On the supposition that one or a few individuals, by the invention of labour-saving machinery, succeed, so as to furnish any particular article much cheaper than it could be done in the ordinary way, in this country where it deprives no one of a living, and goes to forward and hasten the general improvement, it cannot fail to be a benefit to the community. The diminution of price in the articles has been such, that the people have been doubly paid for all the protection granted; and commerce has been benefited by the opening of a foreign market. The failures and fluctuations in the manufacturing establishments have arisen from their weak and incipient state, and the competition of European fabrics. This cause appears greater than want of management and calculation, for the same men have alternately succeeded and failed on the same ground.

Fluctuations, whatever may be the cause, and whether they relate to business, morality, or religion, exert a wide influence on individual and national character. Those to which we are here attending, give currency to monstrous species of swindling, and form a most suitable juncture for unprincipled and unfeeling knavery to grasp with an unsparing hand, while industry and honesty are thrown into the back ground, or kicked out of doors. When such occurrences happen, and the intriguer goes off rewarded and applauded, while the honest man is stripped, despised and neglected, they give a turn to the whole character of the community. The flooding our cities with foreign importations has had this kind of tendency, and produced those evil effects.

Shrewdness and over-reaching are common events. Morality, however much respected in principle, is extremely liable to be set aside in practice. These are some of the bad tendencies of seeking out many useless inventions, and too eager a grasp after traffic and exchange of property, or what is technically called *speculation*. The acquisition and possession of property, are made the main objects of existence, whether it be needed or not. On the other hand, it will be granted, that every objection vanishes, when mechanical inventions acquire permanency, and can be subjected to the regularity of calculations. It may dignify and exalt man to triumph over the known laws of nature, and bring out the hidden treasures of air, earth, and water, in tame submission to his use. For aught we can discern, it would have no injurious effect upon his character, could he extend his journeys and researches further than this globe. One thing is certain, the more he studies and understands the works of nature and Providence, the greater will be his admiration of the display and application of wisdom and goodness. If applied as intended, the more of the resources which have been provided he brings into action, the more he adds to his true dignity and happiness.

Contrivances to favour selfish views and selfish ends are common to the animal creation. The human family are distinguished from the infinity of being, only by a greater possession and cultivation of moral and intellectual faculties. Unlike the most of the animal creation, man is left to provide for himself. Strength and powers are given him, objects are placed before him, and the strongest conceivable motives presented to use this world as not abusing it.

There must be a limit, beyond which refinement will be objectionable. When excessive it is a precursor of a relapse in civilisation.

When wealth and its appearance abound, children are most often brought up in idleness, and indulged in extravagance. Supposing labour a burden, and retrenchment the ruin of happiness, they are made liable to be overtaken by poverty, and with their last energies and ruined characters to be plunged in real misery. Individual calamities of this description, as they accumulate, become national calamities, and foment domestic dissentions. Suffering pride is all the while meditating revenge. It has nothing to lose and will endure any thing to regain what it has lost. Appearances and extravagances are prominent causes of dissention.

when a part are rioting, and a part are suffering. Distinctions of rank are introduced. Individuals and nations who have run into excesses in making and maintaining such distinctions, sooner or later, are wont to be caught in their own snares. Poverty feels the burden of degradation when the power is lost to remove it.

In the present happy condition of the manufacturing districts, there are no advantages enjoyed by the rich, that are not reciprocated with the poor. Labour was never better paid, and the labourer more respected, at any period, or in any part of the world, than it is at present among us. And that man is not a friend to the poor who endeavours to make those dissatisfied with their present condition, who cannot hope, by any possibility of circumstances, to be bettered by a change. This is emphatically *the poor man's country.**

MORAL EFFECTS OF INTERNAL IMPROVEMENT.

In all the efforts that have hitherto been made for the improvement of the country, by means of rail roads and canals, reference has been made to their physical advantage only. In executive recommendations, and the application for chartered companies to construct these works, the enhanced value of lands through which they pass; the importance of establishing communications between commercial cities; the facilities they afford for conveyance of produce to market; the securing the trade of distant regions, to the ports of our own states, are the principal reasons which are urged

* The philanthropist and the political philosopher will enquire, what is the physical and moral condition of the vast population employed in manufactures? The workmen who construct or attend upon all these machines are not to be confounded with the machines themselves, or their wear and tear regarded as a mere arithmetical question. They are men, reasonable, accountable men; they are citizens; they constitute no mean part of the support and strength of the state; on their intelligence and virtue, or their vices and degradation, depend in a considerable measure not only the character of the present age, but of posterity; their interests are as valuable in the eyes of the moralist as those of the classes who occupy higher stations, yet the enquiry should be, not if the manufacturing population are subject to the ills common to humanity, not if there is not much to be lamented, but what is their condition compared with others. It is the destiny of man to earn his bread by the sweat of his brow; idleness, improvidence, and dissoluteness, are found in our large cities, and are invariably the parents of wretchedness; every where, people of all ages and conditions are liable to disease and death. The principal considerations are, the command which the working classes have over the necessaries and comforts of life, their health, their intelligence, and their morals.

upon us why they should be constructed. These indeed are sufficient, if no other could be given, to justify all the expenditures already made to establish such communications, and many more, as soon as the country can bear it. But their moral effects on the community must not be lost sight of by the philanthropist. The effect of an extensive internal commerce, in as large a country as this, on morals and the arts, science and literature, as subservient to morals and religion, are too obvious and important long to escape the notice of an attentive observer. All experience proves that good morals never did, and never can exist, among an indolent people, and people who are poor in consequence of their indolence. "Idleness is the parent of many vices," says an old proverb, and none more true was ever spoken. But in districts far from convenient markets, idleness is inevitable. Never will men labour in any employment if they can avoid it, unless they can foresee some pecuniary advantages sufficient to reward them for their pains-taking. On the contrary, they are too apt, for want of due encouragement to industrious habits, to throw away their time in worse than useless idleness and dissipation. Whoever has experienced the difficulties attendant on almost all efforts for the moral advancement of a poor and scattered population, without this encouragement, and compares them with the facilities afforded by thriving towns and villages, inhabited and surrounded by an industrious and happy people, will see at once that whatever tends to improve the physical condition of man, must, as it renders him more comfortable, conduce, in no small degree, to the improvement of his morals; and that (whatever some may have dreamed otherwise), in real life, poverty, from want of encouragement to industry, is a condition very unfavourable to the practice of virtue. If a people, under these circumstances, are ever moral in their deportment, no credit is due to their condition for it. Let our legislators be assured, that while they are extending towards its completion that system of improvement planned and hitherto carried forward with so much wisdom, they are putting into operation a moral machine which, in proportion as it facilitates a constant and rapid communication between all parts of our land, tends most effectually to perfect the civilisation, and elevate the moral character, of the people.

The general amelioration in the moral condition of communities, by the healthful encouragement of internal industry, and by affording proper aids to the development of national resources, is well worthy of the serious attention of legislators. An idle population is ever vicious and degraded; and perhaps the perpetuity of free institutions and with them a sound state of public morals, cannot exist among a people whose energies are not kept constantly in play by the pursuit of some incessant productive employment. Let us look at the contrast given in the following sketch by a North American resident in South America:—

"It is impossible to look at the present state of our neighbouring republics without a mingled feeling of pity for the weakness, and of contempt for the inefficiency, of their governments. The first out-breaking of the revolution there was hailed by the people of this country with enthusiastic joy, as the grand step towards the formation of other governments equally happy with our own; because based upon like principles, and aspiring to like ends. The success of their undertaking we confidently predicted, for, for them it was not reserved to try the first grand experiment,—that trial had been ours; and when the potentates of Europe, following our example, had come forward and acknowledged the independence of those republics, we felt that we, as a nation, were not alone,—that another, as promising, had risen up to prove the practicability of a new and a distrusted form of government;—we felt that a new light had dawned upon the hitherto benighted half of the great western world, which was to guide them to freedom and happiness, and we exulted in the prospect of the noble contrast about to be presented to the tyranny and despotism of the East. But the day-star of their liberty was the brightest at its dawn. Instead of increasing in splendour as it rose, its rays beamed fainter and fainter, till at length, it is now almost totally obscured in the mists of error, discord, and confusion.

"And we are naturally led to enquire, in view of these facts, into the cause of this. We are at a loss to account for this lamentable failure of reaching that high stand which the world was led to believe the new republics would take,—we compare their first efforts with ours, and we find them equal; indeed, more than equal. While ours were furthered and sustained by petition and remonstrance, and partook more of the character of mild persuasion than of determined opposition, *their* first efforts were accompanied with the heat and the fury of sanguinary conflict; and *their* hopes of redress were founded solely on the extermination of their oppressors.

"How sad is the prospect which, to-day, is presented to our view, in sight of all the nobleness of enterprise and undertaking which characterised the first efforts of our sister republics! There can be no hope of their stability, under their present forms of government. The people have shown themselves unequal to the task of supporting it ; they do not understand, neither can they practise upon, the principles of self-government. And the grand secret ot all this inability lies in the universal propensity of the people to indolence, in their want of enterprise, and in the listlessness which must infallibly spring from such propensity. All the better feelings of that people were called into action in the moment of rebellion ; they were kept alive and nurtured by a constant series of almost unhoped for successes in the grand struggle ; and, at such a time, the men who weighed the most in the scale of popularity, and who were looked up to, by the lower orders, with reverence and respect, were military men,—men who had risen by their valour, or their patriotism, or their zeal in the common cause, to a comparatively high and dignified station. While the struggle lasted, there was no want of energy, or stability, or perseverance among them; the confusion and turmoil of the revolutionary era seemed admirably calculated to give to each and every man an opportunity to display himself in the sphere peculiarly adapted to his powers; and thus all were occupied and satisfied.

"But the contention at last ceased, and the time came when it was found necessary to re-organise the government, and establish it upon the principles for which they had fought. With that moment commenced the troubles and internal divisons which have since brought the country to the verge of ruin. Intriguing and ambitious men had grown up in the midst of them,—hundreds of young officers, whose education had been purely military, and whose views and ambitions were limited to one point, were stopped short in their career, and left, without a single resource in themselves, to plot and plan the means of their own advancement in the sphere of action to which they had so fondly looked forward, and for which they believed themselves solely fitted. Among the more advanced in age and acquirements,—those who had taken a more immediate and active part in the strife just finished,—patriotism, love of country, zeal in the advancement of the national interests, all were buried and forgotten in the all-absorbing consideration of how they might secure to themselves, against the pretensions of the less experienced, those temporary advantages and emoluments of station which were theirs at the close of the revolution.

"Agriculture, commerce, manufactures, and domestic industry, although never much attended to, were now less thought of than ever. They depended entirely upon Europe and North America for the ordinary supplies of the most essential necessaries of life. With a soil the most fertile, and an extent of country sufficient to furnish a supply to half the world, they are still dependent upon

North America for the flour they consume. With their prairies teeming with millions of cattle, they are still dependent, in a great measure, upon foreign countries for their butter and cheese. The mechanic and higher arts are attended to almost exclusively by foreigners; indeed, wherever energy, or enterprise, or industry, is requisite, the native plays but a poor part in competition with the foreigner. This can be easily accounted for: in the first place by their excessive indolence, and in the second by a sort of hereditary pride and loftiness of feeling, which will not suffer them to follow any acknowledged trade or occupation; and which feeling, so far from rendering them superior, either in attainments or appearance, places them actually far below the ordinary standard of mediocrity. Many or most of their young men are living, and must continue to live, upon the scanty resources of their impoverished parents, some of whom, from a state of high affluence, have been reduced to comparative poverty by the destructive internal dissensions, which have laid waste and ravaged the country, and shaken, to their basis, her institutions since the revolution.

"How striking the contrast that our own land, or at least New England, presents! Where, among us, is found the youth, affluent or not, high-bred or low, who acknowledges neither occupation or profession? It is, among us, as deep a stigma as exists, that cast upon him who neglects to adopt *some* means of rendering his natural faculties subservient to one grand end of our being—that of usefulness and assistance to our fellow-men,—and who refuses to occupy that station among them to which he seems called by the particular circumstances and wants of the age, and for which his Creator has fully endowed him, with peculiar faculties and advantages.

"What a striking difference do we perceive in the morals, the feelings, and the habits, of the two people! While the billiard-rooms and the gaming-houses of the one are overflowing with the flower of her young men, and fitting them for any thing save for the performance of their duty in the approaching struggle of life, the workshops and colleges of the other are giving birth to men who are to supply the places and walk in the paths their fathers trod,—who are to further the interests and contribute to the respectability and importance of the nation,—young men who are eminently fitted to enlarge upon and improve the present system of things,—to give force and influence to the virtues, and reform the abuses of those who have gone before them.

"National grandeur and elevation of standing are founded, we may say solely, on the industry and enterprise of the people. The wealth and power of a nation have their existence in them, and the hopes of a nation's prosperity, advancement, and continuance, are, and can be, founded on nothing else. How all-important, then, in view of this, is that great branch of national industry, its manufactures! How evident is the fact that, without them, the noble fabric of our national hopes, and happiness, and freedom, would want, perhaps, the most efficient pillar of its support! The contrast that exists between the moral condition of our own country and that of the South American republics, is too striking to fail of attracting the attention of any one at all conversant with the facts of the case; and we have dwelt thus far on the subject, from the consideration, that thus might be afforded a fresh proof of the superiority, in every point of view, of a nation whose principal resources are in the industry, energy, and enterprise of its people."

DOCUMENTARY TESTIMONY ON THE MORAL INFLUENCE OF MANUFACTURING ESTABLISHMENTS IN NEW ENGLAND.

The following circular was addressed to several heads of manufacturing establishments :—

1. Are there any laws existing in the New England states by which the manufacturers of cotton and wool are prevented from the too constant employment of children? Or from the employment of those of too tender age? Would not such laws prove very salutary?

2. How old are the youngest children usually employed? Are children under fifteen years of age often deprived of opportunities of schooling, by unremitted employment in cotton or woollen factories?

3. Are there not many cotton establishments in which no children under fifteen years are employed? And is this the case with woollen establishments?

4. Are there not many establishments where the proprietors have adopted a regulation, by which children are allowed to work only a portion of the time, with a view that opportunity for schooling may be enjoyed by them? And to what age does this regulation apply?

5. What is the probable proportion of children under fifteen years, to those over fifteen, and adults, employed in cotton factories? What is the proportion in woollen?

6. Are there any factories in New England in which the proprietors employ one set of hands by day and another during the night?

7. How many hours are the operatives employed? Please to specify them. Is there an entire conformity in all the factories?

8. Do the females employed generally live with their parents, or at boarding-houses? And what are the disadvantages attending the system of boarding houses? Are they well regulated, or too large to admit of careful supervision?

9. Are instances of immorality in consequence of the employment of both sexes together, frequent, or otherwise?

10. Do the females employed in these factories generally lay up their earnings, or spend the amount in dress? Are savings banks used by the operatives for depositing their surplus gains?

11. Are first-day or Sunday schools generally established in manufacturing villages, and attended by the children?

12. Are there auxiliary tract societies established generally in these villages, for the purpose of disseminating, at a cheap rate, the excellent moral and religious publications of the American Tract Society? Could not individuals undertake so laudable a work singly?

13. Is it supposed that those persons employed in cotton and woollen manufactories are equally healthy with such as pursue agriculture? If so, can you mention any facts in corroboration?

14. What proportion of the operatives accumulate property? and what classes are generally improvident? Do you not suppose that some of the families who find employment in factories, would, if it were not for such employment, be chargeable to town as paupers?

15. Will you enumerate some of the most striking advantages which have resulted to your town or neighbourhood, by the introduction of manufactures? And also name the prominent disadvantages, if any.

16. What remedies would you propose for those evils which do exist?

17. Do you know of any cotton or woollen factories in which any improved system, or any peculiarly beneficial management, prevails? And will you specify the establishment and give a sketch of its regulations?

18. Are there existing in some manufacturing villages, libraries of useful books which circulate among the operatives?

19. Do you consider the mass of the manufacturing population, equally well educated and intelligent as the mass of agriculturists?

20. Do you know of many instances where families who were in poverty have by their successful industry in the manufactories, made themselves independent? And have you often witnessed the effect of such success in improving their habits and general characters?

21. Is it not the practice in many of the manufacturing villages, for the head of such families as are employed in the mills, to cultivate a small lot of ground, to raise corn, potatoes, and garden vegetables generally and to keep a cow? And is not this productive of much comfort to such families?

From Smith Wilkinson, Esq., Pomfret, Conn. to the author.

"You ask my opinion as to the tendency of manufacturing establishments on the morals of the people. I answer, that my settled opinion is that the natural or consequent influence of all well conducted establishments, is favourable to the promotion of good morals, for the following reasons:— The helps are required to labour all the time, which people can sustain in regular service through the year, consistent with what is necessary to attend to their personal wants,—for meals, sleep and necessary relaxation, and a proper observance of the sabbath. The usual working hours, being twelve, exclusive of meals, six days in the week,—the workmen and children being thus employed, have no time to spend in idleness or vicious amusements. In our village there is not a public house or grog-shop, nor is gaming allowed in any private house, if known by the agent, and very few instances have occurred in twenty-nine years, to my knowledge. In collecting our help, we are obliged to employ poor families, and generally those having the greatest number of children, those who have lived in retired situations on small and poor farms, or in hired houses, where their only means of living has been the labour of the father and the earnings of the mother, while the children spent their time mostly at play. These families are often very ignorant, and too often vicious; but being brought together into a compact village, often into the families, and placed under the restraining influence of example, must conform to the habits and customs of their neighbours, or be despised and neglected by them. Thus it happens sometimes that when it becomes generally known that a family are noted for any vice, they are neglected by the rest, and no person, male or female, will visit or be seen keeping company with them, who is at all concerned to sustain a good name. Another reason is, by being in a way to earn the means, they almost invariably clothe better; and it is a fact of common notoriety, that the females employed in factories clothe better or more expensively than others in similar circumstances as to property, or even than the daughters of our respectable farmers. But this disposition to dress extravagantly soon abates, and the helps contract habits of economy, and lay up their wages by loaning the money at interest.

"I have known a great many, who have laid aside $200 to $300, in from three to four years, and were enabled to fit themselves out decently, when married, for housekeepers. Others, who remained single, laid by four, five, and some seven and eight hundred dollars, and now have it out on interest. As public opinion goes far in regulating the moral habits and behaviour of cities and towns, so it does in manufacturing villages,—by this influence, it is an established fact, that if a female is introduced into a factory of bad or loose character, she must be discharged as soon as her character is fully known, or the rest of the female help will quit the mill. Perhaps I cannot furnish better proof of the practical tendency and effect on female character, than to state, that in twenty-nine years, during which term I have had the sole agency of Pomfret cotton manufacturing establishment, I can assert that but two cases of seduction and bastardy have occurred. One of these was by means which have often proved fatal—where the object was placed in the most disadvantageous circumstances to withstand them.

"The company of the Pomfret establishment, was formed, January 1st, 1806, consisting of,—James Rhodes, Esq., Christie Rhodes, Wm. Rhodes, brothers, all of Pawtucket, R. I.; Oziel Wilkinson, and sons-in-law; Timothy Green, Wm. Wilkinson, of Providence; Abraham Wilkinson, Isaac Wilkinson, David Wilkinson, Daniel Wilkinson, Smith Wilkinson, all of Pawtucket or North Providence, five sons of Oziel Wilkinson.

"The capital stock invested from April 1st, 1806, to October 1808, was sixty thousand dollars—of which, five twelfths was invested in real estate—it was then known by the name of Conger's mills, in Pomfret, Connecticut, on the Quinebaug river, and includes about one thousand acres of land, lying partly in three adjoining towns, namely, Pomfret, Thomson, and Killingly. There was at this time on said lands, a grist mill, saw mill, and blacksmith's shop; two houses, an old gin distillery, then just abandoned; three houses, and some other small buildings of little value. A leading object of this company in buying so much land, was to prevent the introduction of taverns and grog shops, with their usually corrupting, demoralising tendency. Another object was, to be able to give the men employ on the lands, while the children were employed in the factory. The company very early exerted their influence in establishing schools, and introducing public worship on the sabbath. In 1812, they erected a convenient brick building, to answer as a school house, and a place for holding meetings; which is now occupied for those purposes, and has been ever since its first erection."

—

M—— B——, Esq.

TROY, Dec. 26, 1827.

Dear Sir—I fear I have neglected too long to answer your interesting enquiries on the subjects of manufacturing and manufactories; but will now make the attempt, though on several points I have not been enabled to collect the information required. Supposing that you have a copy of the several questions, I will answer them in the order they are put, without repeating them.—(See page 125.)

1. I know of no such restrictive laws in the northern or eastern states, nor can I see any occasion for them. Public opinion, with the independent feelings of the parents and guardians of children, would prevent such abuse should it be attempted; but I never heard of such a practice in our country among manufacturers. Young children are unprofitable in almost every branch of our labour, and so much so, that it is the practice to keep them out of factories as long as the importunities of parents can be resisted.

2. Children under ten years are generally unprofitable at any price, and it is very seldom they are employed, unless their parents work in the mill, and they are brought in to do light chores, or some very light work, such as setting spools in the frame, or piecing rolls. As far as I am acquainted, there is more attention paid to schooling children in manufacturing villages, than in districts of other employments.

3. I do not know of any works where the age is positively limited, nor do I think that it could well be done. There are many boys at fourteen years, who are able, in most employments, to do the work of men; they only want the skill. The heavy work is mostly done by machinery; and there are many girls at fourteen years who are as steady and discreet, as others at sixteen or over. I have no doubt that it would be more profitable to employ young women in our factories generally, except for overseers, if they could be obtained.

4. I do not know of any thing exactly in that shape; it is not consistent with the operations of a mill, that any part of the help should leave their place to spend certain hours in school; but the child is refused employment until it has had its necessary schooling.

5. I have never heard fifteen years referred to, as an age below which employment would be wrong or unprofitable. I should say the proportion might be 10 per cent. There is less young help employed in the woollen than in the cotton manufactures.

6. I never heard of such an instance in our country, though I believe there are those who practise and pursue such a system in England. I do not think it would be tolerated here: public opinion would not suffer it, nor could workmen be procured.

7. An average through the year of twelve hours, is every where under-

stood as factory hours; this is by common consent, nor have I heard of any attempt to increase the number, as a rule of employment.

8. It is customary, in commencing a manufacturing village, to build a boarding-house to begin with: and this is necessary from the nature of the case in most instances; but as soon as families are brought in, the help employed is generally distributed. The custom in most places is, to allow and require every hand to provide for themselves. This is found more satisfactory and best; in this way the price of board is regulated by competition, and labourers choose their associates, and the females in this distribution in families are better protected, and more pleasantly situated.

9. As far as I am acquainted, unfrequent beyond the expectations of any one.

10. There is a disposition to dress among the unmarried females, though many do lay up something, and many help their parents in supporting the younger members of a family. Our factory villages have many widows, who resort there to bring up their families, and are thus enabled to keep them together, and provide for them very comfortably; and here the young women are the stay and support of their mothers, while they receive counsel and protection.

11. Sabbath schools are common to a considerable extent, and are becoming more so in manufacturing villages.

12. In many villages there are tract societies, where from funds of their own, they purchase of the larger institutions, and in others there are auxiliary societies. Something is done, and much more might be done.

13. I have no doubt of the healthiness of the employment. I have been engaged in a cotton factory since 1813, and have employed from sixty to one hundred hands, men, women and children, and do not believe there is a more healthy village any where to be found; and can speak confidently in saying that the farmers in the immediate neighbourhood are not more hardy, nor do I believe they can undergo the same fatigue, because not so accustomed to such constant and regular labour.

14. I cannot say how far they accumulate property; I know that many do, and very many live comfortably and independently, who but for such employment would be paupers. Many families begin in debt and embarrassment, who soon pay their debts, and support their families, and gain property afterwards.

15. This would be to write a volume. The property in the neighbourhood is greatly advanced. It is quite a market for vegetables, fruits, meats, to the farmers around. Industry, education, and morals, are greatly improved. The farmers and mechanics look for the money paid out at the factory store as an unfailing resource for their circulating medium; and depend on furnishing their necessaries, as a sure means of getting money. I not know of, nor can I conceive of, any disadvantages. Our manufactures have greatly increased the commerce of our city, in bringing the raw material and distributing the articles manufactured, and furnish a large market for the product of the farmer. I paid for the last four months $758.63 for the single article of flour for our families.

16. I know of no evils which exist in manufacturing villages as such, which are not increased, and more or less aggravated in other villages, or

which are not to be found in every society. I think any evil is easier remedied in such places than in different society.

17. I will give you our regulations at the close in general terms.

18. I am not acquainted with any where libraries are established, but have no doubt it would be beneficial.

19. I consider them decidedly better educated, more intelligent, of better cultivated manners, higher notions of character, more enterprise, and every way more improved citizens, than the mass of agriculturists. When the latter change to the former there is generally a marked improvement, and when the former to the latter, a deterioration and running down.

20. I do know of many instances where those quite poor have, by their industry and economy, become comparatively independent, and the character of the whole family changed for the better.

21. There are many whose families work in the factories, when the man takes a piece of land on shares, and raises corn and potatoes; but this is a more common practice in the New England states, than with us. When the man cannot be employed to advantage, this may do well, but the leisure hours such an one would have, would be a bad example for the factory hands, and I would prefer giving constant employment at some sacrifice, to having a man of the village seen in the streets or shops on a rainy day at leisure.

M—— B——, Esq.

TROY, Dec. 27, 1827.

Respected Friend—I said, in answering your 17th query, that I would give you our general regulations in our manufacturing establishment. In 1812, five individuals, one of whom was myself, built the establishment which I think you visited with me when at Troy. We were all ignorant of our undertaking, but had very great expectations from what we had been told. I had the principal agency in erecting the buildings, and procuring machinery &c.—but we had one partner who was superintendent, and who professed much, but knew very little. We commenced work in the spring of 1813, but every thing went bad, and we found our superintendent a man of loose, bad notions, bad principles, and he had brought together a bad set of workmen. We dismissed him, and after some time persuaded my brother to come and take charge of it. He was a merchant, and knew nothing of the manufacturing business. Things still went bad; the workmen were deceivers, and my brother had a difficult place to fill; but we dragged along until the peace, and found ourselves very much in debt, and embarrassed, and stopped our works in the fall of 1816. Thus the works remained until the spring of 1817. I then bought eight of the ten shares in which the factory was owned. We had kept a store of groceries, and sold rum to our hands as freely as they required. I have never brought any spirituous liquors to our village since—the hands were all poor and most of them in debt. I bought cotton in April, and started the mill—the hands that chose to stay, and were willing to live without the use of ardent spirits, I kept, and divided their debts into small sums, which they agreed to deduct from their wages weekly—their rents were all payable weekly, that no debts might be suffered to accumulate against the hands, and no one was to ask or expect credit, unless at the beginning of a week, when they could anticipate half the wages of the week if necessary. If they could not live under these regulations, they were at liberty to go; but if they stayed, their old debts must be paid, they must live without spirits, and they were not at liberty to get in debt any where—no liquors could be brought into any workshop under any pretence whatever. Thus I began, now nearly eleven years ago; many of the families are now with me, or those that were young men and girls are now married and have families; they were all poor without exception. I will mention the condition of some of the hands—one young man, an apprentice in the machine shop, is now out of my employ as a steady hand, but does job work for me—he has a large family, but owns a good house, has considerable money at interest, has two buildings for rent, is worth three thousand dollars. Another has two thousand dollars at interest. Another has bought him 100 acres of good land, owns a house in the village, and has money at interest. Another has $1000 at interest—several others have three or four hundred dollars beforehand. Families all above board, with one or two exceptions; we keep a district school the year round, with a competent man teacher—through the season of working in nights, a school goes in at eight o'clock, and out at ten o'clock, which all the young men and women calculate to attend—here are taught writing, arithmetic, and grammar, geography, and history—this is very much encouraged and is a very popular school; we have a very prosperous Sunday school; there is a small house for worship in the village, and one a mile east, and many come into Troy to meeting, it being only about two miles. In order to keep out tippling and grog shops, I have a clause inserted in all the leases given for building lots, that any one selling ardent spirits on the same, forfeits the premises.

A large proportion of our families are hopefully pious, have family prayers daily, and are members of churches in good standing, and a majority of our young people belonging to the cotton factory are professors of religion. Since 1815, there have been three revivals of religion. We have there a bible society, tract society, and domestic missionary society. There are a large number of newspapers taken, and some reviews and quarterlies: and I think a state of society which would be gratifying to the patriot and philanthropist—and the Christian. We have all our hands by the year, which commences on the first of May. We inventory every March, and then engage our help for the year. We seldom have any hands leave us, that we wish to retain. Our young people marry and settle in the same village in many instances. Our contracts are to pay as fast as the individual or family need to live upon, and the balance at the end of the year. To those who will let their balances remain in book we pay interest, but will not give notes, because the advisory influence is in some measure lost if you give notes which can be negotiated; but on our plan, our books become a savings' bank for the hands. If they want a note we pay the balance. We have over five hundred inhabitants, and in 1812 the ground was cleared where our village now stands. Our establishment is very small compared with many of the eastern works, and our buildings and machinery are not after the modern improvements, but we cannot afford to throw them by. We have built a very firm excellent building for the woollen business, and have it well filled with the best machinery that could be procured, and have commenced operation, but it will take time to get such a set of hands as we have at the cotton mill; yet I see no difficulty. The wool business requires more man labour, and this we study to avoid. Women are much more ready to follow good regulations, and are not captious, and do not clan as the men do against their overseers; but I can afford to give a religious man or woman higher wages, than I can one who has no fixed principles of action and government for themselves. It should be the first object of our manufacturing establishments, to have their superintendents, and overseers, and agents, men of religious principles, and let it be felt by the owners that it is always for their interest to support religion, schools, and all those institutions which promote good morals, and diffuse information among the operatives and their families. I feel confident that we have made a sufficient experiment, in the manufacturing business, to see its effect upon those employed and the state of society which it produces, and the influence it has upon a neighbourhood of farmers, and others in the district round about, and have no hesitation in saying, that in every particular it is favourable. It grows up a healthy population, is favourable to early schooling and good education, and early habits of industry; stimulants to enterprise, economy, and frugality in living, and saving the products of their labour—and at the same time the organisation of these establishments in villages, being necessary for their success, they are placed in a more favourable situation for the cultivation of moral and religious character, without which, civilised man is still a savage, and a very limited degree of human happiness attained.

I am, respectfully, your friend and obedient servant,

JEDEDIAH TRACY.

The following remarks are from a correspondent who has paid attention to this subject, and who sincerely wishes well to every branch of useful industry which shall benefit the country:—

"I noted that the legislature of Massachusetts instituted an enquiry some nine or ten years ago, to ascertain the moral influence of manufacturing establishments, which resulted in a favourable report—never published.

"In pursuing thy enquiries upon this deeply interesting subject, I sincerely hope thou wilt state the whole case fairly, so that those points where danger is to be apprehended may be seasonably guarded by the conservators of public morals. The employment of young children of too tender age, should be freely and warmly discouraged; and if at the present moment there should appear to be any increase of this evil, our legislatures should timely adopt such wise and prudent measures as would cure the evil. No patriot could advocate the extension of any branch of national industry which would necessarily bring along with it an ignorant and consequently vicious population.

"We find many men of philanthropic minds who view with alarm the rapid extension in our country of manufacturing industry, under a conviction that it stands opposed to the progress of religion and sound morals—in a word that it is essentially repugnant to the general well being of the community; nor is this surprising, since those whose interests stand opposed to the increase of manufactures on a large scale, have long and vehemently insisted upon its demoralising tendency. A great deal has been said about the sad change this mischievous system has produced among our neighbours of the eastern states—it has been described as a Pandora's box that has filled the land with all sorts of moral plagues. It must be obvious that the subject has been presented to us through a medium somewhat distorted by wrong prejudices, and even the interesting columns of 'The Friend' may have contributed to strengthen these prejudices by the revival of the somewhat trite sentimentality of Goldsmith and Southey—I allude to an article in the second number. I am, however, as little disposed to call in question the motives of our philanthropists in opposing the manufacturing system, as I am to extenuate or defraud any abuses to which it is liable. That abuses do exist, even in this country, I am well aware, and I would be the last person to discourage any well directed effort to remdy them.

"It is certainly an interesting enquiry, whether, as manufactures have advanced in our country, the general character of the operative classes has deteriorated? Have these occupations had an unfavourable influence upon the *intelligence*, the *morals*,* or the *health*, of those engaged in them?

* With reference to this point, we have great satisfaction in adducing the following conclusive testimony :—

WATERFORD, R. I. May 23d, 1835.

Dear Sir,—In reply to yours of 7th inst. will observe, that many persons can give you better views than I can, respecting the condition of the cotton manufacture business in its various stages and fluctuations, since its establishment in this country, and the effect of the tariff laws upon it. Our business has always been seven eighths woollen, and is now exclusively so. We have a woollen mill, eighty feet by thirty-six, and one, three hundred and fifty feet by fifty, both five stories high; for broadcloth principally.

As regards the effects of manufacturing villages on the morals of the people, there can be but one opinion among those who know any thing about the subject. They certainly tend very powerfully to the improvement of morals. In our village, with a population of three hundred to four hundred, not an intemperate person lives. Nearly one hundred females are in the village, and since its establishment, a term of ten years, not a case of illegitimacy has occurred, nor has a rumour of such a nature ever been in the village. No person who has ever resided in the village, has ever become chargeable to the town in any manner. On the first of April last, the people who work in our mills had $10,000 due to them in cash. We have an excellent free school through the year, of about fifty scholars. Yours truly,

WELCOME FARNUM.

"Having had access to authentic information upon this subject, I answer as follows:—

"The cotton manufacture may now be considered permanently established; it is prosperous and rapidly increasing in the New England states, which must remain, as they are at this time, the principal seat of it. For the present, my remarks will be confined to this branch of manufactures.

"A great change has taken place within the last few years, in regard to the proportion of children employed in these factories; the proprietors having found that their interest is promoted by dispensing almost entirely with the labour of children under fifteen years.

"In the factories at Newmarket, N. H., which have been in operation about four years, there are employed, 250 girls, five boys and twenty overseers and assistants—twelve of the overseers have families. Nine only of the girls are under fifteen years of age, six of whom are fourteen. Three of the boys are under fifteen, two of whom are fourteen. In every instance the children under fifteen reside with their parents or guardians in the village, and are admitted into the factories on account of the peculiar circumstances of the families; they are allowed to work only six months in the year—during the other six months, they attend a public school in the village. Besides the operatives mentioned, there are thirty machinists, twenty of whom have families; these, however, are employed in a separate workshop. The relative number of children employed in this establishment, it is believed, will correspond, without much variation, with the proportion to be found in most of the factories east of Providence and its vicinity; in the latter district, the manufactories were established at an earlier period, and still give employment to a larger proportion of children.

"In cases of newly formed villages, it is found necessary to erect at the commencement several boarding-houses, sufficiently spacious to accommodate a large number of the workpeople in each; to this arrangement there are powerful objections. At Newmarket it has been entirely abandoned, and is superseded by the increased number of private families, which have taken up their residence in the village; and not being inconveniently large, are kept under good regulation. A part of the girls whose parents do not live in the village, are distributed as boarders with those families which are disposed to receive them.

"Nearly all of the manufacturing villages are small, and there is very generally attached to each dwelling a lot of ground, which is appropriated to the culture of garden vegetables, and food for a cow and swine; these are considered very essential comforts, and are rarely dispensed with by the industrious operatives.

"It should be borne in mind, that in this country *water-power* is almost exclusively used in manufactures, and, on account of its greater cheapness, the day must be far distant indeed, when steam power will be extensively used; the consequence is, that the manufacturing population must be scattered. We can have no Manchesters on this side the Atlantic, while our thousand rivers and streams afford an inexhaustible supply of unimproved power."

V

Family Labor, "Mill Girls," and Mule Spinners

By far the largest group of Americans to experience firsthand the coming of industry in the mill villages of New England are also by far the most difficult to discover firsthand in the documentary records. These were the women and children and men who were employed to operate the new machines. Unlike the working women of the factory city of Lowell, Massachusetts, who distinguished themselves by writing and publishing a significant amount about their industrial lives, almost nothing of an autobiographical nature survives from the mill village worker. Historians must rely on "unintentional" documents in an effort to reconstruct and understand the experience of labor and domestic economy in the New England mill village.

The section opens with some of the very few surviving private documents in which workers do speak for themselves: several collections of family letters and one published autobiography. The remainder of the section consists of material from company and public records that define workers' lives and their relations to their work.

Help wanted advertisements indicate what kinds of people manufacturers welcomed into their new communities. Company records indicate the varieties of workers employed, the terms of their employment, and in one case the entire economic life of one family over a decade of employment. Factory regulations and individual work contracts help us understand the new forms of authority created by industrial management and offer some insight into the quality of work life in the New England mill village. (MBF)

Hollingworth Family

Letters (1828–1831)

The members of the family who wrote these letters were among hundreds of English immigrants to populate young America's first textile mills. The Hollingworths, unhappy with working conditions in Yorkshire mills, came to the United States intending to become farmers. But after finding jobs in New England's mills, their entrepreneurial instincts emerged, and the laborers soon became owners. In both roles, the Hollingworths were consistently ambivalent in their attitudes toward the new industrial order that they were helping to build. (TWL)

The Hollingworth Letters: Technical Change in the Textile Industry, 1826–1837, ed. Thomas W. Leavitt (Cambridge, Massachusetts: MIT Press, 1969), pp. 24–29, 60–61, 65–66, 80, 87–89, 92–94.

George Hollingworth to William Rawcliff

South Leicester June 28th 1828

Dear Brother

I have long had an intention of writing to you and giving you all the information respecting this Country I possably could, but has hitherto been detered by a perplexed mind, for which neglect I must beg your pardon. I must begin by informing you that we are all at present in good health with the exception of Son John who has not been well for these few weeks past, but perhaps is now a little better. With respect to the Weather they say the past Winter as been verry mild. At any rate it has been verry fare from being insuportable. I have experienced colder Weather in England then I have yet done in America. The truth is that it will freeze ten times as quick and ten times harder then it will do in England and yet will not feel any colder. With respect to our Summer Weather it has hitherto been verry pleasant. It has been uniformly warm with now and then an hotter day. We are almost allways at these times faviouređ with a clear and pure atmosphere and an exhillerating breeze which enables us to sustain the heat of the day with verry little inconveniance. The prospect of the Country at this time is beautifull behond discription, the Woods and Fields ornamented with beautiful Flowering Shrubs of almost every discription such as are considered rare ornaments in your Gentlemen's Gardens in England. Vegetation here growes most rapidly and luxuerantly. You would be surprised to see the uncultivated state of the Land and see the abundance that it will produce conected with the little labour that is bestowed. We have planted about half a Road of Potatoes in verry rough Ground without any manure and they say they will grow verry well. Whether they will or not is yet to prove. Of this I am certain that had they been planted in like manner and with as little labour in your part of England that they would produce nothing. We have made a verry neat little Garden attached to our house in which we have now growing cucumbers Melons, Squash, various kinds of Beans such as is not known in your Country, Carriots Lettuce Parsley Beets Marygolds Sweet Williams Saffron Sage Hysop and various other things all of which we are raising from the Seeds. The uncultivated land even the Road Sides produces abundance of Red and White Clover and are excellant good Pasturage. The horned Cattle are almost universelly of one Kind and Collor either Lighter or Darker

Branded or Red. A great many of them appears to be good ones but very little care taken of them. The Yankee generally takes a Vast more care of his Horse then any thing else, and they are worthy of care For the breed of Horses here are generally the most fleet and Active I ever yet Saw.

*

We all live together in a double House. We have plenty of room. The House contains 8 Rooms besides a Celler under the whole. We pay 60 Dollars a year Rent. Geo. Mellor James Hollingworth and & J. Kenyon Boards with us. Son James is a Filling Spiner viz. a bobing Spiner and can earn 8 or 9 dollars per week. Son John has been a little time a Condenced Spiner. This is a new meserable business for making money. He is now a Slubber which is a fare better Job. He will make better then a dollir a day. Son Joseph has been a Gigger every since he came at only 15½ dollers per Month. As the Job was a new one to him and one I wished him to learn I was not particular about wages at first, but now has he has got an expert hand I am thinking of asking for more wages not less then 20 dollers per month. Edwin is a Warp Winder or Spooler winder for the Warping Machine at 1½ dollers per Week. I am intending to have his wages advanced also. There is not much in America I dislike [excep] ting the too general conduct of Emigrants, and the Factory Sistem which Sistem I hate with a perfect hatered as being only calculated to create bad feelings bad principles and bad practices.

*

Son John's Wife was brought to bed some Months since of a verry fine daughter which they call Elizebeth — We all unite in our Special Regards to you and your Wife — and am Dr Bro.

Yours affectionately
Geo. Hollingworth

If you think proper you may lett my Brother John see this Letter.
I had forgot to mention son Jabez. He is working in the machine Shop at 25 dollors per month.

[To] Mr William Rawcliffe
Oldfield Honley
near Huddersfield
Yorkshire
Old England

George Hollingworth to William Rawcliff

South Leicester Jany 24th 1830

Dear Brother

Since writing to you about a Week ago we have changed our minds respecting Richmond's Factory. In that letter I advised to lett it alone at present fearing we were not able at present to manage it. A few days since Jabez mentioned the case to Mr James Shaw a Saddleworth man who resides here with a large Industerouse Family. This has led to several consultations betwixt us upon the subject and we have finealy agreed and determined to form a Copartnership and engague Richmond's Factory if possoble upon the most reasonable and advantagious Terms. We are aware the greatest difficulty will be to begin or make a start. This over we have not the least doubt of success if we be blessed with Health, for we by our united Families could do all the work and have no wages to pay which is of vital Intrest to the success of a Manufactury. Besides Jas. Shaw is the first man I know who is every way fitted for such an undertaking. He is sober steady Industerous and a good workman, I believe an excellant Carder. He is well informed and has a good knowledge of Men and Things. He has a knowledge or aquaintance with some of the Merchants and Woolstaplers of New York and Albany. This may be of Essential service to our new Establishment. Also he is a man that has a good share of Courage Fortitude and Confidence (viz. what the Yankees call SPUNK) which are necessary requsites for a Trade's-man &c. James Shaw and me has agreed and I hope you will concur, that you and he and me form a Copartnership on equal Shares and engage Richmond's Factory and use our United efforts to sett it a going, for we can perceive no other mode of extercating ourselves from poverty and thraldom. Besides all the members of our respective Families anxiously Concur. Now what we at present want you to do is emediately on the reciept of this letter to see Mr Richmond and tell him that we will Engage his Factory provided he will put it into proper and sufficent repairs. We should wish to have it for some length of time at least 5 Years. We are not exact to a month when we enter to it, provided it be not before the 1st of May. We would rather it was deffered till June or July but you will hear what he says upon these subjects and please to comunicate them to us by letter as soon as possoble. I should like you to get Mr Richmond to employ

Jabez in the repairs because he being upon the Place might not only see what was wanted doing but see that it was done well. If you can succeed in this please to inform us in your letter and we will send Jabez with his tools and every necessary instruction possoble. To prevent a disclosure at this place we shall put this letter into Charlton's Pose Office 3 miles from here, and we request you to direct your letter as follows. –. "Geo: Hollingworth, to be left at Marbles Tavern Charlton, Worcester County, Massachusetts". We at this place are getting into a new Order of things and I have hoped that it might be for the better, but I am not a little afraid that my hopes will be frusterated. They have lowered Weaving to 4 cents per Yd. and it apears to me they intend to have every other thing done as low as possoble. They are posting up a new string of Rules more objectionable than the Old ones. In one of them there is the following "That if any Workman damage any Work or Machinery he shall be liable to pay damage the damage to be assessed by the Superintendant or Agent.

Yours Affectionately
Geo: Hollingworth

Jabez Hollingworth to William Rawcliff

March 14th

Dear Uncle

As Brother Joseph has related to you the loss that has befallen our Family the death of our dear little Sister Hannah I need not say any more on the subject. I shall take up the last mentioned subject. Joseph Haigh requests my Father's advice what is best to do. He says there is plenty of Factorys to sell in the vicinity of Pitsburgh but he says they want as much for them as they cost. But he thinks it is dangerous for him to step into their shoes for if they cannot walk in them he thinks he cannot. My Father is too much troubled to write to him, but he has instructed Brother John to write to him. He also gives him an invitation to meet him at Poughkeepsie, and for them and you to hold a conference on the subject. He also wishes him to write and let us know if he will come and what time he will be there.

With regard to Richmond's Factory it appears to be dropped for very little has been said about it since your letter came, which they did not find any fault with but on the contrary said it was a very sensible one. I have had no work since I left you only what I have done in our own house. I have made a bench to work on which has [cost] me about 6 dollars. I have made a broad loom for Brother John and [. . .] wheelbarrow and some Winterhedges for Sale but I cannot sell them. There is no encouragement for such business here. I am an Englishman amongst Yankees. They want to give me half what they are worth.

March 16 Yesterday morning Father had Notice to Quit as they are going to have all their work done by Girls. Mr. Denny the Agent told Father that they wanted some hands at Southbridge 12 miles from here. Father and I went to see about it but did not make a final agreement. They wished us not to make Application any where else and they would write in a few days. Now you see the Fruits of Large Factorys. Here we are supplanted by Females that is expected to perform the same quantity of work for one half the wages the quality being out of the question. Here we are driven from one Factory to another seeking rest and finding none and when we are in work at what we may call decent wages they have so many different ways to get it all back again that it is impossible to save any thing. The very highest rents fuel Provisions wearing apparrel and every thing else at the very highest prices. The only way to remedy this is unite oursleves I mean our minds and bodily strength together to set about one thing at once and strive to accomplish it. I for my own part has got no money but thank God I am both able and willing to work. I should not be afraid to build houses good enough for any of us to live in.

Yours &c.
Jabez Hollingworth

[To] Mr William Rawcliff
at Wadsworth Factory
Poughkeepsie
Duches County
State of New Yorke

John Hollingworth to William Rawcliff

Woodstock July 4th 1830

Dear Uncle

I write to inform you that we are all in good health at present and hope these lines will find you the same. We have made a genral move this Spring. My Father and family are at Southbridg about the same Distance from here as South Leicester is from Oxford. My Father and James, Joseph, and Edwin are all in work at the above mentioned place. James has got married about 2 Months since. Brother Jabez cousin James and I ha[. . .] at this place to manufacture Sattenette. We came here about the 1st of May and our rent comenced on the 1st ints. We are to pay 500 Dollars per Annum for 3 years. We have Seven houses. The Factory consists of 2 Building connected together 3 storys each. They are about 18 feet wide and 36 or 40 feet long. We have 3 Double Carding Machines and 1 Billy 1 Jenny [a] Picker and Fulling Stock 2 Shearing Frames 1 Press and 1 Dye Kettle.[55] We are to have 6 Power Looms which is to be ready By 10 inst. There is about 2 Acres of Land a Pond of about 100 or 150 Acres which we can draw down 10 feet. It is quiet a pleasant place in fact it realy would be desireable if it was sittuated within 2 Miles of the North River. I forgot to say that there is 15 hand Looms at the place which we can use if we want. There is a Barn and other outbuildings.

*

Aff. Nephew
John Hollingworth

Joseph Hollingworth to William Rawcliff

Muddy Brook Pond Factory,
Woodstock, Connecticut Sept. 5th 1830

D—evolve on me the pleasant task,
E—ach time, to answer what you ask;
A—nd in return for favours done,
R—elate how things are going on.

A—ssisted by a power devine,
U—nveiled before me truth shall shine:
N—ature's grand works may all decay;
T—ruth shall endure to endless day.

A—nd now I take my pen in Hand,
N—ot doubting but you'll understand;
D—esiring, that you wont mistake,
U—nknown, the errors I may make.

N—ow tho' to distant lands we've roved,
C—an we forget those whome Loved;
L—ove, the great source of all our weal,
E—nlightens every mind with Zeal.

I recieved yours of June 20th on the 5th of July. which gave me great sattisfaction. The reason that I did not write sooner is this, Bro. John had written to you Just before I got yours giving you a description of this place, and also desiring you to write to Joseph Haigh, so I concluded not to write then lest you should have letters coming in to thick. Bro. James and Wife lives with Father and family at Southbridge. Mother says she should like to have come over to see you but that she could not well be spared. You want to know what work I have. Well, when I came to Southbrigdde at first there was no work for me at my old buisness, so Mr Sayles set me to work with my Father at the warping macheen. I worked 5 weeks when I thought it time to ask what wages I should have. The reply was NOTHING! that having the chance to learn a fresh trade was thought a Just compensation for my verry valuable services. The result of which was, that I got into a Jackass' fit. Father then took the warping and spooling by the Job. He and Edwin worked at spooling and I at warping untill I got weary of the work. I then came here to work, when Mr Sayles sent for me back, as he wished to hire me to work in the fulling Room for a few days. I went back and worked 21 days for 12 dollars. And finaly I came here again, and am going to do the Napping, Shearing and pressing, when the work is ready. Mary Kenyon has had the misfortun to lose the forefinger of the right Hand. She was weaving on a power loom. She put her finger where it had no buisness, and so the loom in return snapped it of between the first and second Joints.

*

And now for a description of this place. It is situated in the township of Woodstock in the "land of steady habits" alias Con^{ct} about 4 miles south of Southbridge and about 16 from South Leicester alias Clapville. This place contains about 3 acers more or less on which is the Factory, consisting of two buildings Joined together, each 3 stories high, a dye house, a wood shed a Barn, and seven houses or tenements, together with another building divided into 4 sheds with a large chamber over, which may be used as dry house. In the Factory are 3 carding machiens, 2 billys, 4 Jennys, 13 broad hand looms, 4 new satinet power looms, 1 fulling stock called a poacher, 1 picker, 2 broad shears, 1 press, 1 dye kettle, 1 satinet Napper and 2 shears which we have had to buy, and several other things.[59] There is a most excelent watter weel, an over shot weel the best I ever saw.[60] It is suplied with watter from a larg pond called the Muddy Brook pond, although the watter is as clear and as soft as any other. They have hired this place for $500 a year, for 3 years but will have to Quit any time the owners think fit at 12 months notice. The owners are now determined to sell it without delay. They ask 6000 $ and will not take less. There is a party of Yankees wants to buy, but they say they will give our folks the first chance and make the payment easy. Joseph Haig, Father, Brothers John, Jabez, & Jame and Cousin James have determined to buy rather than quit the place. The interest of the money will not be so much as the rent. Joseph Haigh & family arived here last Wedensday and I believe are verry much pleased with the place.

If you could make it convenient to come over, and see us, and the place I should be very glad. Perhaps you would Join our folks in buying of it, or if not sold you might like to buy it youre self. It might be a good place to keep a store, the nearest being 3 miles of. If you come you may come by way of N.Y. and Hartford, from thence by the stage to West Woodstock a place 4 miles from here, and I should like to see Aunt Nancy come along with you.

[To] Mr. William Rawcliff,
Wadsworth's Factory,
Poughkeepsie,
Duches County,
New York.

Joseph Hollingworth to William Rawcliff

Muddy-Brook-Pond Factory, Woodstock Con. Nov 8th 1830

Since Writing the foregoing I have learnt that Father & Joseph Haigh intends to come see you in the course of a few weeks, but have not heard the exact time stated. I should have told you somthing more about our Factory concerns but think it unnesessary as Father is coming. I am Glad you did not come, as I requested in my last, because I now see with different eyes. I like the Place itself, as well as I ever did, but the concern is too much in Yankee fingers for me.

Yankee doodle dandy,
The Yankeys they are handy,
To rogue and cheat,
And make folks sweat—
To smoke Segars, – and drink a glass of Brandy.

I have been informed that Jemmy Anderton, (alias "old Buckram",) South Leicester, (now Clappville) old superintendant, and Slavedriver, is, together with his concubine (Fanny Wilby) residing at a Factory somewhere between Albany and Buffaloe. He is the Superintendant or Slave driver and recieves 5 $ per day. I also understand that you have got Gorge Mellor's Note of 20 £ against Anderton. If so, I would have you look Sharp, and catch the old Mason. I came back here on thursday last, and have finished 200 yds. more of Satinet. If you do take a Factory I should think you will have work for me and I will come any time when you think fit. But dont you think farming the best, and surest way of getting a living? Manufacturing is a very unsteady buisness, somtimes up, and somtimes down, some few gets Rich, and thousands are ruined by it. Rogues, Rascals, Knaves and vagabonds are connected with it. Some persons that you trade with will cheat you in spite of your teeth, and you must cheat others in return to make ends meet and tie. In Short no honnest man can live by it. A Factory too, is liable to be burnt down, but a Farm cannot be easily burnt up. Manufactoring breeds lords and Aristocrats, Poor men and slaves. But the Farmer the American farmer, he, and he alone can be independent, he can be industrious, Healthy and Happy. I am for Agriculture. I am young Just steping into the world. I may probably be married somtime, and have a family, but I cannot bear the idea, that I, or my children (if I should ever have any) should be shut up 16 or 18 hours every day all our life time like

Slaves and that too for a bare subsistence! No, God forbid. If I had the chance to morrow of either a Factory worth 10000 dollars, or a farm worth 5000 dollars, I would take the Farm. But after all, I would say please yourself, you are older than I, and knows the world better. A small Factory with a quantum suficit of land along with it might do pretty well if well managed.

*

View of Millbury, Massachusetts, on the Blackstone Canal. From John Warner Baxter, *Massachusetts Historical Collections . . .*, Worcester, 1839. Old Sturbridge Village Library.

Sally Rice

Letters (1839–1845)

Beginning in the 1820s, when owners of village mills added power looms to their manufacturing operations, they hired mainly young women to operate the new machinery. The weavers were sometimes older members of mill families, but often they were single girls, away from home for the first time. While working as weavers, they often roomed and took their meals with mill families.

Sally (or Sarah) Rice, born in 1821, was the daughter of a Dover, Vermont, farm family. Among the substantial collection of personal letters they wrote are seven written by Sally home to parents and siblings as she made her way out in the world, working first as a domestic servant, then as a weaver, and finally settling into solid working-class respectability as the wife of a Worcester, Massachusetts, railroad engineer. Her letters are valuable evidence of a young woman rationalizing her risky and ethically problematical choices between family and independence, farm and down-country, home and factory. Her own justifications may be balanced against the opinion of her elder brother Hiram, who wrote to their parents that he considered her a "very ungrateful & inconsiderate" child.

All but three of Sally Rice's letters are reprinted here. Two letters, dated March 4, 1840, and July 4, 1842, concern religious anxieties

The letters of Sally Rice are contained in the Hazelton Rice Papers, Dover Free Library, East Dover, Vermont; printed by permission. Hiram Rice's letter, dated April 7, 1843, is in the same collection. For more information, see Nell W. Kull, "'I Can Never Be Happy There in Among So Many Mountains'—the Letters of Sally Rice," *Vermont History,* volume 38, Winter 1970, pp. 49–57. Kull's article includes long, loosely edited excerpts from the letters. The texts printed here are edited from the originals. Periods, terminal question marks, and initial capital letters have been introduced; other punctuation and orthography are as in the originals.

and family trivia. The final surviving letter was written on October 20, 1849, when Sally was settled in Worcester, the mother of a "fat" baby, but still much concerned with poverty. (MBF)

Union Village [New York] Aug. 19th 1838

Honored Parents, Brother Sisters & Friends

It is with pleasure tht I now seat myself to converse a few moments with you, we received your letter the 15th, was glad to here that you was all able to work. My health is very good indeed, & has been ever since I came here except one evening I went to meeting. They preach a great deal of Slavery here and I swalowed so many it made me sick but I went out and puked them up and felt better. I like living here very much indeed. I live with very nice people have enough to eat and drink and enough to do, and I think if I am not contented here I never shall be eny where. There is 7 in the family, Mr. Holmes & wife 3 children there other hired girl & myself. We have 3 cows now, our folks have all been to meting to day but my self I think I shall go this evening. I have the privelage of going to meting half the day sunday, & sunday eve. There has been several deaths of the small pox in this village sinse I came here, but I dont know of any case now. Father, John says he is very much oblige to Mr. Rice for his love. We were some disapointed to hear that [illegible], but just as he can afford. I should like to see y[ou a]ll, but I dont know when. I have been to meting this evening and saw Mary she is well. My best respects to all. It is getting late and I must close,

Sally Rice

I must just [tell?] you see what I have bought: 1 bonnet ready made 1.25, 1 pare seal skin shoes 1.25, 8 yards calico 12 1/2 cts yard 1 dollar. I will send you apeice of my dress, also a peice like a dress Mrs Holmes gave me. So good bye.

Union village 1839

Dear Father & Mother. I am well. I found good crossing the mountain. Got to Arlington about 8 in the eve. Staid over night, and the next morning started for home. Arived at Cambridge about one. Stoped at Comstocks hotel. I found a man there that was going directly to UV. With him I rode here and I never was so glad to see eny place as I was to see my old home and friends, I found [them a]ll well except Mr. Sailsbury. He was very sick and is now with the inflamation of the lungs. It is doubtful whether he ever gets well. His sister is here taking care of him. I found Nancy here and another girl which was not much help. Nancy went away the next day but one and last week the other girl went and so I am alone at present. We expect Maria Gayton will come this week. Elem Knight and I got up monday morning at one oclock to wash. He helpt me some and we got done before light and I should be willing to get up every morning at one if it would make you willing that I should stay here. I can have a home here as long as I will stay and am steady. They are very anxious that I should live with them as long as I work out eny where and no more anxious that I am to stay here. I have one of the best of homes and good society which is a good deal better than I can have there. Not but that I have a good Father & Mother but look at the company that I should be with a profane Sabeth breaking set. I cannot bare the thoughts of going there to live. No one knows how much I suffered the ten weeks that I was at home. I never can be happy there in among so many mountains. I feel as tho I have worn out shoes and I think it would be more consistent to save my strength to raise my boys for if I have got to have 16 I shall need all I have got and as for mayyring and settling in that wilderness I wont, and if a person ever expects to take comfort it is while they are young I feel so. You all know how I was dressed when I came here. I have got so that by next summer if I could stay I could begin to lay up something. It was not for my good that I promis^{ed} to come home nor indeed I dont think it would be for yours. Think of it all around I am now most 19 years old I must of course have something of my own before many more years have passed over my head and where is that something coming from if I go home and earn nothing. What can we [get] of[f] of Rocky farm only 2 or 3 cows. it would be another thing if you kept 9 or 10 cows and could raise corn to sell. It surely would be cheper for you to hire a girl that can do your work one that would be contented to stay in the desert than for me to come home and live in trouble all the time. If you lived within 5 or 10 miles of here I would not say eny thing against living with you but I have lived amonst desent people so long that I dont want to go home. You may think me unkind but how can you blame me for wanting to stay here? I have but one life to live and I want to enjoy my self as well as I can while I live. If I go home I can not have the privelage of going to meting nor eny thing else. Do come away. Dont lay your bones in that place I beg of you

I want you should write me an answer directly and let me know my fate

My love to all who enquire after SR

Mr Holmes raised one hundred and ninety bushels of Rohan potatoes from one bushel planting. Haselton will say I dont believe it.

Sunday Masonville [Connecticut] Feb 23^d 1845

Dear Father

I now take my pen in hand to let you know where I am and how I came here and how my health is. I have been waiting perhaps longer than I ought to without leting you know where I am yet I had a reason for so doing. (My reason was this.) Well knowing that you was dolefully prejudiced against a Cotton Factory, and being no less prejudiced myself I thought it best to wait and see how I prospered & also see whether I were going [to] stay or not. I well knew that if I could not make more in the mill than I can doing house work I should not stay. Now I will tell you how I happened to come. The Saturday after New Years I came to Masonville in Thompson Conneticut with James Alger on a visit to see his sister who weave in the mill. We came Saturday and returned to Millbury on Monday. While here I was asked to come back and learn to weave. I did not fall in with the idea at all because I well knew that I should not like as well as I should housework and knowing you would not approve of my working in the mill. But when I considered that I had got myself to take care off I ought to do that way that I can make the most and save the most. I concluded to come and try promiseing Mrs Waters that if I did not like I would return again the 1st of April. I have wove 4 weeks and have wove 6,89 [689?] yds. We have one dollar and 10 cts for a hundred yards. I wove with Olive Alger one week to learn and I took 2 looms 2 weeks and now I have 3 looms. I get along as well as eny one could expect. I think verry likely that before the year is out I shall be able to tend 4 looms and then I can make more. O & P Alger make 3 dollars a week besides their board. We pay 1,25. cts for our board. We 3 girls board with a Widow Whitemore. She is a first rate homespun woman. I like quite as well as I expected but not as well as I do house work. To be sure it is a noisy [pla]ce and we are confined more than I like to be but I do not wear out my clothes and shoes as I do when I do house work. If I can make 2 dollars per week beside my board and save my clothes and shoes I think it will be better than to do house work for nine shillings [$1.12 1/2] I mean for a year or two. I should not like to spend my days in a mill not by a good deal unless they are short because I like a Farm to well for that. My health is good now. I wrote a letter to Levi and Nancy the week before I came here with a strict comand not to tell eny mortal that I was comeing because if I did not like I should go back to Waters again and if I did not stay I wanted nothing said about it. And I say now if it does not agree with my health I shall give it up at once. I consider that my heatlh is gone I am done at once. I have been blessed with good health always ever since I began to work out. I have not been co[n]fined to my bed but one day since I was sick with mumps at the time Grandmother Rice died. I was very sick one day when I was at Mr Waters.

Dear Father. In my last letter I told you I had morally reformed. Yes I trust I have and I bless God that he unsealed my eyes to see where I was standing. Where have I been since I became a Backslider the name haunts me it all seems like a dream. Pray for me Father that if I ever enjoyed Religion I may enjoy it again and if I never enjoyed it that I may and do as much good as I have hurt in the cause and the great God asisting me I will try to Pray for myself. I feel that I am perfectly will[ing] to give up all into the hands of God and will [try] to lead a better life than I have done. I want [you] should write as soon as you get this. Direct your letter to Masonville Thompson Conn. Give my love to Mother and all our folks tell Brother to write. I have not writen to Hiram yet. I intend to before long.

I want to know where Ephraim is and what he is doing and what you are all about and how you all do. Father Good bye

Sarah Rice

Millbury [Massachusetts] Sept 14th 1845

Dear Father Mother Brother & Sister

I have waited a long time for a letter from some of you. I have not heard from eny of my friends for a good many weeks—and now I ask the reason. I fear you are sick—if you are why do you not let me know it. I have writen you 2 letters before this since I have heard from you.

My health is very good indeed. My work is very hard and I get some tired. Mr. Waters is building a house this summer which makes the family much larger than usual. You surely cannot blame me for leaving the factory so long as I realised that it was killing me to work in it. I went to the factory because I expected to earn more than I can at housework. To be sure I might if I had my health. Could you have seen me att the time or a week before I came away you would advised me as many others did to leave immediately. I realise that if I lose my health which is all I possess on earth or have eny reason to expect to posess that I shall be in a sad condition. I want to see you all and proberbly shall in the course of a month or two. I want you should write immediately and tell all the news you can think of. What is Haselton up to? Henry told me that Daniel is married. There is one case of the small pox in this part of the town and we shall think it very strange if there is no more. We have very dry weather here. It rains to day and is very cold. There was some frost on the ground Saturday morning. [The rest of this letter is taken up with overwrought reflections upon her supposed moral lapses, which she concludes hopefully.—MBF] New light sprang into my soul. I felt that that promise was for me and that there was hope for me yet, that I had something to do. If I was not a penitent child I guess there never was one. My conviction of sin was not a faint one. Sometimes it seemed that I must die. God grant that I may be steadfast, unmoveable, and meet you all in heaven.

H. Rice — Sarah

Bennett Family

Letters (1839–1846)

Among the collected letters of the Bennett family in the Haverhill, Massachusetts, Public Library are a group of early ones in which several women—cousins, aunts, and nieces—discuss the migration from farm to factory work. Unlike Sally Rice, who perceived her occupational choice to be between the mill and domestic service, these women consider possible alternatives in schoolkeeping and "the trade"—dressmaking and millinery. Like Sally Rice, they are preoccupied with problems of health and religion and assume that a woman must make her way in the world at gainful employment. (MBF)

Bennett Family Letters, Haverhill, Massachusetts, Public Library; printed by permission. Punctuation has been normalized, orthography has not. A much larger selection of Bennett family letters appears in Thomas Dublin ed., *Farm and Factory: The Mill Experience and Women's Lives in New England, 1830–1860* (New York: Columbia University Press, 1981). The transcriptions printed here differ from Dublin's in small matters.

To Sabrina Bennett

Nashua April 4 1839

MY DEAR COUSIN,

Doubtless you will be suprised to hear from me as I am not in the habit of holding correspondence with you. It is not because I have forgotten you no, Dear S[abrina]. I often think of you & could as often wish to see you. I suppose I need not apolygize for past negligence as you are guilty of the same yourself. I hope we shall let past neglig[en]ce suffice & for the future commence a correspondence. There shall be no lack on my part. I have often thought of you since I came to this place & especially since I heard that you was obliged to give up your shop on account of ill health. I left Vermont last july came to Bristol & stayed until October when I came to this place. I work in the mill like very well enjoy myself much better than I expected am very confined could wish to have my liberty a little more but however I can put up with that as I am favored with other priveleges. I think I shall visit you this Summer. I think if nothing in providence prevents I shall stay here untill fall. It seems now a long time since I left home am almost homesick sometimes. I heard from home last week & likewise from Bristol. Our folks all & Grandfarthers are all in good health except Aunt Bryant she has not been able to work this winter.

I will just say I hope you will answer this soon. Give my love to uncle & Aunt & all friends. If you do not think of coming here to work I hope y[ou come] & visit us. I want to see you very much hope to soon. Write us all the news you have & believe [me] to be [your] undeviating Friend & Cousin

Persis L. Edwards

[Persis Edwards's mother appended a letter of her own on the same page.—MBF]

DEAR SABRINA,

I have nothing special to write you but Persis has commenced a letter [and] I will try and think of something [even] if it is not so very interesting. You have been informed I supose that I am a factory girl and that I am at Nashua and I have wished you were here too but I suppose your mother would think it far beneith your dignity to be a factory girl. Their are very many young Ladies at work in the factories that have given up milinary d[r]essmaking & s[c]hool keeping for to work in the mill. But I would not advise any one to do it for I was so sick of it at first I wished a factory had never been thought of. But the longer I stay the better I like and I think if nothing unforesene calls me away I shall stay here till fall. Persis has told you that the folks at Bristol were all well but sister Bryant and I fear if she does not get help soon she never will be any better. Your uncle Frisbies folks have moved to New York where his Brother lives. Your uncle [Daniel] Sandborn has buried his father he died the 4 of March. Give my respects to your fathers folks and except much love your self from me. Write soon and write me all the news you can think. I want to hear from Haverhill. Write too where you are and what you are doing and what you intend to do this summer. My health is very poor indeed but it is better than it was when I left home. If you should have any idea of working in the factory I will do the best I can to get you a place with us. We have an excelent boarding place. We board with a family with whome I was acquainted with when I lived at Haverhill. Pleas to write us soon and believe your affectionate Aunt

M[alenda] M. Edwards

To Sabrina Bennett

Barnet [Vermont] April the 18 1840

MY VERY DEAR COUSSIN

I received your letter dated jan 24 after a long time it layed in the office. Be assured it met with the most hearty welcome was read over & over again & again. It brought to mind the many social hours we have spent together which are now past. [. . .] When I came home last fall found Sister E[liza] confined with the Fever. She recovered in a few weeks then Mother & Alcemena were confined at a time & myself likewise. After we got better Almira & John they get about. Then James was sick about three weeks. The first day of january Father froze his foot was not able to work for five or six weeks. The first day he went to work, the boys went into the woods with him to chop wood James cut his foot was not able to go to school for three weeks & so we have had one trouble after another ever since I came home till this Spring. We are now all enjoying good health, which above every thing else we should be thankful for. I feel as though it was through the Goodness & Mercy of God that we are spared. It can surely be nothing that we merit from our own goodness. [. . .] I do not know what my employment will be this Summer. Mother is not willing I should go to the Factory. I thought some of learning the Milleners & Dressmakers trade but have failed in the attempt. I wrote to Uncle Bryants folks to know if I could get in there to Haverhill can not under four or five months [and? p]ay my own board. Cannot do that. They thought it would be rather inconvenient for them to board me if I could get in. You may well know the reason I am not popular. Cousin Ann [Blake] is the top of Haverhill Corner. She had a Broadcloth cloak last Winter cost over 30 dollars.

If I could learn the trade there is a very pleasant village in this town which would be a good place to work. There is no one in the place that keeps shop. Hope you will try to visit us this Summer. Come & spend a long time with us. Write to me as soon as you get this tell me of your Prosperity & how you are employed. Dont delay. If you work at your trade I should be glad to work with you. I wish you were here in a shop. Could you come we should enjoy all the pleasures imaginable. Father & Mother send love to your Parents. Wish them to visit us as soon as convenient. Give them my love. My Brothers & Sisters all send love to you all, you all have our best wishes for your Prosperity. Cousin there shall be no lack on my part aboutt keeping up a correspondence. Answer this as soon as possible. Direct your letter to Peacham. Barnet Post Office is five miles from us. Believe me to remain your very affectionate Cousin Persis L Edwards.

To Richard and Ruth Bennett

Nashville [Nashua, N.H.] May 14th 1843

DEAR BROTHER & SISTER BENNETT,

I thought I would jest say a word to you if it is not quite so brite. Our famely is all in good health except myself. I have been q[u]ite out of health this spring but am much better now. The Doctor says I have the liver complaint. You will probely want to know the cause of our moveing here which are many. I will menshion afew of them. One of them is the hard times to get aliving off the farm for so large famely so we have devided our famely. For this year we have left Plummer and Luther to care on the farm with granmarm and aunt Polly. The rest of us have moved here to Nashvill thinking the girls and Charles they would probely worke in the Mill but we have had bad luck in giting them in. Only Jane has got in yet. Ann has the promis of going in the mill next week. Hannah is going to school. We are in hopes to take a few borders but have not got any yet. We live on canall street vary nere Indian head factory. We heard from father, folks last week. They were all well. They had lately heard from Mary. She wrote she was well fat sausy and happy and had got a little girl, the prittyest little babe you ever see. She sayd they ware agoing to move to Indinia in April. They wrote they had bought a farm there and ware agoing to farming. They did not write the name of the town so we do not know whar to derict a letter to her tell we here from them. We have not herd from Brother Frisbye folks scence last winter. He was vary low and feeble then. They did not expect he could live but alittle while. Brother [John] Bryant has got a new office. He is postmaster and Deputy Marshal of this state. David [Edwards] has moved to Nuberry about nine miles from Bryants and taken afarm. I think Eliza would like to come down here and work in Mill. There is a grate many more trying to git in than can git in. It is quite apleasant place hear but it dont seeme much like home. It would seeme more like home if any of my folks lived here. You know I never was weaned from fathers house before. It is rather a hard case but I suppose I must try and bare with it. You must come and see us as soon as you can. It is only 20 miles. You can take the cars and come in a few minits. I have some good news to tell you about father. He became q[u]ite pious last fall. & O Ruth it would have affected your heart to have seen our aged farther agoing fored to the anxious seats and bending the knee for prayers. He is vary particular to crave a blessing before eating. You know that is a grate undertaking for him. There was a grate many that professed to have meet with achaing [a change] last fall, mostly old people. Sence we moved here we have hered that Luther and aunt Polly had experinced religion. I think Plummer will be vary much rejoiced to have so many of the familey join withe him in praising his maker. Sabryna I was vary sorrow to hear of your sickness. I hope you are fast againeng your health. When you git well enough to ride abroad come and make us a good longe visit. I should have been glad if Ann could have gone to Haverhil and lernt the trade but she thinks she must try the mill aspell first for the want of clothes that is fit to wear.

I think I shall not have to make any appology, only say that Daniel has gone to Brystol and you will not think strange of my bad spelling and interlineng. I am as lonesome as you can think here among all strangers. You must all come and see us as soon as you can.

I must draw to acloas by subcrybing you loveing Sister

Jemima W Sanborn

To Sabrina Bennett

[May 14, 1843]

DEAR COUSIN

It is with pleasure I sit my self to write to you informing you of my good helth &c. I feel as well contented as could be expected concidering all things. I think it would be best for me to work in the mill a year and then I should be better prepared to learn a traid. I should like to have gone [to Haverhill] but our folks moving to Nashville I thought I should like to try the Mill and see how I like it. I think I shall like very mutch for I go in all moast every day to see Jane and have all most stole the t[ ]. I think I shall go in to work next weak. It is [ ] imposable for eny one to get in to the Mill. They do not engage only half the help they did before they reduced the spead. Ann and George Bryant was down this last winter. George said he should come to Nashvill this summer he tho[u]ght in June to go to Haverhill. He was taken sick with his old complaint. His father came down and caried him home. His grandmother was down this spring she said he had got well so I think he will come. I think if I have good luck I shall go to Haverhill but do not wait for me. Come if you can conviently. Father is gon to Bristol so we are very loanly. We received your letter and was very glad to hear from you.

You must excuse all bad mistakes as I am in a grate hurry. Give my love to all the good folks you know.

Ann M B[lake]

To Sabrina Bennett

Friday Eve Nashua Sept 25th 1846

DEAR SISTER BENNET

It is with much pleasure I seat myself at my table to converse with you by the silent language of the pen. How my Dear S. are you these days and all the good friends of H[averhill]. I assure you I wold give all the bright cents I have to see you and some others in H tonight but as that will do no good I will tell something of my times and health since I saw you. I could not get a chance to suit me, so I came here to work in the Mill. The work was much harder than I expected and quite new to me. After I had been there a number of days I was obliged to stay out sick but I did not mean to give it up so and tried it again but was obliged to give it up altogether. I have now been out about one week and am some better then when I left but not verry well. I think myself cured of my Mill fever as I cannot stand it to work there. The people that I board with have been verry kind to me and want me to stay hear and work in a shop. There are a number of chances and I think some of stopping but have not decided yet. My friend in Boston wished me to come there. If I worked at my trade and if I thought I could make more I would go and think some of going next week and see what I can do. I like the place and the people hear verry much but wish to work whare I can make the most. Will you pleas to write so I can get your letter next Monday and write me whare to find Dr. Gleason as I wish to see him when I go to Boston. My head has been considerably affected since I went into the Mill and you must excuse my writeing much this time as I have a few lines to write to Abby but will you pleas to ask Miss Forbs to excuse me for not paying my bill. I called a day or two before I left to pay her and she was engaged and the day I left it entirely sliped my mind. Pleas tell her I if I do not come to H soon I shal send to you when I pay my assesments. My love if you pleas to all our good Sisters your family and all our friends. Pleas write so that I can get it on Monday if you can as I intend to go to B on tuesday and I wish to see the Dr verry much. Next time I write I hope my head will feel better and I will write more and I hope better but write all the news you can think of will you.

Yours truly
L[ucy] M Davis

Hiram Munger

Life and Religious Experience (1856)

In addition to business records and public documents, some firsthand accounts survive to help us understand family life and work in early mill villages. Manuscript memoirs, diaries, published autobiographies, and letter collections are by no means common, but enough of them do exist to suggest that the responses of American workers to the new industrial work were quite varied.

As a young child of impoverished parents, Hiram Munger (born in 1806) was one of the early factory workers in Massachusetts. Like most, he worked in a textile mill for only a short time, but the experience remained vivid forty years later when he composed his autobiography. He spent most of his life in various manual occupations and became an energetic itinerant Methodist lay preacher in central Massachusetts. (RP)

Hiram Munger, *The Life and Religious Experience of Hiram Munger* (Chicopee Falls, Massachusetts, 1856), pp. 10–15.

I was born in Monson, Mass., September 27, 1806, of poor parents. I was the oldest son of Stillman and Susan Munger, who were the parents of five sons and six daughters, who have all, except one, lived up to the present time, this 9th day of August, 1855. I am consequently nearly 49 years of age. There is nothing remarkable in my experience of early life any more than in that of many others. But I can recollect so distinctly circumstances that took place when I was very young, that it may refresh my memory concerning later dates to note a few things as I passed from childhood up to where I now am ; and as memory is the most I have to depend upon, it needs refreshing, and this I offer as a reason to my friends for commencing my narration previous to what they or I expected at first. I recollect a number of circumstances that took place when I was less than two and a half years of age, while living in Monson. My father moved to Ludlow in the year 1809, and tended a grist-mill for a Mr. Putman, in the place then called "Put's Bridge," since called Jenksville. While there I tended the toll-gate on the bridge. I recollect demanding the *two cents* of a colored man, who refused to pay me, and threatened me if I did not open the gate. I went for help, or to inform my father in the mill : when we came out in sight, he was on the gate (which was very high) getting over—my father shook him off, which so enraged him, that he cursed and swore at a great rate, which scared me for the first time in my life that I recollect. The same hour, and a short distance from that place, he committed a crime worthy of death, and was executed in Northampton. His name was *Piner.*—Many will recollect this circumstance as well as I do, for there was much excitement in that place at the time of his capture and trial.

The next work I remember doing was going into the small cotton factory over the grist-mill, started by Benjamin Jenks & Co., who came from Rhode Island. This was the first factory of that kind in Massachusetts. The help necessary to carry it on was about twelve or fifteen hands. Here was where I was first made acquainted with American slavery in the *second degree.* The treatment of the help in those days was cruel, especially to poor children, of whom I was one. Although I was young, I recollect of thinking that life must be a burden if I was obliged to work in a factory under such tyrants as the Jenks' were *then,* and they never improved, unless it was when they failed and cheated the community out of $100,000, or more, and then left the parts.

In a few years, we moved to another mill three miles north, but in the same town, and lived there three years. Here I began my education with tending grist-mill. There being few inhabitants in the place, my mother was sent for when there was any sickness, and I, being the oldest of her four children, had *all* the care when my father was absent. I remember that my second sister was at play around the fire, and her dress took fire ; father and mother being gone, I tried in vain to put it out, 'till she was very badly burned,—her screams terrified the rest of the children, and no neighbors being near, I was in a straight place sure enough. I thought of the brook, and in an instant took the child, and amid the screams, confusion and fire, hastened down the bank a number of rods through bushes and weeds, and threw her in. The brook being large and high at the time, she went down some distance before I could get her out. This operation put the *fire out* and stopped her crying, for she had strangled by rolling over so many times while going down to a place where I could get her out. She soon revived, to my joy, for I was afraid that my sudden remedy was fatal. But she got well, sooner probably by having the cold water bath. I must have been at that time about ten years of age. We next removed to Wilbraham, and lived a year or so. I worked that summer for Abner Cady, on a farm, for three dollars per month.—This was the cold summer of 1816. My summer wages bought my father a cow, which we kept until we moved to Chicopee, the town where I now reside. I was now large enough to help in the mills, and was subject to my father for a number of years : with him I struggled with poverty, the family now being large.

My second brother and myself were all the

help he had, to carry on a grist-mill, and some of the time two saw-mills ; and we were so poor that I had not clothes that were comfortable for winter or decent for summer much of the time. This was the misfortune of being very poor ; it was not caused by indolence nor intemperance of my father, for there is hardly a man that *lives*, or ever *did live* or ever *will*, that worked harder and more hours to support a family than he did, and my mother too. I was old enough to know that it was out of their power to do any better by their children. But, like other boys, I was often dissatisfied with staying at home without clothes to go to school or meeting but very little. I was nearly 16 years old before I could write, or read in a paper ; and I could not cipher at all. I was ashamed to go to school there then, and at last got rather headstrong and unruly, and determined to run away. I recollect setting a time to start : got my little all done up in a cotton handkerchief, and about 8 o'clock in the evening I started for Monson, to my uncle's—about fifteen miles. It looked like a great undertaking in those days. But I started, and had got about half a mile, when my attention was arrested by hearing some one praying up the river about one and a half miles from where I then was. I could hear distinctly what was said, and I staid nearly an hour and listened, until I concluded to go back home and put my goods in at the chamber window where

I got out. I went to bed thinking about that praying up the river: *that* turned my mind from running away. I staid at home peaceably for a year.

Manufacturers' and Farmers' Journal
Massachusetts Spy

Help Wanted Advertisements (1821–1832)

The unskilled workers, including families, who tended machinery in New England textile villages frequently came from nearby towns. Many moved from one local mill to another. Skilled and supervisory workers often were hired from greater distances, as were some of the young women who tended power looms. Some hiring was done through newspaper advertisements, and the *Manufacturers' and Farmers' Journal* of Providence was a prime source of help wanted ads, as it also was for information about factory properties for sale. Ads appeared in other regional papers, as well, among them the *Massachusetts Spy* of Worcester, from which the last two selections here are taken. (RP)

[June 28, 1821]

WANTED,
A mule and Throstle Spinner, and a few Girls from fifteen to twenty years of age, who have been accustomed to tending spinning frames, and can come well recommended. . . .

John C. Dodge, Agent
(for Tyler Manufacturing
Co., Attleboro, Mass.)

[July 2, 1821]

A family who wishes employment in a Cotton Factory, may find suitable encouragement, if they can be recommended to be of good habits, and willing to work, say one reeler, one or two to tend spinning, and one or two to tend drawing and roping. . . .

[November 1, 1821]

The subscriber is in want of a mule Spinner—one who is well versed in the business and can give satisfactory recommendations of his skill and habits, will find liberal encouragement by calling on the subscriber.

IRA P. EVANS

N.B. One with family would be preferred.

Chepachet (R.I.), Nov. 1

[April 15, 1822]

The subscribers are in want of eight or ten Girls to weave on Water looms. Those who understand the business may have constant employ, good treatment and liberal wages, by applying to Mr. Kent, at the Hope Factory, Scituate, or to EPHRAIM TALBOT & CO., Providence.

N.B. None need apply but those who are willing to work twelve hours a day. One or two good families are also wanted at the above mill. . . .

[April 25, 1822]

Wanted, a family consisting of six or eight persons to work in a Cotton mill, near this town. Two of them must be Spinners; and the remainder work in a Carding Room. None need apply unless well recommended, and are willing to comply to good and necessary regulations: to such a Family liberal wages will be paid, either in cash or otherwise as may be agreed upon. . . .

[April 29, 1822]

Wanted, A man to take charge of a carding room in a cotton factory in Walpole, N.H. One very well acquainted with the business and that could bring good recommendations would meet with good encouragement.

Sampson Drury
Anan Evans

Drewsville, Walpole, April 29

[September 19, 1822]

WANTED—At the Middlesex Factory in Hopkinton [Massachusetts], a man that understands carding, spinning, weaving and repairing a cotton mill. No one need apply without he is master of the business. To such a man good encouragement will be given. Also, a girl that understands taking care of a dresser may meet with good encouragement by applying at the factory in Hopkinton.

[October 24, 1822]

WANTED At the Plainfield Union Factory, in Plainfield [Connecticut], two good machine makers. Inquire of the agent at the Factory, or of David Anthony, agent at Providence.

[December 19, 1822]

WANTED to Hire, a good workman, to repair Cotton Machinery at a Cotton Mill near this town—one with a family to work in a mill, would be preferred.

Philip Allen

[June 28, 1824]

Wanted by the Framingham Manufacturing Company, at Framingham, Mass. 30 miles from Providence, twenty girls, acquainted with spinning and weaving. Also a good Mechanick, acquainted with the operation of machinery. . . .

Those answering the above description, will find constant employment and receive good wages, in cash, for their services.

Samuel Murdock, Agent
Framingham, June 28

[October 1, 1829]

The subscribers are in want of Twenty-four good [female] Water Loom Weavers, to whom good wages will be given, part cash or all, if particularly required. None need apply unless they are willing to work twelve hours perday, and be subject to good rules and regulations while weaving.

J. Underwood & Co.

[July 31, 1832]

Wanted at the new Cotton Manufactory in Windham [Connecticut], two or three large Families, to whom liberal wages and constant employment will be given (as there is no lack of water) and full employment for men's labour if wanted.

ALSO A Machine Maker who understands forging.

Perez O. Richmond

[November 2, 1831]

Wanted, 10 to 15 Families, at the New England Village, in Grafton, to work in a Cotton Factory. Also, two or three Mule Spinners. None need apply but those with steady habits.

[March 14, 1832]

Wanted, at the Cotton Mills in Phillipston, as Overseer in the Card Room. Also, a Machinist who is well experienced in iron work and acquainted with repairing as well as building machinery. Likewise a Family who would take 12 to 18 Boarders. Also, a young man who is acquainted with warp spinning, and a man well acquainted with finishing Satinetts, may find good encouragement by applying soon and bringing good testimonials of character and experience.

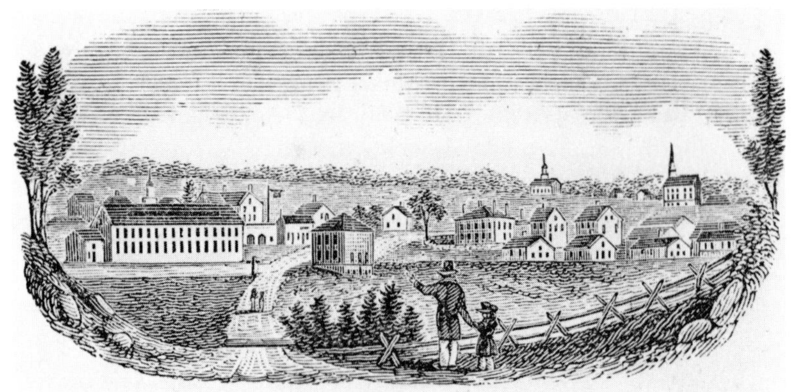

View of a manufacturing village, Webster, Massachusetts. From John Warner Baxter, *Massachusetts Historical Collections . . .*, Worcester, 1839. Old Sturbridge Village Library.

William Davis Family

Records (1803–1827)

Families were the backbone of the work force in most of New England's early cotton mill villages. Many of the jobs in cotton factories could be performed by unskilled youngsters, who were promoted from one manufacturing operation to another, according to age and sex. Rural heads of households like William Davis, many of them marginal farmers or even propertyless laborers, sometimes chose to move their families into company-owned tenements in a nearby town and put their children, both boys and girls, to work in the mill. Had they stayed on the farm, such parents might have had to send some of their youngsters to work for and live with relatives or neighboring farmers who were better off. In the mills, such children could contribute to their families' support without being separated from their parents.

Although examples can be found of children starting work at younger ages, twelve or thirteen seems to have been a common age at which to begin. Most young people left the mills by the time they had reached their early twenties. Their fathers, who usually lacked factory skills for which manufacturers were willing to pay adult wages, sometimes worked at odd jobs around the mill or as day laborers on farms in the surrounding community.

Between 1812 and 1827 the William Davis family moved to the mills, working for the Slater-Tiffany company in what is now Webster, Massachusetts. The following data, compiled by Old Sturbridge Village from company records as well as from public records like the Federal Manuscript Census and local vital statistics and registry of

The Slater-Tiffany records are in the Baker Library, Harvard University.

deeds, suggest the kinds of information about family life in a mill village that can be gleaned from such sources. Often this is all that can be learned about otherwise anonymous mill families. (RP)

Biographical Data, 1803–1827

September 1803

William Davis married Rachel Humphrey, in Oxford, Massachusetts. They will have seven children:

Jemina	b. October 10, 1803
John	b. March 9, 1806
Ebenezer	b. May 4, 1808
Ruth	b. December 9, 1811
Mary	b. December 5, 1814
Alonzo	b. July 3, 1816
William	b. February 1, 1820

February 1812

William Davis living on his own farm and working irregularly as a home weaver in Oxford, begins purchasing items at the Mill Store of the newly organized Slater-Tiffany Cotton Mill. Before he has done a day's work in the mill he has built up a debt of $24.23 by purchasing the following:

Snuff	1 box	Brandy	2 1/2 qts.
Tobacco	1 1/4 lbs.	Gin	1 3/4 gallons
Tea	1 lb.	Rum	3 3/4 gallons
Brown Sugar	4 1/2 lbs.	Bottle	1 pint size
Molasses	3 qts.	Hair Comb	1
Rice	7 lbs.		

September 1812

Credit is noted in William Davis account for providing 168 feet of 2 inch plank.

June 1813

William Davis begins work in mill as an operator. His work record is irregular:

June	6 days
July	12 1/2 days
August	24 1/2 days
September	6 days
October	0 days
November	4 1/2 days
December	1 1/2 days

Throughout 1813 he continues to make purchases like those listed above at the mill store.

January 1814

William Davis issues a promissory note to the Slater-Tiffany Mills for amount he owes at the mill store.

April 1814

William Davis works for 4 1/2 days. This is his only entry in the mill's time books for this year.

April 1815

The William Davis family moves into a mill-owned tenement house. William Davis and his three oldest children, Jemina, John, and Ebenezer, are working in the mill. They work regularly, from fifteen to twenty-two days a month for the remainder of the year.

June 1815

By June 30, the Davis family owes the mill store for the following:

Beef	90 1/2 lbs.	Tobacco	1 3/4 lbs.
Pork	47 1/4 lbs.	Tea	1 lb.
Veal	12 1/2 lbs.	Biscuits	2 doz.
Rye	2 bu.	Chocolate	2 cakes
Flour	49 lbs.	Wood	16 ft.
Rice	28 lbs.	Shoes	1 pr.
Sugar	2 lbs.	Thread	5 lbs. 2 oz.
Molasses	2 7/8 gallons	Penknife	2
Butter	5 lbs.	Keeping Davis Family cow	

February–June 1816

William, Jemina, John, and Ebenezer Davis continue to work at the mill.

July–September 1816

The number of employees at the Slater-Tiffany Mill is reduced from 101 in March to 21 in July to 14 in September. By September, William is the only Davis family member working at the mill.

September–October 1816

The Davis family makes the following purchases at the mill store:

Salt Pork	53 3/4 lbs.	Cheese	5 1/2 lbs.
Salt Fish	2 lbs.	Tea	1 1/4 lbs.
Mutton	4 lbs.	Tobacco	2 1/4 lbs.
Rye	1 bu.	Biscuits	3 1/2 doz.
Corn	9 bu.	Wood	6 1/4 ft.
Potatoes	2 1/2 bu.	Candles	1 lb.
Sugar	3 lbs.	Material	2 1/2 yards
Molasses	1 1/2 gallons	Spelling book	

March 1817

Jemina, John, and Ebenezer Davis return to regular work in the mill, working twenty-two to twenty-four days per month.

June 1817

Ruth Davis begins work in the mill.

July–August 1817

The Davis family makes the following purchases at the mill store:

Beef	64 13/16 lbs.	Tobacco	5 lbs.
Pork	40 3/8 lbs.	Tea	1 1/2 lbs.
Veal	13 1/2 lbs.	Coffee	2 lbs.
Mutton	19 1/2 lbs.	Crackers	3 doz.
Salt Fish	38 3/4 lbs.	Peppers	1/4 lb.
Codfish	5 lbs.	Ginger	1/2 lb.

Rye	4 3/4 bu.	Pearlash	2 1/2 lbs.
Potatoes	11 1/4 bu.	Salt	1/2 package
Sugar	6 lbs.	Pimento	1/4 lb.
Molasses	4 3/4 gallons	Wood	15 ft.
Butter	19 lbs.	Candles	1 lb.
Cheese	55 3/8 lbs.	Thread material	
Creampt	1	shirting 20 yards	
Crockery	1	other kinds 11 yards	
Primer	1		

1818

William, Jemina, John, Ebenezer, and Ruth Davis continue to work regularly in the mill.

June 1819

The Davis family account in the records of the Slater-Tiffany Mill is closed.

April 1820

William, Jemina, John, Ebenezer, and Ruth Davis return to work in the mill.

October 1821

Mary Davis begins work in the mill.

1822

Jemina Davis is given a separate page in the mill account books and charged separately for her room and board.

August 1823

Ebenezer Davis leaves employment in the mill.

December 1823

Alonzo Davis begins work in the mill.

April 1825

John Davis leaves employment in the mill.

1826

William, Ruth, Mary, and Alonzo Davis continue to work in the mill. Jemina Davis continues work in the mill and is listed separately in the account books.

1827

There is a change in the organization and management of the Slater-Tiffany Mills. The work force of 145 employees is reduced to 24. None of the Davis family members is listed. Elliot Mansfield, Jemina Davis's husband-to-be is listed among the new workers.

Earnings, August 5, 1820–December 2, 1820

	Debit Samuel Slater	to William Davis		1820
1	August 5	for Balance due on Settlement		48.10
	Dec. 2	" my work in full to date 74 days 8 1/2 hours @ $17 per mo.		48.86
		" son John the same 16 weeks 3/4 days		20.16
2		@ 7/6 [per week]		
		" son Ebenezer H. the same 15 weeks 1 1/2 @ 6/ [per week]		15.25
		" Daut. Jemima the same 15 weeks 3 5/12 days @ 12/ [per week]		31.14
		" Daut. Ruth the same 14 weeks 4 3/4 days @ 4/3 [per week]		10.48
				$173.99
3	Dec. 2	By Amt. of his own rendering	150.11	$173.99
		" Balance credited in new acct	23.88	

Errors Excepted Oxford Dec. 2nd 1820 [signed] William Davis

1. That is money owed by Davis to Slater from the previous settling of accounts. Note that during the period covered by this settlement the Davis family improved its position, from a $48.10 debt to a $23.88 credit.

2. "7/6" means 7 shillings and 6 pence. The English form of computing currency was still used interchangeably with the American decimal system, figuring 6 shillings to the dollar. The work week consisted of six twelve-hour days.

3. The "Amt. of his own rendering" is the sum of Davis family purchases at the company store.

Analysis of Expenses and Income, 1820–1821

Expenses

Item	Cost	Percentage of total expenses
Food	$183.73	49.6
Clothing	71.27	19.0
Shelter	27.52	8.0
Heat (wood)	22.59	6.0
Light (tallow, candles, wicks)	7.83	2.0
Household utensils (plates, knives, barrels and brooms)	6.40	1.6
Miscellaneous (highway tax, soap, Baptist Society fee, books, paper, combs, and "goods")	9.99	2.8
Unaccounted ("Dist. To . . ." or cash)	41.98	11.0
Total	$371.31	100.0

Earnings

Person	Cash	Percentage of total family income
William Davis	$170.21	42.5
John (age 14)	61.93	15.5
Ebenezer (age 12)	48.21	12.0
Jemina (age 17)	87.81	22.0
Ruth (age 9)	32.52	8.0
Total Earnings	$400.68	100.0
Total Family Expenses	$371.31	
Family Income After Expenses	$ 29.37	

Purchases, August 5, 1820–December 2, 1820

1820—Debtor		
8/8	to 1 3/4 lb. Butter	.19
8/9	to 1 Peck Beans	.31
	to 1 Bu. Grain	.94
8/10	to 1 Load old wood	.36
	to 1 1/4# Butter	.20
8/11	to 7# Rice	.38
	to 2 qts. molasses	.26
8/14	to 1 11/16# Butter	.27
	to 2 1/4 yds. Sattinet 8/3	3.09
	to vesting	.75
8/15	to 1/4# Tobacco	.07
	to 4 3/4# Codfish	.21
	to 3/4# S. Tea	.59
	to 1# Candles	.23
8/16	to 1 Bu. Grain	.94
	to 1 1/8# Butter	.18
8/18	to pd. yr. highway tax, 1820	.54
8/19	to 1 Thread	.06
	to 1 Comb	.03
	to 1 yd. ribbon	.10
	to 1# Sugar	.15
	to 1 10/16# Butter	.26
8/21	to 2 qts. molasses	.26
	to 1 yd. shirting	.28
	to 1 skn. thread	.05
8/22	to 7# Lamb	.38
	to 1/4# Tobacco	.07
8/23	to 1 6/16# Butter	.22
	to 1/4# Pepper	.12
8/24	to 1 Bu. Grain	.94
8/25	to 2 3/16# Poulters [Poultry]	.35
	to 1/4# Tea	.12

1820—Debtor		
8/26	to 1/4# S. Tea	.22
	to 1# Sugar	.15
	to 1/2# Pearlash	.11
	to 1/2 Barrel Flour	3.10
8/28	to 1 1/2# Butter	.24
8/29	to 4# Lard	.60
	to 5 11/16 Lamb	.29
	to 6# Cheese	.36
	to 1/4# Tobacco	.07
	to 2 qts. molasses	.26
	to 2 Wattermellons	.12
8/31	to 2 qts. Vinegar	.13
	to 2 needles	.01
	to 1 Square cambric	.36
9/1	to 8# Pork @ .12	.96
9/2	to 1/2 Bu. corn	.42
9/4	to 1 1/16# Butter	.17
	to 5 1/2# Cheese	.33
9/5	to 6 crackers	.06
	to Goods pd. Wm. Young	.26
9/6	to 7 1/4 yds. Bombazet	4.57
	to 2 skeins Silk	.14
	to 1/4# Tobacco	.07
	to 7 1/4 Pork 11 C	.80
	to 1 Bu. Grain	.92
9/8	to Remnant Bombazet	.37
9/9	to 1/2# S. Tea	.49
	to 5 10/16# lamb	.28
	to head [tobacco]	.05
	to 2 qts. molasses	.26
9/12	to 3 1/2 yds. shirting 28	.98
	to 1 skein warp	.07
	to 1 7/16# Butter	.23

1820—Debtor

	to 7# Rice	.35
9/13	to 1 1/2 yd. Ribbon	.18
9/14	to 1/2# Pimento	.25
9/16	to 1 1/2# Butter	.24
9/19	to dist with J. Bates	.87
	to 1/2 Gall. Molasses	.26
	to 7# Rice 5	.35
9/20	to 1 14/16 Butter	.30
	to 2 yds. Sattinet 8/	2.67
	to 35 1/4# Beef 4 1/2	1.59
	to 23 1/2# Tallow 10	2.35
	to 1/2 Pint Oil	.10
	to 4′ 9″ wood 11/6	1.14
	to Good Pd. R. Davis	.25
9/21	to Cash on acct.	.20
	to Balance on Plates	.04
	to 1 Comb	.10
9/22	to 15 oz. candle wick	.67
	to 1 oz No. 8 Filling	.72
9/23	to 9 crackers	.09
	to 1/2 Peck F. Salt	.19
9/24	to 1 14/16# Butter	.30
	to Goods Pd. R. Davis	1.00
	to Bu. Grain	.82
9/25	to 6# Lamb	.30
	to 1/4# Tobacco	.08
	to 13 1/4# Pork 12	1.59
9/27	to 1 E Phamber	.18
	to 1 skn. Thread	.05
	to 1 Pint oil	.20
9/28	to 1 Comb	.12
9/29	to 1# Pearlash	.20
	to 6 1/4 yds. cloth 9	4.23
	to 7 1/2# Sole leather	1.88
9/29	to 1/2 yd shirting	.14

1820—Debtor

	to 1 very old vase	.06
	to 1/4# Tobacco	.08
	to Dist. with Wm. Bigelow	3.75
10/2	to 1 Bu. Grain	.82
	to 7 14/16# Beef	.42
	to Cash pd. C. Dunham	.06
	to do John 21st.	.30
	to 2 qts. molasses	.26
	to 1 Handkercheif	.60
	to 1 Pair Shoes	1.25
10/3	to 2 hdkf.	.90
	to 1 penknife	.15
	to 1 hdkf.	.60
10/4	to 2# Candles	.46
10/5	to 14 oz. soap	.15
	to yarn pd Wm. Mown	1.62
10/6	to 8 1/4# Lamb	.41
10/7	to 28 Bu. Potatoes	6.16
	to 14# Cheese	.84
	to 2 3/6# Butter	.44
	to 1/4# Tobacco	.08
	to 1 Bu. Grain	.82
	to 1/2 side neat leather	2.38
	to 1/2# S. Tea	.39
10/10	to 4 1/2′ Wood	1.08
10/11	to 11 lb. lamb	.55
	to 1 Plate	.05
	to 2 qts. molasses	.26
10/13	to 6 1/4# Pork	.78
	to 3 3/4# Mackeral	.21
10/14	to dist. with A. Ballard	.77
10/15	to 1/4 lb. Tobacco	.08
10/16	to 7# Rice	.35
	to 1 Penknife	.25

1820—Debtor

10/17	to 2# Butter	.36
10/18	to 1 Bu. Grain	.82
	to 4 lb. Mackeral	.22
10/19	to Dist. with J. Bates Sen.	.87
	to 1/4# Tobacco	.08
	to 1 oz. Filling	.72
	to 2 [?]	1.50
10/20	to 2 qts. molasses	.26
10/21	to 2# Butter	.36
10/23	to 7# Rice	.35
	to 2 yd. woolen cloth	2.50
	to 5 1/4 Mackeral	.24
10/24	to dist. with A. Ross	3.47
10/25	to 66# Beef 5	3.30
	to 1 skein thread	.05
	to 24# Beef 6	1.44
	to 1/4# Tobacco	.08
10/26	to 1/4# H. S. Tea	.22
	to 1 Chestnut Barrel	.84
10/27	to 1# Butter	.18
	to 1/2 G. Molasses	.26
	to 1 Bu. Grain	.82
	to 2# Wool 84	.80
	to deduct for dump .4	
	to 4 skein 13 D. Blu. Twist	.36
10/28	to 1 candlestick	.20
	to 1 load wood	.75
10/31	to lamb	.39
11/2	to 1 Ivory comb	.16
	to Bal. on hdf	.08
	to 1 1/8# Soup	.19
11/3	to 1# Butter	.18
11/4	to 2 Bu. Grain	.14

1820—Debtor

	to Goods pd. for cabbages	.24
	to 1# Sugar	.15
11/6	to 1/4# Tobacco	.08
	to 2# Butter	.36
	to 1/2 Gal. Molasses	.26
	to dist. with J. Bates Sr.	.82
	to 4 yrds Calicoe 36	1.44
	to 1/2 skein thread	.03
11/7	to 3 1/2 yds. cassimere	6.50
	to 1 3/4 yds. shirting	.52
	to (Wound 7) thread 6	.13
11/8	to 1/2 pk. coarse salt	.17
	to 1/2 yd. Calicoe	.18
	to 4 1/2 ft. wood	.85
11/9	to 7# Rice	.35
	to 2 qts. Molasses	.26
11/10	to 1/4# Tobacco	.08
	to 1 Bu. corn	.67
11/11	to 2# Butter	.40
	to 1 Doz. Crackers	.17
11/14	to 1 Hat for Ebenezar	.83
11/15	to 1 Comb	.06
	to 1/4# Tobacco	.08
11/16	to 1/4# S. Tea	.20
11/17	to 2# Butter	.40
	to 1 sk. D. Brown yarn 12	.10
11/17	to 13# Cheese 5 1/2	.72
11/18	to 6 yds Ian for 1820 to J. Larned	1.57
	to 1 Comb	.10
	to 2 Bus. corn	.34
	to 7# Codfish	.35
11/20	to 1/4# Tobacco	.08
11/21	to 2 qts. molasses	.26

1820—Debtor

11/22	to 1 Bus. Grain	.74
	to 1/4# Ginger	.05
	to 1 1/16# Soap	.17
	to 1/2 Bus. apples	.12
	to 2# Butter	.40
	to 12 10/16# Beef 6	.76
11/24	to beef shank	.33
	to pins	.16
11/25	to 1/4# Tobacco	.08
11/27	to 1 Bus. Grain	.74
	to 1 Comb	.10
	to Salt @ Mr. Cortess	.70
	to 1/2 Gal. Molasses	.26
11/28	to 4 1/2′ Wood 10/	.85
11/29	to 2# Butter	.40
	to 19 1/2# Pork 3 hud	.59
11/30	to 1 Bus. corn	.67
	to 1 3/4# wool	.70
	to an order on E. Howard	1.29

1820—Debtor

	to 12 1/2# Fowls 6 1/2	.81
	to 8 14/16# Sole leather 1/6	2.22
	to 1 spelling book	.22
12/1	to 1/4# Tobacco	.08
12/2	to 1 writing book	.12
	to 12# N. M. Cheese 8 1/2	1.02
	to quills	.03
	to 1 writing book	.12
	to 1/4# H. S. Tea	.22
	to 1/4# Ginger	.05
	to Rent of tenement from Aug. 5 to Dec. 2, 1820, 3 mo. 27 days @ 25$ per annum	8.12
		$150.11

List of Employees (1816)

One of the few glimpses we have inside an early cotton mill is provided by the agent for the Ramapo New York Cotton Factory who drew up a list of his employees, indicating the "prices" paid per week for labor and the age and sex requirements for the various tasks in the mill. This small spinning mill used both "mules" and "throstles" to produce the relatively coarse "Number 20" yarn. The agent, whom we know only by his initials, "J.S.S.," indicated that these fifty-four employees, performing all the operations in a mill with 3,600 spindles, could, in a week, work up 1,917 1/2 pounds of cotton into 38,352 hanks (32,315,680 feet) of this yarn, at an average labor cost of 12 1/2 cents a hank. Some of the wages are computed in shillings (for example, 16/), each shilling being worth 12 1/2 cents. (TZP)

Ramapo Cotton Factory, Estimate of the cost of spinning No. 20 yarn . . . , September 3, 1816, Peirson & Co. Collection, case 6, Baker Library, Harvard Business School.

		$	cts.
1	Carder (the same as 6000 spindles would require)	10	50
1	Boy to strip cards &c.	3.	—
3	Girls to feed cards	6.	—
4	Girls to attend 2 Drawing Frames	8	—
1	Girl to attend the Bk [back] of 2 Roving Frames	2.	—
1	Girl to attend the Front of 2 " "	2.	25
3	Girls to attend the Bk of 3 streching Frames	6	75
3	Strechers	26	—
1	Boy to attend the front of 9 cards	2	—
1	spare hand	2	—
19	Work people in Card Room & amt. of wages $	68	50
14	Throstle Piecers at 16/	28	—
4	Bobbin changers at 16/ (& spare hands)	8	—
1	Master Spinner	9	—
19	Work people in the throstle Room; amt. of wages $	45	—
4	Mule Spinners on 4 Pr. Mules of 276 spindles each at 24/100 [pr.] Hundred Skeins and each man 9 (hks [hanks] pr spindle pr. wk.)	47	69
8	Mule piecers 16/	16	—
12	Work people in the Mule room; amt of wages	63	69
1	woman attending the Picking Machine	3	—
1	Boy to assist to make up yarn	3	—
1	Man as an overseer of the mules & stretchers to make Banding & spin on mules	10	50
1	clerk	7	15
	amt. of Reeling Mule yarn	22	08
	" " " Throstle yarn	17	46
	total amt. of wages $	240	38

Asenath Maria Townsley

Samuel Newell

Death Records (1829, 1823)

Although it is a simple matter to determine the number of men, women, and children working in a manufacturing village by reading census records and other statistical sources, it is a much more difficult task to identify and study particular individuals associated with early factories. The higher the socioeconomic status, the easier it is to learn about the life of a specific person because persons with substantial property leave more behind them in land, tax, and probate records. And men are easier to trace through the historical record than women, for the latter seldom appear on tax lists or census records (before 1850) unless they have been widowed.

Asenath Maria Townsley and Samuel Newell lived and worked in the mill villages of Southbridge, Massachusetts. Townsley was killed June 3, 1829, when she was "caught in the machinery of a factory." Her forebears had joined in settling the town of Brimfield, Massachusetts, in the early 1700s. A century later, when industry came to the nearby town of Southbridge, Asenath Maria, the young daughter of the farming family of Abisha and Eleanor Townsley went to work in the mills. Nothing more is known about Asenath Maria than is carved into her tombstone or written into the vital statistics of the Massachusetts towns of Brimfield and Southbridge. She died at the age of eighteen and probably owned little more than her workclothes, a few books, a silk bonnet, and perhaps a calico dress. Her tombstone stands

The basic information on the lives of Asenath Townsley and Samuel Newell was obtained from the vital statistics of the Massachusetts towns of Sturbridge, Southbridge, and Brimfield. The Sturbridge records were published by the New England Historic Genealogical Society in 1906, the Brimfield records in 1931. Vital statistics for Southbridge have not been published and are available only in manuscript form at the town hall.

in the Brimfield cemetery as solitary testament to the short working life and untimely death of this minor female factory worker.

Six years earlier, Samuel Louis Newell, part-owner and superintendent of the Columbian Manufacturing Company in Southbridge, was killed when he was caught in a belt and carried up through the shafting overhead. At his death, Newell was twenty-nine years old. His one-fifth share in the cotton mill was valued at $3,178, and his personal and business property covered many pages of the Worcester County probate records in contrast to Asenath Maria Townsley's absence in them.

How did someone like Samuel, born in agricultural Sturbridge in 1793 with no particular formal education, become part-owner of the Columbian Manufacturing Company? Very likely kinship was most important. Samuel's uncle Stephen Newell was a partner in the Sturbridge Cotton Company with Moses Fiske and John and Ziba Plimpton; their factory was built on the Quinebaug River in 1812. Just downstream, two other Fiskes and Gershom Plimpton constructed the first mill in Globe Village two years later. Gershom Plimpton's wife was Kezia Fiske, Samuel Newell's aunt. Their son, Moses Plimpton, was one of the partners in the Columbian Manufacturing Company with his cousin Samuel. Although it remains to be determined how these young men were set up in business or how the shares in the company were divided, it is obvious that factory ownership and management in Southbridge was very much a family affair.

Whereas Asenath Maria Townsley left nothing behind other than her tombstone, Samuel Louis Newell left a clear track through the historical record. His probate included a detailed inventory of the goods in the company store as well as in the cotton factory and its machine shop. His personal effects are included here as a measure of the wealth and position of a male head of household and millowner. (TZP)

Inventory of the estate of Samuel L. Newell, deceased, taken by the appraisers Feb. 15, 1823.

Wearing Apparel

1	Castor hat 33/ & 1 water proof do 3/	6.00
1	Blue Broadcloth surtout 24/ & 1 old do 6/	5.00
1	Great coat	2.00
1	Under coat 36/- 1 do worn 9/ & 1 do 6/	8.50
1	pair pantaloons 12/ & 3 pair sattinett do @ 4/	4.00
1	Cotton roundabout & pantaloons	1.50
2	pair woolen drawers & pantaloons	1.50
1	silk vest	.50
1	woolen do @ 6/	2.00
1	White Marseilles do	1.50
1	valencia do	2.00
6	Cotton & 3 linnen shirts @ 6/	9.00
1	Fine shirt & shirtee	2.00
5	pair woolen & 3 pair cotton socks	1.30
2	" woolen mittens & 1 pair Beaver Gloves	.42
9	cravats	2.00
2	pair boots 24/ & 2 pair shoes 6/	5.00

Beds & bedding

1	Bed, Bolster, pillows, underbed, bedstead & cord	20.50
1	do do do do do do	22.50
1	do do do do – –	20.00
1	Canopy Bedstead & cord	6.00
3	pair woolen Blankets	18.00
4	Cotton bed quilts, (viz) 24/ 24/ 15/ 12/	12.50
1	woolen do	4.00
1	cotton comforter	2.50
1	woolen coverlid	5.00
3	pair flannel sheets @ 21/	10.50
3	do cotton 12/ & 8 pair linnen do 15/	26.00
12	pair linnen pillow cases @ 3/	6.00
1	First bed curtains	4.00

Furniture	
1 Bureau	15.00
2 circle tables 16 dolls & 1 pine do 2	18.00
1 cherry & 1 pine candle stand	3.50
1/2 doz. dining chairs 6.00 & 1 doz kitchen chairs 15/	8.50
1 Rocking do & cushion	1.50
9 linnen & cotton table spreads @ 12/	18.00
12 Towels & napkins @ 2/3	4.50
2 pair shovel & tongs 9/ & 7/6	2.75
1 looking glass	1.50
2 pine chests 6/ & 12/	3.00
Bellows & cricket	.34
1 pair brass candlesticks 4/16—1 pair iron do 1/2	.95
1 Brass Kettle	8.00
1 tea kettle 6/ & 2 small dish kettles 6/	2.00
1 pail pot & cover 6/, 1 spider 2/6	1.42
1 Bake pan 6/ & 1 pair sad Iron 5/	1.84
1 Grid Iron 4/6 & 1 Toast Iron 2/	1.09
1 Chicken boiler	.25
5 Treys & Salt mortar	1.00
3 pails, 1 wash bowl, & 1 B.P. Chamber	1.25
1 carpet	15.00
1 doz. B.P. plates 8/	1.34
1 doz B.P. do each 6/- 4/6 & 3/	2.25
2 B.P. oval dishes @ 2/ each & 1 do @ 2/6	1.08
1 B.P. Larett, 15/	2.50
1 pair B.P. butter boats 3/	.50
1 Fancy pint pitcher /9 & 1 Quart Bowl, /9	.25
1 Edgd. nappie 2/ & 1 do oval dish 1/	.50
1/2 doz. Edgd. plates	.25
3 C C oval dishes @ /9^{d}	.38
4 " " nappies @ 1/	.67
1/2 doz. C. C. plates	.25
1 pair pint decanters 2/6 & 1/2 doz. wines 1/6	.67
4 tumblers @ /4 1/2	.25
1 pair vinegar cruits	.17
1/2 doz. Glass Bottles @ 6/	.50
1/2 doz. phials @ 2/	.17

2 jugs 2/ & 2 milk pans 1/2 4 Earthen pots 3/	1.04
2 Silver Table spoons	2.50
1/2 doz. do. Tea spoon	3.00
1/2 doz. Britannia do	.38
2 cases knives & forks	2.00
1 Brittania Tea pot	.75
1 small cannister & creamer	.17
4 Tin Baisins	.25
1 small tin pail & dipper	.25
1 pewter porringer & ladle	.25
1 coffee pot 1/6 & 1 small tin pail & cover 1/6	.50
1 sett sugar boxes	.50
1 chopping knife & tin dipper	.25
1 japanned waiter	1.25
1 Silver watch	7.00
1 Second hand waggon & harness	15.00
2 Meat tubs	2.00
2 Wash tubs	1.50
6 Bushel Potatoes @ 1/6	1.50
1/2 Bbl. Flour	4.50
1 Bbl. Cider	1.00
	$378.51

James Wolcott Jun.
Lement Bacon
Ira Carpenter

One fifth part of the Property belonging to the Columbian Manufacturing Company $3178.03 1/5

Samuel Fiske

Worcester . . . , March 29, 1823. Personally appeared Samuel Fiske, Administrator of the Estate of Samuel L. Newell late of Southbridge decd. and made oath that this is a just and true

Inventory of all the estate of which the said Newell died seized or possessed excepting however, the Estate which is common with the Columbian Manufacturing Company, and that if any other Estate should be by them found belonging to said Estate that he would exhibit a just and true acct. of the same at the Probate Office.

Before me
Nath. Paine. J. Prob.

Photo by Donald F. Eaton.

Pomfret Manufacturing Company

Workers' Contracts (1807–1830)

Although too few company records survive to give a clear indication of how many workers were hired under the provisions of written contracts, some companies are known to have made widespread use of them. Most of the selections here, drawn from the Pomfret Manufacturing Company's Contract and Memorandum books, are memoranda written by Smith Wilkinson, the company's agent, stating the terms under which he had hired people. Occasional references to a "written contract on file" suggest that there were often more formal documents signed by both parties. These selections cover the period from 1807, when the company's first mill was still under construction, to 1830, when it had two mills in operation and was hiring a number of single, female weavers in addition to traditional family labor. Some of Wilkinson's workers remained on the job for a number of years; others left after a short time. (RP)

Pomfret Manufacturing Company Records, Connecticut State Library, Hartford, Contract and Memorandum Books 1 and 2.

March 28, 1807
[John Henry of Woodstock, Conn., to work seven months, beginning April 6] at farming, building wall, tending saw mill, or any out door work as they may want and board himself for nineteen dollars per month. [He is to put four of his children to work in the mill for a year: Sally, age 14, at six shillings per week; Sukey, 12, at five shillings; Fanny, 10, at 4 shillings; Harriet, 7, at 3 shillings, 6 pence. The company was to rent the family two rooms in the Stall House at $20 per year and pay wages] as he may want, when due.

April 3, 1807
Salmon Caswell hath agreed to work for the Pomfret Mg. Compy. one year from date for one hundred and ninety dollars and board himself, and that his oldest son shall work in the Cotton manufactory for four shillings pr week and his second son for three shillings & six pence Pr week for one year. [. . .] He is to Board with us at 5½ dollars pr mo. (except Sundays) untill we get a Tenement done for him to move. . . .

April 16, 1807
Caleb Kelton [a mason] hath agreed pr. advise of O[ziel] Wilkinson to work when we want him for 8/ [8 shillings] Pr Day and be found [have room and board provided] . [. . .] We are to consult him about his work when we want him & how long at a time, so that he may undertake Jobs for others.

May 7, 1807
Sukey Blois, a girl about 14 years old, began to work at the Boarding House and if she suits is to have 3/ pr week. Quit May 10.

June 2, 1807
Resolved Sabin agrees to mix the Paint, Prime the Sash and Glaze the Windows for the Still House [a former distillery that had been converted into a tenement] complete, say about 360 Lights or Pains of Glass at two cents pr Square. . . .

June 13, 1807
Asa White hath agreed to build the wood work of a doubling and

Twisting Frame with two reels complete for Sixteen Dollars & pay his own Board.

November 19, 1807
Agreed with Asa White to allow him nine Dollars for his Brother Salem's Clothes this season in addition to what he has had in full for his services. It is also agreed that the Co. send said Salem to school half the time this winter on their Right in the Town School and employ him in the mill the Rest of the Time—say every other week & on the Saturday of the week that he goes to school, said Asa White paying his board in Proportion of Six Days in fourteen.

December 8, 1807
Agreed to give Zebulon Cady 5/9 [five shillings, 9 pence] pr day for what he works for us by the day this winter from this date. [. . .] Gave him to understand that we should want him to move by the 1st of February next, as we shall want the house room for familys who have children to work in the factory.

February 1, 1808
Abigail Dexter began to Reel by agreement with Mr. Congdon. She is to have as much as others according to what she does. Sally Obrien began to Reel also & is to have by understanding with Mr. Titus Adams 7/6 pr week if she Reels 6 doffs pr day.

February 23, 1808
Rachael Adams of Killingly hath agreed to Remove here with her family and Put the following Children to work in the manufactory, health Permitting, on the Fourth day of March next—

Say, Shubell Adams at 6/ pr week
Samuel Adams — 6/ " "
Laura Adams — 3/ " "
& Geo. Adams — 3/ " "

Making three dollars for the whole, to continue one year from the time they begin and we the P. Mg. Co. hereby let to said Rachael a Tenement in the Cargill House of two Rooms with a Privilige in the Cellar & Garret and a garden spot for twelve dollars. They will keep her a cow at 12 cents pr week in summer & 3/ in Winter, makg $18–20 pr year, also fuel at 11/ & 12/ pr cord as the price may vary. Provisions of all kinds & W[est] I[ndia] Goods at Cash Price.

In Witness of the above written agreement we have hereto sign'd our names.

her
Rachael X Adams
mark

Dexter Tiffany

Smith Wilkinson,
Agent for said Company

February 24, 1808
Be it Remembered that we the Pomfret Mg. Company and Benj. Greene & James Wild have made & Entered into the following agreement—Vizt. Said Company Let to said Greene & Wild the use of their Work Shop, Engines, Lathes, Grindstone & other Tools in the factory (They Doing their Proportion of Keeping them in Repair), also the use of the Brass furnace (they to keep it in Repair) and the Blacksmith Shop to Repair & fix Tools, all for the Consideration of fifty cents pr Day, and when they have a man to Work in the Blacksmith Shop, forging machinery, they are to allow   cents pr day for the use of Shop & Tools.

The above agreement to continue untill said Greene & Wild complete a Job of four Throssles of 66 spindles Each, & the Preparation, say 1 Roping, a Sett of Drawing and five Carding Machines.

And said Greene is to Work for said Co. when Required in Setting & Grinding Cards & other Jobs which their hands do not understand, and Oversee & give Directions to their hands for Two Dollars pr day, and said Wild is to spin on the mules at the Old Contract, when we can furnish Roping, for four Days and is to have Two Piecers. But said Company have the Liberty of Hiring a Person to spin on the mules when they please.

March 1, 1808
Be it Remembered that I, Stephen Chapman of Gloucester (R.I.) Have this day agreed and by these Presents do agree to move my

family to the Manufactory & Put four of my children at Work in the same, of the following ages & Prices—

Vizt.,	Phebe Chapman,	at 6/9	pr Week —	12	years old
	Thomas Chapman	" 6/6	do. [ditto]	15	do.
	Stephen Chapman	" 6/	do.	10	do.
	Sally Chapman	" 3/	do.	6	do.

for the full term of one year from the day they may begin.

And further that I will work for said Co. at carpentering when they want for one dollar pr day & find myself.

And the Pomfret Mg. Co. on their Part agree to employ his children as aforesaid and him one half the year at the rate mentioned (having the Preference in his services, I.E. he is to work for us when we have business in preference to any other Employ he may find during the year) and also to Let him the Tenement where Mr. Cady now lives, consisting of Three Rooms with a Privilege in the Cellar for Twenty dollars Pr. year.

And said Co. will keep his cow both summer & winter as they do for others, the Present Prices being 3/ pr week winters and twenty cents pr week for Pasturing—Will furnish him fire wood at not exceeding Two Dollars pr Cord, when call'd on, He to buy of others when he can at a fair price, and will allow him when is in our Employ the same drink that we do others.

It is now understood that said Chapman is to move on or before the 20th of March, weather permitting, and that said Co. are to build an oven in the Tenement they have Let to him, and also that the usual Quantity of [apple] sauce ground go with the house at the aforesaid Rent of Twenty dollars pr year.

March 5, 1808

I Benajiah Matthews have agreed . . . to serve the Pomfret Manufacturing Company as an Overseer in the Spinning Room and Doing up yarn (being the usual duty of that station in addition to Governing the Children) for the Term of two years from the first Day of March Present, for and at the Rate of Two hundred and Twenty dollars pr year and find myself.

And said Co. by their agent agree to pay said Matthews the wages aforesaid as he may Want from time to Time, and Let him the same Tenement & Privileges he occupied the year past at Seventeen dollars pr year, and allow him 10/6 pr Week for board, Lodging, Washing & mending and the use of a bed he has with the room it now stands in, untill we want the Room for another family.

I Henry Greene have contracted and agreed, and by thse Presents do agree, to Serve the Pomfret Mg. Company as an Overseer or Superintendent of the Carding Room in their factory for the Term of Two years from the day my Last Agreement will be out, for the following Wages, say for the first year six shillings & six pence pr day, and six shillings and nine pence pr day for the second year, and said Company by their agent agree to pay said Greene the Prices above stated for every day he shall serve in their business, and also agree to Let him the Tenement he now Lives in for Twenty dollars pr year, and to keep him a Cow for Eighteen Dollars and forty cents pr year. Other things to be Regulated by the Current prices as heretofore.

March 22, 1808
Isaac Russell hath agreed to work for us for one hundred & Eighty dollars pr year and Put Three children into the factory, say Abigail at 6/, Sally at 4/6 & Francis at 4/ pr week Each, making $305.67 [per year] for the whole. And we have agreed to let him the Two Rooms where N. Wade mov'd out & the small unplaistered Room in the Garret with [a] Garden at sixteen dollars pr year, also to assist him in buying a cow, to pasture her for 20 cts pr week &c & the balance in cash at the years End. He is to come on the 6th of April next.

April 9, 1808
Mr. John Henry Hath agreed to Work one year on the farm & out-doors work as usual for two hundred & Ten Dollars, or Seven Months for Twenty dollars pr Mo., and to put his four Girls into the factory at the following Prices,

Say, Jane to Reel by the Doff, Sally at seven shillings Pr week, Sukey at six shillings & six pence per week, and Fanny at four shillings & six pence pr Week.

His House Rent to be same as Last year, with a garden spot, other things at usual Prices as Heretofore. . . .

January 4, 1810

This agreement made & concluded by & between Malachi Greene of Providence, state of Rhode Island on one part and Smith Wilkinson of Pomfret as agent of the Pomfret Mg. Company on the other part, Witnesseth—That said Wilkinson . . . doth Lett unto said Greene the dwelling house, shop, shed, Privilege in the Barn and garden, as lately occupied by Mr. Nathl. Pearce, for the term of one year, from the fifteenth Instant, for one hundred and ten dollars, and will furnish him with Boarders—from six to ten, as may be convenient in their business, and will [provide?] beds sufficient to lodge them . . . and will pay said Greene one dollar and seventy five cents pr week for boarding, lodging & washing in a decent & suitable manner, and will further pay him eight cents pr week for such as he shall mend for. . . . Said Wilkinson will pay the taxes on said Tavern, and will not sell Liquor or oats to travelers, neighbors or townspeople by small measures, to Injure said Tavern, and further agree to lett said Greene the Privilege of Pasturing Travelers' Horses or Cattle for one half what he shall take therefor, an accurate account to be kept by him. And Malachi Greene . . . also agrees to put two children into the factory, on the same terms as they give others of the same ages, sexes & station to work steady during said time . . . and said Greene doth further agree to take of us his West India & Dry goods, grain & other farm produce, as we have to sell, (if as low as he can buy of others) and in preference to buying of others, and also agrees to pay the expence of a licence for said House. And it is further agreed by said Parties that the board for the Partners of the Co. when here shall be two shillings pr day, and extras in addition, transient custom when the Co. agrees to pay [shall be] twenty cents pr meal. . . .

March 3, 1810
James Arnold, son of Moses, came this noon on trial & if he will do, his father agrees that he shall serve until he is of age, say a year from Octr. next at eight dollars pr. month.

March 5, 1810
Mr. Willard Arnold hath quit, having served 2 apprenticeships, one of 2 years & one of a year in the short space of seven months, besides getting married, &c. Settled with him in full, Amen.

March 17, 1810
I hereby agree to work for the P. Mg. Co at [?] and doing up yarn & other work in the factory, ten months from the first of April next for seven dollar pr [month?] being boarded.

Shubel Adams

June 22, 1810
A widow, alias single woman, Daughter to Isaiah Brown in the edge of Foster, Rhode Island, formerly wife to Resolved Walling, moved into east end of Still House this day & occupies one room, at one shilling per week rent, and puts 3 children at work in the factory, say Alice at 6/, Amy 5/6, and Anna 4/6, makg 16/ pr week for the whole. They began 19th Instant.

August 23, 1810
Mr. Coady of Dudley hath begun on trial at forging machinery & calls himself a Workman at turning, filing &c, and is to work on trial a week at 6/ pr day and longer if he suits & we can agree.

October 12, 1810
A Mr. Asa Todd or some name I have forgot was here from Chaplin and talks of moving here with his family. I showed him the Tenement when J. Lawrence lived last, informed him what the Rent would be, & cow keepg &c., & told him what I would give him for 4 children, 1—7 yrs old, 1—9 yrs old, 1—10, & 1—13: Say 17/6 for the whole. If he concludes to come will come next week. Also agreed to employ 1 girl at reeling.

January 28, 1812
Jesse Daggett & Andrew Leavens have agreed to serve as a watch for the safety of the factory & Village against fire &c &c. To

commence at 8 o'clock in the evening & patrole from the Hill near Joseph Heaths over to the old barn and into every Room in the factory, once in a half hour. One is to serve from 8 till one oclock & the other from one till six in the morning, and are to give the usual watch word. They are to have 2/6 each pr. night.

March 24, 1812
I hereby agree to work for Pomfret Mg. Co. one year at bundling yarn and other Business connected with finishing cotton goods ready for market as required by the Agent of said Co., for nine dollars Pr. month, being found, time to commence March 17th past, the day on which I began, and am to take half my pay in clothing at fair prices & the other half in cash, as witness my hand, Henry P. Pierce Green. . . .
Settled March 17th, 1813.

December 14, 1812
Mr. Congdon agreed to send his Daughter Catharine into the factory to Work after breakfast. Price Proposed by me was 7/ pr week, or as much as we allow others of her size.

March 15, 1813
Mr. Ephraim Rice came this morn and began work on Trial as a Blacksmith. Began board at Timo. Lamsons at Dinner. . . . Settled with said Rice March 27th. He did not suit.

March 15, 1813
Miss Mary M. Munroe, Niece to Cyrus Gorton, began as a Reeler and boards at Thos. Richmonds. . . . She is to have usual Prices & pay pr doff. and is to Pay her own board.

March 22, 1813
Mr. Stephen Stone, son of John Stone, came this day & began in factory store, on a Bargain for Two years, & is to work at doing up yarn, bundling & Papering, & helpg Collendar & do up Goods, make out & Put out webbs & Take in Clothes & all Parts of the Business of said Store, being found. Is to have 8 dolls pr mo. the first and Ten Dollars pr mo. the Second year. The bargain was Talked over between his father & myself and also with said Stephen.

May 8, 1813

Luther Bates hath agreed to Lathe & Plaister the House we are now Building on the hill near the Mill Pond and Lay all the Hearths complete & find his own Grog . . . and being boarded & tended, from the 1st floor to the Roof, Including stair ways, clossetts, & cupboards, for thirty-six dollars, Payable same as the rest of our contract. Vizt., half cash and half in our way of Trade.

December 12, 1814

Mr. Leml. H. Elliott of Woodstock began the winter school for District no. 14 this morn & is to Teach 3 months after the Rate of 13 dollars pr Mo. Boarded with S. Wilkinson this day. Ephm. Congdon has agreed to board him the Quarter at 12/ pr week including Washing. He is to begin tomorrow morn.

September 11, 1815

Saml Hernton Junr. hath agreed to work one year at mule spinning for twenty dollars pr month, being boarded or allowed two dollars pr week therefor and is to take the mule now used by Joseph in the garret and spin twenty two hundred pr week for six months and twenty-four-hundred pr week the six successive months beginning March next.

March 25, 1816

Be it remembered that Amos Hammond hath agreed . . . to work for Pomfret Mfg Co. seven months from the 1st day of April next, on their Bundy farm under Wm Martin, being boarded &c for fourteen dollars & 25 cents pr month and boarding self on Sundays & do my own washing & mending and find my own tools, as follows, say, scythe, rake, hoe, shovel & axe—and also to husk seven evenings during the season of corn harvest. . . .

August 26, 1816

[Agreement with Green Capron] We hereby agree to consider the old contract at an end, and said Capron agrees to spin at two shillings pr hundred & find himself (being found a piecer), and to take pay in such articles as he wants for his family, & for the rest, such pay as will pay his debts. This agreement to run six months if the mill continues to run, but if [it] stops the contract at an end.

[The company, like many New England manufacturers, was in financial trouble in this postwar period. A number of workers were laid off.—RP]

March 4, 1819
Mr. Daniel Sprague of Thompson, son of Jona Sprague, Decd., applied for Employ at the Blacksmith business and thought he ought to have 20 Dolls pr mo by the year, finding himself. We gave him encouragement of employ, provided he answered our purpose, and with the view he agrees to come on trial, 8 or 10 days, about the 25th Instant. March 29th, said Sprague came on trial about 10 o'clock. April 10, settled with & discharged him, Judging he would not answer our wishes as to workmanship.

November 7, 1821
Mr. Eden Leavens agreed to board his son John Leavens at work in Co. service for 28 cents pr day, to be reckoned on the time, on actual days service which is equal to $1.68 pr week. If he loses any time is to pay his father for all board when not in Co. service.

May 1, 1823
Capt. Eleazer Sabin agreed to work for P. Mg. Co. at digging stone at W. Cutlers ledge & find himself, board & drink at one dollar pr day, co. to find tools. Said he would work 5 days in each week as long as we wanted him.

March 15, 1824
I hereby agree to serve Pomfret Mg. Co at mason work on their new factory to begin May 3d & work till the stone work is done at six shillings pr day being found. Pardon Bennet.

July 2, 1824
Mr. Rufus Davison hath agreed to do all the forging for 4 wide Power Looms now begun (agreeable to the work of 6 looms just finished) & find himself for seven dollars each to be credited in a/c payable same as his contract, say half cash & half goods.

October 3, 1824
Be it remembered that I Joseph Bartholick of Coventry, R.I., have agreed to go to work for Pomfret Mg. Co. for two weeks on trial

& if I suit & am pleased with my employ, board &c am to serve them one year from date at mule spinning or in the factory, machine shop or store at such business as they may have to do for twelve dollars pr month, being found &c . . . and will spin 2800 skeins pr week and have 20 cts pr hundred for all I spin over & will deduct after the same rate for what I fall short of that.

October 18, 1824
Miss Lucy Green began work for the Pomfret Mg. Co. the 7 Instant at weaving on water looms, to have the same price as other weavers & begin boarding with Augustus Howe at noon.

November 4, 1824
This day Nancy Torrey began work for the Pomfret Mg. Co. at weaving on water looms and is to Reel or do any other work in the weaving shop that is wanted. Is to have the price pr week when employed at Reeling or spooling that she averages on the looms. Pay to be the same as others, 1/4 cash. Began boarding with Rufus Davis on this morning.

December 1, 1824
Miss Martha Pettiface of Burrillville agreed to come on Tuesday next to weave on water looms at our usual terms, say 3/4 of a cent pr yd. payable one-quarter cash & residue in board & goods. Miss Amy Anne Matthewson of Burrillville agreed to come in a week from Monday next to weave on our usual terms as above & is to work at [the] spooler or beam or other work untill looms are ready to start.

February 9, 1826
Mr. Martic Pierce recommended by Mr. George Danielson experience mule spinner and a person of exemplary moral character applied for employ . . . agreed to come when we get our new mill ready to start and run one of our mules for one year at 18 cents per hundred and find himself and pay his own piecer and spin shuttle cops. Either party to be at liberty to quit by giving 2 weeks notice.

February 13, 1826
Miss Eliza Warren came this day to work in weaving room—is to weave at the usual price. Went to board at E. Congdon's at noon. Quit February 25 without leave or consultation.

March 11, 1826
Mr. Ephraim Congdon has agreed to move into the tenement where T. Chapman lives, rent $40 and put his son Welcome to work in the mill and take 7 or 8 girls to board at 6/6 per week and his wife to make harnisses as in years past. To pay as heretofore.

July 30, 1828
Mr John Kimball agreed to learn to tend a dresser in new mill. The company are to pay his board and the girl that runs the dresser until he learns so as to be able to run it alone and then he is to have 3 cents per cut and board himself for one year. Worked 8 1/2 days till August 13 at noon and quite. We are to pay Sayles for his board.

October 3, 1828
Chloe Bugbee agrees to recommence work at weaving in the old mill tomorrow morning and her sister Rhoda Bugbee agrees to commence at the same time to weave in the old mill. Both on the customary terms. Board at Thomas Richmonds'. Began as agreed. Rhoda chooses to weave 2 wks by the wk at 12/—then by the yard.

November 15, 1828
Miss Louisa Packard agreed to go to weaving in the new mill week after visit on usual terms, say 7 mills pr yard payable in board, 1/3 cash & residue in goods. To work by the yard from commencement & to stay if agreeable on six months at least, extras excepted.

November 17, 1828
I Wm. A Brown of Killingly hereby agree to serve Pomfret mg co as assistant in the weaving room of their new mill under Mr. Stephen Comer for the term of three months to fill Mr. David Comer's place for his board, lodging and usual mending & to enter on said duties in two weeks from date, health permitting.

April 15, 1829

I Lucy Pooler hereby agree to tend spooler for P Mg Co in new mill at 12/ pr week & to pay 7/ pr week for my board & to take what goods I want & to have cash for balance to work til April 1, 1830, sickness only excepted. To come on Monday next, to be sent for by co.

her
Lucy X Pooler
mark

September 9, 1829

Augustus Buck, mule spinner, came this day & commenced work at spinning on 2 mules in new mill. See contract on file. Went to board at A. Elliott at noon.

October 6, 1829

James Young son of widow Patience began in weaving room this morning as assistant under Wm Smith to bring wood, water, trim and fill Lamps & any work in & connected with said business. To have $11 pr mo incl. board. . . . Settled Nov. 9.

December 23, 1829

Agreement made & resolved following day: Betsey Barrett allowed 1/6 extra to do her work well & dispense with a seat, is to have pay for her extra work Saturday night.

Smith Wilkinson

Contract and Memorandum Books (1824)

While we do not have any figures on the mill N. B. Gordon managed in Mansfield, Massachusetts, it was undoubtedly smaller than Smith Wilkinson's Pomfret Manufacturing Company, which by 1831 operated 3,000 spindles and had some 150 employees. (An average company of the time operated about 1,000 spindles.) Wilkinson probably also had more people to whom he could delegate responsibilities, and he also had a more complex operation, including a tavern and a company-owned farm. His "Contract and Memorandum" books show him involved in a greater variety of activities than Gordon was.

The period covered here is from April through September 1824. April was always a busy month, with many employee contracts expiring, some families moving away, and others moving into more desirable tenements that had become vacant. During this particular period the company also was building a new stone mill, and Wilkinson here makes notations concerning the hiring of masons and the making of contracts to dig stone and have new machinery built. As always, there was also business relating to the farm and company store. Most of the entries here relate to activities outside of the mill, and we can only assume that Wilkinson must have involved himself to some extent with technical and production problems, as did Gordon. But it is obvious that he also spent much of his time tending to other kinds of work. (RP)

Pomfret Manufacturing Company Records, Connecticut State Library, Hartford, Contract and Memorandum Book 2.

April 1

Widow Selinda Phipps removed from this place to Ashford factory on date. Had no cow.

Mr. Alpheus Chaffee moved into the tenement left by William Andrews this day.

Mr. B. Mathers moved from this place onto his farm in Killingly on date and drove away his cow.

Mr. Thomas Dyke moved into the tenement in house #6 left by Armin Bolles this day and put a cow in our keeping.

Mr. David Hall moved into the tenement left by Wid. S. Phipps this day.

Amos Hammond moved into the tenement in Harris House left by Chad Carey.

Mr. William Leach moved into the tenement in Stall House last occupied by Alpheus Chaffee on date.

April 5

Mr. John W. Field moved from this place on date into John Spencer formerly the Olney Place a little south of E. Gleasons. Has not had a cow since he came here.

Mr. Cyrus Burlingame moved into the tenement last occupied by Amos Hammond. Rent 30 dollars per year.

Mr. Leonard Bugbee moved into the tenement left by B. Mathers on date.

April 6

Mr. Ephraim Congdon hath agreed and doth hereby agree to work for Pomfret Mg. Co., the year ensuing and find himself for four shillings per day and make fair weather and to tend the water on the meadow and the rest of the time to work at carpentering or tend the sawmill or at farming as said company may wish me to do and receive pay as verbally agreed with S. Wilkinsons, agent. . . .

Mr. Jedidiah Leavens moved into the tenement last occupied by John W. Field in Blanchard House on date, and put a cow in our keeping.

Mr. Johathan Clough moved into the tenement left by David Hall in Stall House and is to pay twenty dollars per year rent.

April 8

Mr. John M. Sabin on the Aqueduct ground and . . . agreed to put up all the breaches made in his wall and fences in laying our aqueduct across the Sabin farm, being 5 places and also to pick and cart off all the stones left on his land to his own liking as an offset for all the damage done by him in carting wood across our lot on the plain no. of the red barn the winter past, I made the offer and he accepted it. I then set B. Warren and 3 hands to mend and finish the covering of the aqueduct and to seed the land over the same, on the meadow next to the river and the pasture east of the road, the residue being plowland did not require it—all was done to Sabin's satisfaction.

Sent D. Chaffee to put up the fence adjoining Joseph Wheaton's land over which our hands took down the winter past to fall the trees which stood near the line. S. A. Stone helped do it.

April 15

I hereby agree to serve Pomfret Mg. Co. at mason work on their new stone factory the present season from May 3rd next until the stone work of said building is done, being found, and find my own tools, said company keeping them in repair as last season, for seven shillings per day. Also that my brother Levi shall serve them for the same term and wages health permitting. Nathan Wood. settled with Sept. 8th, 1824.

April 17

Memo of families who have moved on the place, and families who have removed into other tenements on the place this spring whose tenements were whitewashed. Viz.

Amos Hammond in tenement left by C. Cary					
Penuel Weld	"	"	"	"	J. Heath
Andrew Willard		"	"	"	L. Lawrence
Alpheus Chaffee		"	"	"	W. Andrews
Wm. Leach	"	"	"	"	A. Chaffee
Jonathan Clough (white washed himself)		"	"	"	D. Hall
Perez Park	"	"	"	"	D. Starr
Thomas Dike	"	"	"	"	A. Bolles
Arther Tripp (white washed by company)	"	"	"	"	Wd. [widow] Stanley
Cyrus Burlingame		"	"	"	A. Hammond
Jedidiah Leavens		"	"	"	J.W. Field
David Hall (white washed himself)		"	"	"	Wd. S. Phipps
Charles Buck (new—not white washed)					
Armin Bolles in tenement left by T. W. Chapman					
T. W. Chapman	"	"	"	"	Wd. Cleveland
Benjamin Warren		"	"	"	N.K. Aldrich
Leonard Bugbee (white washed by company)		"	"	"	B. Mathers

Jacob Mann's rent and cow keep settled to date. Settled May 8th.

April 19
Miss Sukey Bishop began the day school in this village on date to keep at $1 per week. Hired by Mrs. Converses committee.

April 20
Examined and marked the gardens for the newcomers this day as follows—B. Warren takes that left by N. K. Aldrich and Mrs. Nancy Stanley that left by O. Greenhill, Leonard Bugbee that left by B. Mathers, E. Congdon that last occupied by Asa White and Armin Bolles that which T. Chapman occupied last year and all

the rest on Pomfret side the same as last year. 11 gardens in all. And on east side—P. Park takes Starr's, C. Burlingame overtakes Hammond's, A. Tripp takes late S. Stanley. Thomas Dyke takes Mrs. S. Phipps. A. Hammond takes C. Cary's. P. Weld takes Joseph Heath's. A. Willard takes L. Lawrence's. A. Chaffee takes W. Andrews. Jona Clough takes D. Hall's. Wm. Leach takes A. Chaffee's. Jedidiah Leavens takes J.W. Field's. David Hall takes Selinda Phipps. C. Buck a new garden. Thomas Chapman takes that left by Widow Mary Cleveland. All the other families same as they occupied last year. 27 gardens in all on the Thompson side. Vizt. 7 in nine partners, 10 in barn lot, 2 west of new barn, 2 by the Cady house, and 6 in Rock field. . . .

April 21
This day Benjamin Warren took the bed and bedding from the retail store, leaving the bedstead and cord, and J. Leavens put a bed &c in the store and let Warren have a bedstead for which the company found a cord.

April 27
Memo. Went to Killingly yesterday to view the stone on the Wilson's and Isaac Ballards land and ageed with Mes^{rs} Aaron and Atwell Wilsons to board, lodge and work for four hands while digging stone on their or Ballard's land for 10/per week cash and for liberty to dig stone anywhere in their woodland for 20¢ per cord, but if they prove very hard to dig and poor are to allow but 1/per cord or load. Then went to Isaac Ballard's and agreed with him for the priveledge to dig stone at any of his ledges in the woodlands at the same as we agreed to allow the Wilsons . . . and to allow him for what we dug last year in his pastures 25¢ per load, but are not to dig any more there on a/c of damaging his pasture and mowing. Charles Buck was present and heard the above bargains.

May 9
The cows were turned out on date.

May 10

This day agreed with Capt. Noah Sabin for 4-40 feet sticks for factory beams to be sawed 10 × 9 to be delivered at the saw mill in all this month—at 151 each—to be credited in a/c. Also went to Calvin Warren's and picked out 7 trees standing for 36 feet beams and 2 do. for 40 feet do, (1 of these is felled), which he agrees to cut, trim, and cut off, say 37 and 41 feet and we are to draw them and allow him 11/ each for the 36s and 12/ each for the 40s, to be cut by the 24th inst.

May 11

Mr. William Hopkins of Plainfield wishes to furnish us with one of his patent whipping machines, will bring it here and start it and warrant it to perform well for 50 dollars. I have agreed to view one erected by him at the Swamp Factory and if I think it worth buing am to write him as soon as may be conveinient. . . . July 10 agreed to have him bring one. . . .

May 24

Widow Betsy Cundall's a/c was settled up to 22nd inst. She is now to have weekly duebill.

Mr. William Chamberlain, near the Hearthstone Ledge west [of] Quosset agreed to get out 32 corner stones ready to trim as per memo. Gave him ten cents per square foot area. . . .

May 28

Bought of Mr. John Brown of Thompson a bay of good hay in his barn—12 feet square by 15 feet high, estimated at 5 tons for forty seven dollars and fifty cents—payable in cash next week, to take the hay as it lies and if he helps load it we are to pay him for the firm.

Mr. Webb Cutler agreed to let us dig what stone we can find at his ledges for 1/per load. Also agreed with Stephen A. Stone some weeks since to board our hands at 9/9 per week cash while digging stone at Cutler's.

June 1
Master Henrey Wilkinson, son of Mr. William Wilkinson began to board with Smith Wilkinson on date and went into company's store to learn to be useful if he shall prove steady and active.

June 9
The brindled ox of the yoke bought of Willard Arnold in April 1823 died last night being overheated and by Mr. Hammond in drawing load stone from Wilsons. The weather was extreme hot and the ox appeared to fail and lose his appetite in the morning.

Settled up with Simon Davies Leavens this morning in full. See a/c at retail store.

June 10
Bought of B. Warren a lot on quantity of hay in the leanto of the barn on the farm he used to live on, estimated by him at five tons for thirty dollars and also agreed to allow him for carting said hay 10/6 per ton and to call the lot 5 1/2 tons . . . .57/9 . . .

June 16
Harvey Chamberlain began to work this day at getting out stone at Cutler's and is to have four shillings and three pence per day being found and make fair weather.

June 21
William Alton sold Pomfret Mg. Co. this day all the stone in a half wall in his pasture a little east of Webb Cutler's which will answer their purpose for 2/per load or cord. Also agreed to let us dig what stone we can find . . . in said pasture at 20¢ per load.

Mr. Calvin Warren agreed to relinquish the contract for 2 beams made May 10th, being for 7-36's, of which we have only recd. 5. The 2 remaining not proving very good, we concluded to get them of Mr. J. Perrin.

June 25
This day swapped the yoke of oxen we had of B. Warren last April called 6 years old, to Eleazer Keith for a yoke of red oxen, also

called 6 years old—and some and paid him eight dollars for both. Bought a large dun-colored ox of Benjamin Brown of Chestnut Hill on the 10th inst. to replace the ox which died on the 9th inst. and gave 30 dollars for him. This company now own 8 yoke oxen.

July 1

Albigence Warren agreed to begin work this morning to dig stones at Wilsons ledge with our hands at 5/6 or 6/per day, being found. I agreed if he worked faithfull to allow him the last named price.

Agreed to allow N. Profit five shillings per day and find him to work at haying this season if he sticks steady and makes fair weather.

July 2

Mr. Rufus Davison hath agreed to do all the forging for 4 wide Power looms now begun (agreeable to the work of 6 looms just finished) and find himself for seven dollars each to be credited in a/c payable same as his contract, say half cash and half goods. Began yesterday noon.

Settled George Glasko's account to 1st inst. including services, house rent and cow keeping. Agreed with him to work at haying and cradling the season and make fair weather at $1.20 per day and find himself board and drink.

July 22

Francis Pierce began to work at haying this morning and is to have two shillings per day in addition to his yearly wages. J. H. Morris is to tend mule at 4/per day while Pierce works out.

August 4

Contracted to receive of the firm of Henry and Cary, card makers of Brookfield, 2 sets machine cards, vizt, 1 breaker of no. 32 wire and 1 finisher of no. 33 wire at 35 dollars per set, to be done in about 60 days and to be sent to Thompson—payable half in cotton goods at our cash prices and half in cash on receipt of the cards.

Agreed to receive of Leffingwell—Pedler—120 to 150 lbs. good geese feathers at 55¢ and pay him in cotton goods at fair prices, say sheetings 16¢ and plaids 17¢ &c &c.

August 16
Settled with George Glasko in full for his services to date and agreed to allow him 5/per day for his services in this month, he boarding himself and 4/6 per day for his services in Sept. with our hands in drink.

August 17
Turned out the breachy Keith ox and Scurvy Sabin ox to fat in lower meadow yesterday.

August 21
Sold Phillip Dean the Brown Arnold ox mate to the one that died being lame for four and half hundred good beef payable when said ox is well fatted and one dollar in cash. . . .

August 24
Zera Clough son of David began work this morning on trial. Price to be agreed on after he has worked 5 days. Boards with Jona Clough. Talked of allowing him 7 dollars for board.

September 7
Daniel F. Ellis was taken sick on the 2nd inst. and was carried this morning to Mr. Sparrow's in Killingly Boarded with Morris since he left work 5 1/3 days—chd. at settlement of account March 31, 1825 with 2 days more in October, say 7 1/2 cents.

September 9
. . . this day finished the outside walls on mason's work of the new stone factory, of which the foundation was laid June 9, 1823.

September 11
Settled with David Chaffee for tending mason 4 Mos. @ 9/ (per agreement on file), gave him a duebill.

September 13

Richard Gay of Thompson has agreed to get the glass for our new factory about 300 squares. We are to furnish the materials for making putty, say oil, whiting, and old tin, which he is to receive here, with the sash and glass, and return the windows done in a good workman like manner, the glass to be well cut in or beded for 1 1/2 cents per square. 1/4 cash and residue in goods, or 1 1/4 cents per square all cash at his option. N. B. The glass not to be beded in putty.

Cyrus Burlingame has agreed to work what time we want him from now until winter for seven shillings and nine pence per day and board himself.

September 14

Asa Gary hath engaged to us a good, fat cow. To be deld in one month, at 2 1/per hundred, payable half cash and half store pay. . . .

September 20

Memo. This day turned out the other ox had of Eleazer Keith and Mr. Sabin's stag on the lower meadow to fat, being mates of the two turned out Aug. 16, they were bled.

September 21

Settled with Jonathan Clough for boarding Zera to the 11th inst. 18 days @ 7 dollars per month. See memo of 24th ult—gave him due bill with 3 weeks rent a/c to the 25th inst. Zera began on the 11th to board at Thomas Dike's and stayed until the 20th when he went to Cyrus Burlingame's. Zera lost while boarding with Mr. Dike 2 3/8 days by sickness for which board is to be charged him. See time book Sept. 15-16-&17. charged Zera the above board at retail store Dec. 28, 1824.

September 27

Agreed to set Rufus Spink to work at such business as we have to do on trial, to employ him at small wages until we start our new looms and then to employ him in the weaving room at what may appear to be fair wages according to his ability. . . . Oct. 3 came and boarded at B. Warren's.

September 29
William H. Wilkinson went home this day after dinner in company with Jedediah Leavens, who went on company business. He returned Oct. 1 at night.

Factory Rules and Regulations (n.d., 1843)

The fact that owners and managers of textile mills published formal work rules at all is even more interesting than the actual content of those rules. Handwritten notices, of which an unusual example survives in the papers of Zachariah Allen, were superseded by printed ones, which made even more clear that the relationship between the workers and management was not personal but a formal contractual arrangement in which disembodied and mathematically exact rules governed all. Proprietors, however, were not of one mind what the precise regulations should be, as a comparison of the two sets included in this collection indicate. Selections from the diary of N. B. Gordon, which are reprinted in this collection, indicate that punctuality and absenteeism remained intractable problems for management and suggest that the appearance of formal work rules should not be interpreted as hard evidence that workers necessarily obeyed them. (MBF)

The manuscript "Rules" of an unidentified mill and the "Regulations" of the Columbian Manufacturing Company are both found in the Zachariah Allen Papers, Rhode Island Historical Society.

RULES

From 20th Sept[r] to 20th March the Wheel Starts at Sunrise, Stops at half past Seven O Clock—One half hour is allowed for Breakfast—Three Quarters for Dinner—From 20th March to 20th Sept[r] Starts at Sunrise & Stops at Sunset—One half hour is allowed for breakfast—One Hour for Dinner—It is expected every person will be at their places ready to work at the Starting of the Wheel, which will be at the above mentioned hours—Any Person being absent 15 Minutes after the Wheel is Started will have ¼ of a day deducted from their wages—

No person will leave the room under any pretence without permission from the Overseer

The Overseer will pay particular attention to the hands under his charge, that they perform the Work allot[d] them—He will not permit them to leave the room without sufficient reason—Nor allow any Person not employed in the Establishment to Visit his room without permission—

REGULATIONS

TO BE OBSERVED BY ALL PERSONS EMPLOYED BY THE

COLUMBIAN MANUFACTURING COMPANY,

IN THEIR FACTORY.

Every Overseer is required to be punctual himself, and to see that those employed under him are so.

The Overseers may at their discretion, grant leave of absence to those employed under them, when there is sufficient help in the room to run all the machinery; but when short of this, they are not to grant leave of absence except in cases of absolute necessity.

All persons are required to observe the regulations of the room in which they are employed.—They are not allowed to be absent from their work without the consent of their overseer, except in case of sickness, and then they are required to send him word of the cause of their absence.

All persons while in the employment of the company are requested to attend public worship on the Sabbath, and are required to conduct themselves in a respectable and becoming manner, both in and out of the Factory.

All persons intending to leave the employment of the company, are required to give notice of the same to their overseer, at least two weeks previous to the time of leaving.

The above regulations are considered as a part of the contract with all persons entering the employment of the Columbian Manufacturing Company.

S. Smith ~~J. B. BACON~~, Agent.

Mason Village, ~~May 1, 1840.~~ (71.4.) June 13. 1843

Hand loom weaver, early nineteenth century. London, Tibart and Company, 1804. Slater Mill Historic Site.

Weaving Instructions (1815)

The expansion of waterpowered spinning mills, particularly in the years from 1807 to 1815, fostered the growth of hand weaving. (See the Albert Gallatin documents in this volume for statistics on weaving during this period.) Most of this weaving was done by farm families as a part-time trade, though a small number of full-time weavers worked on the finer and more complex fabrics. Spinning mills frequently "put-out" (gave on consignment) yarn to individual weavers who worked in their own homes at their own pace. Merchant weavers, who served as intermediaries between the mills and individual weavers, became important during the years of embargo and war (1807 to 1815), though some mill owners continued to deal directly with their weavers. Both mill owners and merchant weavers customarily transmitted detailed directions specifying the pattern to be woven and the type of cloth to be produced. Such directions eliminated a basic and traditional element of choice, as the weaver's independence was eroded by the dictates of mill owners and merchant weavers.

New England handloom weavers, however, did not respond to assaults upon their skills and prerogatives in the same fashion as their English counterparts. Because most American weavers were only part-time workers, they were only slightly affected by the fluctuating world textile market and never experienced the full plight of the English hand weavers, who were driven to the edges of destitution by declining prices for woven goods. Relatively few in number, American

The source of the Greenville poem is the Mowry Papers, Rhode Island Historical Society. It is undated, but was probably written around 1815 by "Rambler." Greenville is a town in north central Rhode Island. The weaving ticket is from the Zachariah Allen Papers, Rhode Island Historical Society.

weavers did not mount the spirited and large-scale resistance to degradation that marked the behavior of their English counterparts. American weavers, however, did register resistance in quiet and subtle ways.

The following lines, written by an anonymous Rhode Island weaver-poet, poke fun at the pretensions of the merchant weavers. At the same time, the language of the poem marks out a major dimension of historic change. The title, "The Price of Weavers in Greenville," suggests that the poet had grasped the fact that human labor had become just another commodity to be bought and sold on the market, but his language lags behind this new reality. The term "price" is, of course, the traditional term referring not to the cost of labor but to the cost of labor's product—in this case woolen cloth. The poem begins:

All ye that have woolen weavers for sale,
Can sell them in Greenville wholesale or retail;
That great man of traffic, that deals in the fleece,
Will readily buy them at nine-pence apiece.

A famed speculator is this Mr. Steere,
When wool is so cheap, and cloth is so dear,
And knowing that laborers always were geese,
He thinks he can buy them at nine-pence apiece.

For "Mr. Steere" and other mill owners and merchant weavers, the "weaving ticket" was the crucial link in their transactions with hand-weavers in the "putting out system." It was the means by which entrepreneurs controlled the work. Such weaving tickets indicate shifting relations of production. The weaver worked not for an individual customer or for a generalized market, the tastes and needs of which he or she had to judge, but for a manufacturer-middleman who dealt with the market and contracted by means of these tickets for the kinds of cloth he assumed he could sell. (GK)

Prices for Weaving Cotton Shirtings 7-8ths yd. wide, and Stripes and Chambrays 3-4ths yd. wide.

No. 8 & 9	is 5 cents per yard.
10, 11 & 12	6
13 & 14,	7
15 & 16,	8
17 & 18,	9
19 & 20,	10
21 & 22,	11
23 & 24,	12

Tickings, 7 8ths wide, at 12½ cts
do 3-4ths wide, 10
Sheetings, 4-4ths wide, add 2 cts. per yard to above prices, and 1 cent more for every additional eighth.

Where there is more than one shuttle used in a piece, one cent per yard is added to the weaving for every shuttle over one; and when the warp and woof do not correspond with the above list, the price will be proportioned.

☞ Weavers must return the Yarn left of a piece with the cloth.

Cloth must be trimmed and returned free from stains and dirt; and if it is made too sleazy, or damaged in any way, a deduction will be made from the weaving.

This Piece is calculated to make 41 yards in a 40 slaie. 3/4 Wide

Weight, lbs. oz.

No. 56 *Pattern,* C

			No.
24 Sks.	D Blue	Warp,	11
12 do.	Yellow	do.	12
24 do.	Lt Blue	do.	–
~~do.~~		~~do.~~	
~~do.~~		~~do.~~	
~~do.~~		~~do.~~	
33 Sks.	D Blue	Woof,	–
33 do.	Lt. Blue	do.	–
~~do.~~		~~do.~~	
~~do.~~		~~do.~~	

To be Warped.

8 Deep Blue
2 Yellow
8 Lt Blue
2 Yellow

To be Filled.

10 Deep Blue
10 Lt. Blue

The weaving will be 7 cents per yard if well wove and trimmed.

20 Spools, 3 Skeins on a spool, will warp the piece.

Return this with the cloth.

N. B. Cotton Yarn for sale.

Joseph France

The Weaver's Guide (1814)

Joseph France, a weaver from Burrillville, Rhode Island, wrote a weaver's guide in 1814. The preface, reprinted here, sketches his reasons for doing so. France believed that in instructing ordinary weavers to make their own "drafts" they might weave their own goods and no longer have to rely on the instructions of master weavers and mill owners. "Drafts" or "direction papers" were instructions for weaving, usually drawn on sectioned paper, giving the correct plan for arranging the warp threads in order to create a desired pattern in the finished cloth. France recognized that he would "incur much censure and blame" for serving his "own class, the laborious and manufacturing part of the community." But he was prepared to weather the storm, believing that his instructions would only hurt a "few selfish, narrow-minded draught-makers, and little upstart beings of manufacturers." (GK)

Joseph France, *The Weaver's Complete Guide; or The Webb Analyzed* (Burrillville, Rhode Island, 1814), pp. iii–vi. This is one of five "weaver's guides" published in the United States between 1792 and 1818. See Rita J. Adrosko's introduction to the earliest guide, John Hargrove's *The Weaver's Draft Book and Clothier's Assistant* (Baltimore, 1792), reprinted by the American Antiquarian Society, Worcester, Massachusetts, 1979.

THE WEAVER'S

COMPLETE GUIDE,

OR

THE WEBB ANALYZED.

TO WHICH IS ANNEXED,

THE WEAVER'S

Complete DRAUGHT-BOOK,

CONTAINING

Seventy-Three different Draughts,

From Two to Sixteen Treadles,

Neatly Engraved on Copperplate,

WITH INSTRUCTIONS ADAPTED TO ANY CAPACITY.

By Joseph France.

Preface.

THO' the generality of mankind may look on the labors of the *loom*, and indeed on all other kind of handicraft work, as destitute of contrivance and ingenuity, and as things which require no degree of skill to attain to a perfect knowledge of the same; my design, (in part) therefore, by writing this little Treatise, was to shew the fallacy of such an opinion, and to convince the world, that both skill and ingenuity are alike requisite to attain to a knowledge of the simplest kind of webb, so as to be able to manufacture the same, merely by the simple means of inspecting the stuff itself; for I dare be bold enough to affirm, that, out of a thousand tradesmen who manufacture goods of various kinds, there is scarce one individual amongst the number, who knows the *internal structure*, (if I may be granted the expression) or, if you please, the real texture or make of those goods, which he daily handles and inspects, and of which he is supposed to have a perfect knowledge; nay more, there is scarce one in a hundred of the weavers themselves, who frame those goods, that knows the method of proceeding in forming a Draught, or Direction Paper, in order to enable him to work any kind of goods, by barely examining a *Patch*, (as weavers express it) or a remnant of the cloth itself; and, though strange and singular as this may appear to one unacquainted with trade or manufacture, 'tis an absolute and notorious fact.

But, however, though this is real matter of fact, with respect to the ignorance of the generality of tradesmen; yet, if any reasonable and thinking man amongst them will reflect but for a moment, he will quickly perceive what an advantage it gives to him, who has this knowledge over the rest of his fellow tradesmen; since by this means he is enabled *immediately* to fit out an order, and while the rest of his neighbors are running with the patch or sample of the goods wanted, from one draughtmaker to another, his goods are gone to market, and he has the advantage of reaping the highest prices, which are the constant reward of an early sale. With equal truth too, this may apply to the laborer likewise, for the highest wages are given during the heat of a demand.

I have in this little Treatise (the first attempt of the kind ever made) endeavored at great things, no less than the establishment of a set of fundamental rules for weaving to a system; and though my endeavors may not be crowned with any great success, I have at least shewn my good intentions; therefore, I hope it will excuse my defects in learning, and the keen eye of criticism will be induced to overlook the errors he may find, simply on that score; for, from the pen of a poor illiterate country weaver, he cannot expect much.

I beg leave to recommend this little Treatise, as an accompaniment to a work of mine, entitled,—The Manufacturer's and Weaver's Complete Draught-Book, as the one may tend in some measure to illustrate the other. The reader of the following pages, perhaps, may blame me for not laying down more examples for his practice, than I have here thought proper to give; but I must confess, I saw no real use they could be of, and that any more than what are here given, would only have swelled the work to a greater extent than was necessary. I have therefore intentionally omitted the insertion of too many. Some objections too may possibly be made to those examples which are given, as selected chiefly from the draughts in use amongst fustian manufacturers; this I ingenuously confess was owing to my want of others, and not having opportunity or power to procure such as I could have wished; but these objections are in themselves of no weight, for if any one will adhere to the great and general rule, which I have thought proper to lay down at the beginning of this work, he cannot fail to accomplish his design, in laying down any draught whatever, merely by unravelling the stuffs, and therefore, any kind of draughts may serve equally well to illustrate the same, for that is all here required to be done.

I am not unapprized or ignorant, that I shall incur much censure and blame, by thus endeavoring to serve my countrymen, especially my own class, the laborious and manufacturing part of the community; but while I am endeavoring to assist thousands of poor laborers and mechanics, and only hurting (and that but in imagination) a few selfish, narrow-minded draught-makers, and little upstart beings of manufacturers; I am sure my reason and conscience will not condemn me, and I persuade myself I shall not be condemned by that tradesman or mechanic, whose great and liberal mind connects the happiness of every being with his own, and wishes to see the blessings arising from trade and commerce diffused abroad, so that every one may partake of their benefits : such an one I am sensible will not fail to encourage that laudable ambition amongst his countrymen to rival the rest of the world, in the beauty and goodness of their manufactures ; since he is conscious to himself, that as long as he can excel the rest of the world, so long he shall have a brisk and constant demand, and of course, can be enabled to give his laborers or mechanics higher wages than they can procure elsewhere ; thus much, however, I can say, that what I have here done, was done with a design PRO BONO PUBLICO, and as such. I submit it to the world.

THE AUTHOR.

BURRILLVILLE, RHODE-ISLAND STATE,
December 28, 1814.

Labor Contract (1823)

The following is a contract signed by the Greene Manufacturing Company of Warwick, Rhode Island, and the mule spinner Samuel G. Niles. There are a number of provisions in the contract worth noting. First, Niles is to spin on only one mule rather than two, which was then customary in England and would soon be true in the United States. Second, like all mule spinners in this period, he is to pay his own piecers, a form of subcontracting customary only among skilled workers. Third, he is to receive a "fured hat" for good work, a striking form of nonmonetary compensation, unusual even among highly skilled workers. Fourth, he is to have three weeks notice before dismissal, again uncommon except among skilled workers. This is, of course, still a period when nonstandardized contracts are the rule and mill owners have not yet developed systematic standards of pay and performance for their workers. (GK)

Greene Mfg. Co. Papers, Rhode Island Historical Society; published by permission.

Warwick Feb. 8. 1823

I agree to Spin for Greene M. Co.
on Small Mule One year
from April 1.— Shuttle Cops
25 cts. pr hund. Sks. & Said Co to
find back side peicer & I agree
to have fore side peicer & pay
him myself— I also agree to
be steady to my work & make
good Cops for weaving— pay
One quarter Cash the remainder
goods— If said Niles makes as
good Cops as Thos. Warner, Said
Co. at the end of the year agree
to give him a fine hat—
If he does not make good Cops
Said Co. have liberty to dismiss him
by giving 3 weeks notice

Saml. G. Niles

Warwick Feb. 8, 1823

I agree to spin for Greene M. C°. on Small Mule one year from April 1. -- Shuttle Cops 25 cts. p^r hund. Sks. & Said C° to find back side peicer. & I agree to have fore side peicer & pay him myself -- I also agree to be steady to my work & make Good Cops for Weaving -- pay One quarter cash the remainder goods -- If said Niles makes as good Cops as Thos. Warner, Said C°. at the end of the year agree to give him a fured hat -- If he does not make good Cops Said C°. have liberty to dismiss him by giving 3 weeks notice

Saml G Niles

VI

Labor Organization and Protest

Strikes did not occur frequently in early nineteenth century mill villages, but when they did they carried more than just particular and local meanings. They crystalized all the resentments that first-generation mill workers felt toward the new modes of work and authority and the new methods of discipline common to textile mills. Often precipitated by the efforts of mill owners to reduce wages, these early strikes were not so much "economic" acts as they were eloquent statements that a "market" for wages had not yet been accepted. The resentments that fed strikes were far more pervasive than strikes themselves, for the small size and isolation of mill villages, the absence of public meeting places, and the power of local mill owners all militated against them. Typically, the first strikes—those at Pawtucket and Taunton—occurred in or near urban centers.

Labor organizations in the mill villages were even less common, though many mill villages were nominally represented in the counsels of the New England Association of Farmers, Mechanics and Other Workingmen—the region's first inclusive labor organization—which for a time drew attention to the problems of child labor and the abuses of mill village life. Skilled workers, particularly those with British trade-union experience, like the Scottish carpet weavers of Thompsonville, Connecticut, could be expected to organize. But mill workers in general did not join trade unions. Unlike many urban artisans—shoemakers in particular—textile mill workers did not experience the industrial revolution as an assault upon their craft skills. Freshly recruited from nearby farms, many mill workers were entering the world of paid work for the first time, without the customary sense of craft independence so central to artisan life.

Women were, of course, central to the textile mill labor force and even more important in the early strikes. Despite their reputation that they were more easily managed than men, women were in the foreground of every important textile mill strike in the early nineteenth century. Two selections in this part suggest that. Coming from roughly similar backgrounds and working alongside each other, the young women weavers came to symbolize the new nature of industrial work. Drawing on a work-culture marked by mutuality and collective self-help, they responded forcefully and collectively when they felt their rights threatened.

The following documents, drawn from a number of different kinds of sources, are illustrative of a larger—but by no means voluminous—body of evidence useful for probing the meaning and extent of labor unrest in the mill village. (GK)

A State of Excitement and Disorder (1824)

In response to mill owners' efforts to lower piece rates and increase hours, Pawtucket's textile workers "turned-out" in late May 1824. Young women weavers, demonstrating resolve and organizing ability, played a critical role in the turnout. They did not act alone, however. Filling the streets, numerous workers and artisans carried the conflict to the mill owners' houses, subjecting the owners to verbal abuse but doing no real damage. A few days later, an unknown arsonist attempted to burn one of the village mills. With their mills threatened and the village against them, mill owners successfully sought a compromise settlement to end the week-long strike.

Unlike the newly created mill villages of southern New England, Pawtucket had been a settled artisan community prior to the introduction of textile mills. Here there were no rows of company housing, no company store with a monopoly on local business. No single mill owner exercised political and cultural control. Pawtucket's workers, therefore, had more breathing space both to create their own culture and to nourish opposition to the growing power of textile mill owners.

The following passage is the first printed account of the strike. The *Manufacturers' and Farmers' Journal* in which it appeared was founded by mill owners and takes a characteristically disdainful view

Manufacturers' and Farmers' Journal, May 31, 1824. See also Gary Kulik, "Pawtucket Village and the Strike of 1824, Origins of Class Conflict in Rhode Island," *Radical History Review,* volume 17, Spring 1978, pp. 5–37. Another version of this article appears in Milton Cantor, ed., *American Workingclass Culture: Explorations in American Labor and Social History* (Westport, Connecticut: Greenwood Press, 1979), pp. 209–239.

of the strike, the village, and the young weavers. Despite this, the passage still clearly suggests the collective resolve of the women weavers and underlines the level of community support. (GK)

The citizens of Pawtucket have, for a few days past, been in a state of excitement and disorder, which reminds us of the accounts we frequently read of the tumults of manufacturing places in England, though unattended with the destruction and damage usually accompanying these riots.

The present depressed state of the cotton manufacture, on which the village principally depends for its support, has occasioned much anxiety in the minds of the manufacturers, and, the prospect being gloomy and unpromising, they concluded that something must be done to compete successfully with the country manufacturers who, generally, by working their mills more hours in a day, than has heretofore been usual in this village, and procuring weavers at lower wages combined with other advantages, have been able to manufacture at a less price. On Monday last there was a meeting of the manufacturers, which was generally attended, and an agreement made to run the mills about an hour longer, and to reduce the wages of those who worked by the piece, after the 1st of June next, about 20 per cent. When the laboring part of the community learned the results of the meeting, they very generally determined to work only the usual hours; and when the bell rang to call them to their employment, they assembled in great numbers, accompanied by many who were not interested in the affair, round the doors of the mills, apparently for the purpose of hindering or preventing the entrance of those who were disposed to accede to the resolution of the master manufacturers—no force, however, was used. The female weavers assembled in parliament to the number, it is stated, of *one hundred and two*—one of the most active, and most talkative, was placed in the chair, and the meeting, it is understood, was conducted, however strange it may appear, without noise, or scarcely a single speech. The result of the meeting was a resolution to abandon their looms, unless allowed the old prices.

On Wednesday evening a tumultuous crowd filled the streets, led by the most unprincipled and disorderly part of the village, and made an excessive noise—they visited successively the houses of the manufacturers, shouting, exclaiming, and using every imaginable term of abuse and insult. The window in the yellow mill was broken in—but the riot, considering the characters of those who led, and the apparent want of all reflection in those that followed, was not so injurious to property and personal security, as might have been reasonably apprehended. The next day the manufacturers shut their gates and the mills have not run since—a comparative stillness now reigns—and it is to be hoped that the people will return to a sense of propriety—satisfied, as they must be, on reflection, that the prosperity of a community can never be promoted by riot and tumult.

Notice to Mule Spinners (1824)

Mule spinners were the first English textile factory workers to organize in great numbers. Their skill in manipulating their partially hand powered machinery to produce finely spun yarn elicited the grudging respect of mill owners. In the United States mule spinners were less powerful because they were less numerous and less concentrated. After 1820, they worked in largely integrated factories, whereas their British counterparts worked in factories exclusively devoted to fine spinning. Still American mule spinners were not without some influence, as evidenced by their relatively high wages, favorable labor contracts, efforts to organize, and independent work behavior. One American mill owner claimed that the skilled spinners had "brought with them the disorderly habits of English workmen" and demonstrated high levels of absenteeism. The mill owners' response, originating in Manchester, England, after the spinners' strike in 1824, was to develop an automatic mule that could be run by unskilled workers. For reasons that still require close investigation, the automatic mule did not completely displace skilled male labor. It was not until the late nineteenth and early twentieth century, when it became practical to replace mules with high-speed ring frames, that the American mule-spinners' fate was sealed.

The following notice appeared several times in more than one Providence-area newspaper throughout June 1824. It closely followed

Providence *Patriot,* June 16, 1824. For further reading, see Harold Catling, *The Spinning Mule* (Newton Abbott, Devon, England: David and Charles, 1970). The most thorough study of the introduction of automatic mule spinning machinery in relation to spinners' independent power is William Lazonick, "Industrial Relations and Technological Change: The Case of the Self-Acting Mule," *Cambridge Journal of Economic History* volume 3 (1979), pp. 231–262.

the partially successful Pawtucket strike of the same month (see previous document). It is noteworthy that the meeting of area mule spinners is called for mid-day on Monday, July 5 when it might have easily been called for July 4, a holiday. The mule spinners were clearly bent on disrupting production in the middle of a work day and apparently did not doubt that they had the unchallenged power to do so. No detailed evidence has yet emerged on the Rhode Island Mule Spinners' Society. (GK)

Notice to Mule-Spinners.

Notice is hereby given, that the meeting of the Mule-Spinners in the State of Rhode-Island and the vicinity, will be held at Joseph Randall's, innholder in Smithfield, about a mile from Pawtucket, on Monday, the 5th day of July next, at 11 o'clock, A.M. for the purpose of organizing a Society, and establishing rules and regulations for the government thereof.
By request of the Clerk and Committee.

Mule spinning room as it may have appeared in Samuel Slater's mill. Note gas lighting fixtures over machines. From George S. White, *Memoir of Samuel Slater,* Philadelphia, 1836.

Pawtucket *Chronicle*

Interesting Law Case (1829)

The following article relates the story of an unsuccessful mule spinners' strike in Valley Falls, Rhode Island. The strike occurred in the midst of a serious depression, and the local justice of the peace sided with the mill agent when he resorted to forcing the mule spinners from the mill. (GK)

Pawtucket *Chronicle,* April 15, 1829.

A question, interesting to the owners of factories, was decided in a justices court in this place, on Wednesday last. An action for assault and battery, was brought against Mr. Francis C. Gardner, agent of a mill at Valley Falls, for expelling by force some individuals from the spinning-room. It appeared, on trial, that some difficulty had appeared between the agent and four or five mule-spinners, with regard to wages, and the number of the yarn to be spun on the mules. The due-bills which were offered them for the last week's work, was for a less sum than those they had been in the habit of receiving.—When they came to work, on Monday morning, they entered into an agreement among themselves, not to work any longer than the noon of that day, unless the bills were altered, and proceeded in a body to the store to inform the agent of this resolution, and to get them altered, but could not find him. After dinner, persuant to their resolution, they did not return to the mill, but went to a bowling alley some miles off, where they spent several hours. While there they agreed to return to the mill, and should they find any person employed on the mules, compel them to leave off work. They did return to the mill and found other persons there who had been employed by the agent after they had left—all of these persons, except one, were driven from the mules—several hundred threads were broken in a dresser—the belts were thrown off, and one mule was thrown out of gear. During these proceedings the agent was absent; his brother who was employed there ordered them to leave the mill, which order they refused to obey. On the return of Mr. Gardner to the store, he was informed of the proceedings in the factory; he armed himself with a heavy stick and repaired there; he ordered them to leave his premises instantly, at the same time making use of threats in case they did not comply; they went towards the door, where they made a stand, some saying they would not go out at all, and others saying they would not be driven out. Mr. Gardner put his hand on the collar of one of the men and attempted to lead him to the stairs, when he was seized by several; at this juncture he was induced to make his stick rather more intimately acquainted with their heads—he dealt several severe blows among the crowd, and finally succeeded in driving them from the mill. It was for the infliction of these blows, that the action was brought.

The Court, after a minute investigation of the matter, acquitted Mr. Gardner, on the ground that he had a right to expel the complainants from his premises, and that the means made use of were not unreasonable.

New England Association of Farmers, Mechanics and Other Workingmen

Report of the Committee on Education (1832)

The New England Association of Farmers, Mechanics and Other Workingmen, the area's first labor organization, was dominated by urban artisans deeply committed to democratic values, the amelioration of the worst features of factory labor, public education, mechanics' lien laws, a broadened franchise, and an end to the imprisonment of debtors. Some artisans carried their critique of the new industrial order further by suggesting the outlines of a cooperative society. As a political force, they fielded a group of workingmen's candidates for public office in the early 1830s. Never able to organize deeply in the new factory villages, they were strongest in port cities with sizable artisan communities, such as Boston, Providence, New Bedford, and Salem. They also had representatives in numerous mill villages, such as Pawtucket, Coventry, and Woonsocket Falls, Rhode Island; Chicopee, Massachusetts; Norwich, Connecticut; and Saco and Biddeford, Maine. Their preference for electoral politics over organizing in the work place or building cooperative industries, however, limited their effectiveness, for the majority of mill workers were legally barred from voting. The Constitution prohibited women from voting, of course, and many male workers were too poor to meet property qualifications imposed by state and local authorities.

The legacy of the New England Association was not simply failure, however. During its years of influence in the early 1830s, it succeeded in forcefully raising such issues as the abuse of child labor and contributed to the self-education of countless workers through its newspaper, *The New England Artisan.* It contested mill owners and other

A[mos] G[ilbert], "The Working Men," *The Free Enquirer,* June 14, 1832. The report was published in other labor papers as well.

"mushroom aristocrats" over the meaning of the revolutionary heritage.

The following report of the association's Committee on Education suggests the association's concern with child labor and public education while it also provides evidence on the length of the work day. Note the reference to continued conflict in Pawtucket over the mill owners' definition of time. "A. G.," who introduces the association's report, was Amos Gilbert, co-editor of *The Free Enquirer,* a reform newspaper; the other editors were Robert Dale Owen and Frances Wright. (GK)

THE WORKING MEN.

There was a convention of delegates from the working men, in various parts of New England, held in Boston in last February. Part of the business transacted at the convention, was, the reading of the following reports. No lover of his species can avoid regret that there are facts justifying such reports, and still more so, that they exist in New England, where it has been thought, ample provision was made for the instruction of every one; but it must be gratifying to him to see that there is zeal and industry sufficient to ferret out the causes, and enough of moral courage to expose them. It is to the promptitude of such persons as constituted the convention that the United States will owe their preservation from the trammels and toils of the monied aristocracy if they are preserved; and their advancement, in useful knowledge if they do advance.

"Like master, like man." In what principle does the labour statement in the second report, differ from one made in *Old England*.

A. G.

Your Committee, to whom was referred the subject of Morals and Education, have attended to the duty assigned them, and beg leave to report the following Resolutions.

Resolved, That in the opinion of this Convention, the want of education is the great and original cause of the present comparative degradation of Mechanics and Working Men in this country; and that the diffusion of knowledge is the best and most direct means to ameliorate the condition of this class of the community.

Therefore, Resolved, That this Convention recommend that Associations for mutual improvement, consisting of both sexes, be formed in every town in New-England, to be dependent on their own members for advancement in knowledge.

Resolved, That the several delegates composing this Convention, will use their influence for the formation of these institutions.

The Committee appointed to take into consideration the subject of the education of children in manufacturing districts, have attended to that duty, and beg leave to report:

That, from statements of facts made to your committee, by delegates to this body, the number of youth and children of both sexes, under sixteen years of age, employed in manufactories, constitute about two fifths of the whole number of persons employed. From the returns from a number of manufactories, your committee have made up the following summary, which, with some few exceptions and slight variations, they are fully persuaded will serve as a fair specimen of the general state of things. The regular returns made include the establishments in Massachusetts, New Hampshire and Rhode Island; which employ altogether, something more than four thousand hands. Of these, sixteen hundred are between the ages of seven and sixteen years. In the return from Hope Factory, R. I. it is stated shat the practice is, to ring the first bell in the morning at ten minutes after the break of day, the second bell at ten minutes after the first, in five minutes after which, or in twenty-five minutes after the break of day, all hands are to be at their labor. The time for shutting the gates at night, as the signal for labor to cease, is eight o'clock, by the *factory time*, which is from twenty to twenty-five minutes *behind* the *true* time. And the only respite from labor during the day is twenty-five minutes at breakfast, and the same number at dinner. From the village of Nashau, in the town of Dunstable, N. H. we learn that the time of labor is from the break of day in the morning, until eight o'clock in the evening; and that the *factory time* is twenty-five minutes *behind* the *true Solar* time. From the Arkwright and Harris Mills in Coventry, R. I. it is stated that the last bell in the morning rings and the wheel starts as early as the help can see to work; and that, a great part of the year, as early as four o'clock. Labor ceases at eight o'clock at night, *factory time*, and one hour in the day is allowed for meals. From the Rock-land Factory in Scituate R. I. the Richmond Factory in the same town, the various establishments at Fall River Mass. and those at Somerworth, N. H. we collect similar details. At the numerous establishments in the village of Pawtucket, the state of things is very similar, with the exception of the fact, that within a few weeks, public opinion has had the effect to reduce the *factory time* to the *true Solar* standard. And in fact, we believe these details to serve very nearly to illustrate the *general* practice.

From these facts, your committee gather the following conclusions. 1. That on a general average, the youth and children that are employed in the Cotton Mills are compelled to labor at least thirteen and a half, perhaps fourteen hours per day, factory time: and 2. That in addition to this, there are about twenty or twenty-five minutes added, by reason of that time being so much slower than the true Solar time; thus making a day of labor to consist of at least fourteen hours, winter and summer, out of which is allowed, on an average, not to exceed one hour, for rest and refreshment. Your Committee also learn, that in general, no child can be taken from a Cotton Mill to be placed at school, for any length of time however short, without certain loss of employ; as, with very few exceptions, no provision is made by manufacturers to obtain temporary help of this description, in order that one class may enjoy the advantages of the school while the other class is employed in the mill. Nor are parents, having a number of children in a mill, allowed to withdraw one or more, without withdrawing the whole; and for which reason, as such children are generally the offspring of parents whose poverty has made them entirely dependant on the will of their employers, and are very seldom taken from the mills to be placed at school.

From all the facts in the case, it is with regret that your Committee are absolutely forced to the conclusion, that the only opportunities allowed to children generally, employed in manufactories, to obtain an education, are on the Sabbath, and after half past eight o'clock of the evening of other days. To these facts, however, your Committee take pleasure in adding two or three others of a more honorable character. It is believed, that in the town of Lowell, no children are admitted to the labors of the mills under twelve years of age; and that the various corporations provide and support a sufficient number of good schools, for the education of those that have not attained that age. In the Chicopee Factory Village, Springfield, Mass. and also in the town of New Market, N. H. we also learn that schools are provided, and the children actually employed in mills allowed the privilege of attending school during a portion, say about one quarter of the year. Your Committee mention these facts as honorable exceptions to the general rule, with a desire to do justice to all concerned, and the hope that others may be inspired by their example to go much farther still in their efforts to remove their existing evils. A few more instances of the above character may exist; but if so, they have not come to the knowledge of your Committee, and they have every reason to believe them to be extremely rare.

Your Committee cannot, therefore, without the violation of a solemn trust, withhold their unanimous opinion, that the opportunities allowed to children and youth employed in manufactories, to obtain an education suitable to the character of American freemen, and the *wives* and *mothers* of such, are altogether inadequate to the purpose: that the evils complained of, are unjust and cruel; and are no less than the sacrifice of the dearest interests of thousands of the rising generation of our country, to the cupidity and avarice of their employers. And they can see no other result in prospect, as likely to eventuate from such practices, than generation on generation, reared up in profound ignorance, and the final prostration of their liberties at the shrine of a powerful aristocracy. Deeply deploring the existing evils, and deprecating the dreadful abuses that may be hereafter practiced, your committee respectfully recommend the adoption of the following resolutions:

Resolved, That a committee of vigilance be appointed in each State, represented in this Convention, whose duty it shall be to collect and publish facts respecting the condition of laboring men, women, and children, and abuses practiced on them by their employers: that it shall also be the duty of said committee, as soon as may be, to get up memorials to the Legislatures of their respective states, praying for the regulation of the hours of labor, according to the standard adopted by this Association, and for some wholesome regulations with regard to the education of children and youth employed in manufactories; and to make report of their doings at the meeting of this body, on the first Thursday of September next.

Voted, That in accordance with the foregoing resolution, the delegates constituting this body, also constitute a committee of vigilance, for the purpose therein specified, and forward the statements of facts they may obtain from time to time, for publication in the "New England Artisan."

John Adams

Testimony, Thompsonville Manufacturing Company *versus* William Taylor et al. *(1836)*

In July 1834, the Thompsonville Carpet Manufacturing Company of Thompsonville, Connecticut, charged its weavers with a "Conspiracy for being Concerned in a Strike for Higher Wages," naming William Taylor, Edward Gorman, and Thomas Norton as principal defendants. The company, one of the two largest carpet weaving workshops in the United States, brought the case to court almost exactly a year after the strike. The case, prosecuted under the English common-law doctrine of criminal conspiracy, was similar to many others brought against workers' organizations in the early nineteenth century. Unlike most of the earlier cases, this one was won by the defendants. Two trials, one in August 1834 and a second in January 1836, resulted in a verdict favoring the weavers.

Many of the Thompsonville weavers were Scottish immigrants, and it was a skilled Scotsman who installed the company's intricate carpet looms in 1828. Carpet weaving in this period was still done on hand looms. Requiring both strength and skill, it was a trade dominated by European immigrants. The Thompsonville weavers, drawing on a tradition of Scottish militance and trade union organization, were one

Adams's testimony in the case of *Thompsonville Carpet Manufacturing Company* v. *William Taylor, Edward Gorman, and Thomas Norton,* tried before the Superior Court for Hartford County, Connecticut, January term 1836, has been reprinted, along with the entire transcript of the second trial, in John Commons et al., *Documentary History of American Industrial Society,* volume IV, Supplement (Cleveland, 1916), pp. 29–37. The text reprinted here has been lightly edited, and footnotes have been added. For further reading, see Arthur H. Cole and Harold F. Williamson, *The American Carpet Manufacture* (Cambridge, Massachusetts, 1941); H. A. Turner, *Trade Union Growth, Structure and Policy* (London, 1962); and John Commons et al., *History of Labour in the United States,* volume I (New York, 1918).

of the few groups of European textile craftsmen to successfully import their union to the new nation.

The following testimony of John Adams, the president of the local weaver's union in 1833, provides a cogent outline of the strike. Adams's testimony suggests the strength of the Scottish weavers' network in the United States. The weavers successfully wrote to friends in other manufactories and to a tavern in New York frequented by Scottish weavers to warn them against coming to work in Thompsonville. The unfortunate Adams, electing to return to work when threatened with prison, became contemptible in the eyes of his fellow weavers. Despite the resolve of the weavers, the six-week strike was lost. (GK)

John Adams. Upon the 6th of July 1833, I was
requested to call the shop together of which I was
President. Mr. Taylor thought the opinion of the
shop ought to be taken in relation to a grievance of
which one of the workmen complained. One Boyle
had complained that he had finished his piece within
the time and could get no premium. William Keys 1

also complained to me on the same day that he could
not get a ticket, and that he had been much hindered
in his work by poor filling. I called a meeting at 4
o'clock on Saturday and sent to Mr. Thompson. The
reason of the refusal to pay Keys was that he had not
2 complied with the rule requiring a webb's notice of
his intention to leave the employment of the company.
It was always understood at Thompsonville, and all
other Factories, that a webb's notice was to be
given by the employer or employed, if the one intended
to discharge a workman, or he intended to leave. There
is a great danger that if two workmen, even good workmen, are employed upon the same piece, that the figure
will not match, and that the piece will be thus destroyed. At the meeting on Saturday before mentioned
Mr. Henry Thompson attended and consented to pay
Keys, and that he might go away, but said he wished
it distinctly understood that a webb's notice would be
required thereafter. Mr. Boyle's complaint about not
receiving the premium was also considered at this
meeting, and Mr. Thompson said the premiums should
be given up, and upon being inquired of if fines were
to be given up also, he answered yes. Before the agent
Mr. T. left the meeting, he was requested to raise the
prices upon some difficult fabrics, to which he replied
that he would refer them to the company who would
soon meet. A committee of six were appointed to
draw a petition, which was presented to the company
at their next meeting. An answer was returned by the
company on the following day. A meeting was then
called in the High Shop and the answer was read and
was unsatisfactory. And the question was then put,
"shall we make a stand or not?" It was determined
to ask for an increase of wages on all kinds of fabrics,

and the terms were agreed upon. The three ply 3
weavers had never before made any complaint. We concluded to ascertain the views of those present by a personal vote, and each individual was asked "stand or not." Seventy said "stand," and but one made any objection, and he a slight one. Resolutions were then adopted and a committee of which John Elder was one, was appointed to present them to the company. The meeting remained organized until the committee returned with a report that the company were not prepared to make any other alterations. Toward evening the Foreman (Mr. Ronald) came to the shop where we had assembled, and told us he was directed by Mr. Thompson to require us to leave the shop. I requested that we might be permitted to remain thirty minutes to finish our business, but he refused, saying he was ordered to shut the shops. We left the shop and went into Mrs. Metcalf's lot, where I was chosen President, and George Black Secretary. A committee of management consisting of nine was also appointed. The committee was composed of the President and Secretary, with William Liddle, James Alexander, William Taylor, John Connor, John Flood, Samuel Sturgeon and Henry White. A tax of six cents was laid to pay postages. We were then instructed to repair to some convenient place, and not to separate until we had written to all the principal Carpet Manufactories in the United States. We did so, and a letter was written which was approved. It was then determined that each of the committee should make a copy & address it to some individual in some other Factory with whom he was acquainted. One read from the letter first written and the committee wrote. I wrote to James Wells, West Farms, N.Y., George Black wrote to

Alexander Winkie, at New Haven, James Alexander to Hugh Torrence, Baltimore, William Taylor to Robert Young, Lowell, Mass. John Conner to William Gibbs, Tariffville, Samuel Sturgeon to Robert Wilson, New Jersey, Henry White to John Park, Rochester, N.Y., John Flood to his brother, at Saxonville, Mass. There were eleven letters in all written. Liddle did not write but read the first letter while others wrote. Two were written on the following day. It was thought advisable to write to Mr. Miller of New York, who keeps the Blue Bonnet Tavern, and request him to give information to Charles Cunningham of the Rob Roy House in that city, that they might give notice to weavers of the strike. The letter to Miller differed from the rest in the request that he would give information to Cunningham. It was proposed that the letters written should be lodged in the Post Office at Suffield, and I delivered them to Samuel Sturgeon for that purpose, and gave him the money to pay the postage. Nothing more was done that evening.

On the next day Mr. Thompson requested me to call the men together in the yard at 2 o'clock p.m. which I did. They assembled and Mr. Thompson read a paper. (It will be found in note marked D.) We went into the field and upon consultation refused to accede. On the morning following which was Saturday, at about 6 o'clock, Mr. T. handed me another paper, to which we were required to agree by Monday or to go in afterwards, if at all, at reduced prices. It was determined to call a meeting that forenoon. I saw Thompson before the meeting and suggested some alterations, which I informed him I believed would render the proposition acceptable. Mr. Thompson would not agree to any alterations. A meeting was held on that day in the field. The resolutions were again publicly read, and I explained what I had said to Mr. Thompson. Several present, among whom was Richard McDowell and James Taylor, reprimanded me severely, and said among other things that I was their servant and not their director. The question was then taken shall we accept the offer or stand permanently. The question was taken by personal vote, each being asked individually, and they unanimously said stand. I was then requested to invite those who belonged to the support society to remain after the meeting was dissolved. The meeting was then dismissed. In the Society's meeting it was proposed to dissolve that Society which had been in existence for several years, and draw the money from the Saving's Bank and divide it amongst the members. The proposition was agreed to. George Simpson and William Taylor were in favor of this measure. On Thursday following, one week from the day of the strike, a meeting was holden on the River bank, and the money divided & two letters read from Tariffville, approving of our conduct. Early next week a committee meeting was held at William Taylor's room, at which about six of the committee were present. The object was, to provide funds for the support of those who had families and others in need. William Liddle, William Taylor and Henry White offered to furnish $5 each, and Black offered to lend $5 more, if needed, and act as Treasurer. The money was to be loaned and returned afterwards if the person borrowing had ability, and was to be apportioned among the needy as follows. To a family of more than two children $3.00 – of two children $2.50 – and single persons $2.00 per week. On the 10th day of the strike there was a meeting held in Smith's lot. The committee reported their views to the meeting, which were approved. The committee were then authorized to contract for $60 in any manner for the support of the indigent. At this meeting James Taylor, Thomas Green and James Law were added to the committee. On my way home from the meeting, I met James Borland who handed me a letter from Tariffville which contained $54.00. The letter was addressed to me. Mr. Law soon came down and I delivered the money to him, and informed him he had been elected Treasurer, and that the money must not be paid out without my order.

On Monday after the strike the committee met and prepared an article for the *Old Countryman*, a paper published in New York. We soon began to receive letters of approval and promises of money to aid us. The first letters which we received did not contain money. We wrote them a second time informing them that we were still standing out, and intimating that we might have need of aid. About two hundred dollars were received during the first three weeks. During the third week Mr. Thompson asked us what we intended to do, whether we should go back and work or remain as we were. I told him our wages had been reduced & we wanted them increased. He said we should be ruined if we persisted. He then requested me to call six of the men to the office whom he named, and they came. Charles McGill, one of the six, told him he might make him work in the shop with those who worked at an under price, but he could not make him associate with them. Wm. Dixon, Esq. was present at this meeting and proposed to meet six of the men at the office in the afternoon, and read them the laws of the country about turn-outs. It was with difficulty that I could procure six to attend, and they were young men. Mr. Dixon read the law about turn-outs, but the men went away as much dissatisfied as ever. Mr. Thompson told us they should bring men from New York and fill the looms if we did not go back. George Simpson said they might put hell on the looms if they wished, they had ruined other Factories, and could ruin them, and send their Factory down the river.

At a general meeting on Saturday a resolution was passed, or an agreement made that each one should be as active as possible in preventing weavers from coming there, and if any did come to notify them of the strike before Mr. Thompson knew it. One James Latty came there from Scotland. I asked him if he had any knowledge of the turn-out. He said after he had been requested to come, he had been informed of it at the Blue Bonnet tavern in New York. We informed him of the facts, and he said he would do us no harm – offered to furnish him money if he needed it. Charles Stewart, William Taylor and David Gibson were present at the conversation. Latty went that evening by our advice to David Gibson's and I did not see him again. Latty came up in the Boat, and some few always attended upon the Boat on its arrival.

A weaver by the name of Clayton came there also, and William Taylor told me he gave him a dollar, and sent him to the Shakers to seek for work. Taylor said he (Clayton) was out of money, and he gave him the dollar to bear expenses. He wished the money refunded. Law, Sturgeon, and Liddle were present.

On Saturday the 15th of August, the defendants were arrested on the first suit. It was agreed among the

weavers that one or two should go and see them each day, to see that they were supplied with every thing necessary, to consult with counsel and if possible to procure bail. Mr. Martin the Clerk told me that the company were determined to fill up the works, and that I had better go in. The company threatened to arrest me and did so, and on my way to prison I gave up and consented if I could be discharged and protected, to return to work. I signed a paper and was discharged accordingly and again went to my loom. Some young men had already gone in when I returned, which was on Monday the 19th of August. From the time I went in, the old weavers hissed and spit at me and shouted at me as I passed and called me skunk. John Hanson spit at me from the Tavern window. They would stand about in clusters and insult me
4 daily – called me a traitor and a Monteith. Some one in a house said it would be well to throw a pail of water on me, but I can't say whether it was a man or a woman. On Saturday of that week the Foreman of the Norwich Factory came there, and I took a letter from him. On Monday following several came to the loom where I worked, three or four at a time. James Buckridge, one of the weavers engaged in the strike said, "can this be the same John Adams that wrought here a few months ago?" John Lamont said, look at him again and you will find that it is the same man. They called me a traitor, &c. One said to Thomas Little, I thought you knew better about strikes, & had
5 been in the Aberdeen strike. Others shouted and hissed at me that afternoon. I went to Mr. Thompson's office but did not find him, and the men came again and insulted me. Next morning I went into the shop but went away that day – went by the way of New Haven.

At one of the meetings spoken of, I think it was said, if any one went back he would be put on the black catalogue. The defendants were all present at the first meeting, but do not know that they were at every other. No new fabrics were introduced before the strike – no fabric is new which does not require an alteration of the reed. In the resolutions adopted at the strike, we claimed ½ a cent more on the whole per yard than we claimed in our petition. The object of the three ply weavers in joining us was to help us as I supposed, as they had not complained. Don't know
that double shot abouts had been woven in this coun- 6
try until introduced at Thompsonville.

1. It was the practice of the company to pay a premium if a piece was completed within the time set by the company. A fine was imposed if it was not.

2. Some hand weavers still refer to the finished piece as a webb, or web.

3. Three-ply refers to a type of ingrain carpet using three sets of warp and filling yarns.

4. Likely refers to John Montieth, a textile manufacturer who introduced power looms to the Glasgow area in 1796, thus undercutting the livelihood of hand loom weavers. But there is another, more obscure, possibility. William Montieth (1790–1864) was a Lieutenant General in the Indian Army, born in Scotland, who campaigned both with and against the Russian Army in a two-year period in the late 1820s. The epithet, "a Montieth," connotes a traitor and turncoat and may have drawn its force from reference to both Montieths.

5. The Aberdeen strike probably refers to the great Scottish weavers' strike of 1812 involving 40,000 weavers from Carlisle to Aberdeen. The Scottish weavers had the strongest and most extensive textile union in the world during the early nineteenth century.

6. A form of carpet using four shuttles of different color, creating the possibility of multiple variations in design. The Thompsonville weavers sought higher prices for this type of carpet.

Salome Lincoln

Almond Davis

Memoir of Salome Lincoln (1843)

Salome Lincoln was a leader of the Taunton, Massachusetts, strike of 1829, and is one of the few women active in the textile strikes of the 1820s and 1830s about whom anything is known. Her public commitment to evangelical religion—the reason for her biography—makes her all the more interesting. Born in 1807 in Raynham, in southeastern Massachusetts, Lincoln was the eldest of six children. In 1821, at the age of fifteen, she went to work in the Hopewell Mill, one mile from Taunton green. Troubled about her salvation, she embraced evangelical religion in the following year. In 1828 Lincoln became a preacher herself, and she continued to preach while she remained a weaver. A tall, dark-haired woman with a deep voice, she was reportedly an excellent speaker, carrying herself in a simple yet solemn and imposing fashion.

In May 1829, Lincoln found another use for her preaching skills. Her biographer describes the Taunton strike and her role in it. Two aspects of that strike bear comment. First, its causes were the same as those which produced the 1824 Pawtucket strike, and, in both cases, the young women weavers assumed the central role. Far more decorous than the Pawtucket "turn-out," the Taunton strike was marked by its "procession"—a strike form that seems to have appeared first at Dover, New Hampshire, in 1828 and was repeated at subsequent

Almond H. Davis, *The Female Preacher, or, Memoir of Salome Lincoln, Afterwards the Wife of Elder Junia S. Mowry* (Providence, R.I., 1843; New York: Arno Press, 1972), pp. 49–53. See also Anthony F. C. Wallace, *Rockdale: The Growth of an American Village in the Early Industrial Revolution* (New York: Knopf, 1978); Nancy Cott, *The Bonds of Womanhood, "Woman's Sphere" in New England, 1780–1835* (New Haven: Yale University Press, 1977); and A. D. Williams, *The Rhode Island Free-Will Baptist Pulpit* (Boston, 1852).

strikes in the 1830s. Second, after Lincoln "manfully" refused to break her word and go back to the Taunton Mill, she apparently had no difficulty in finding another mill job nearby. She even boarded in her new employer's house—evidence that her disagreement was with a particular owner rather than with the system of cotton mill life. But clearly no "black-list," of a type even then being elaborated in Lowell, existed in the Taunton area to punish labor activists. Labor in southern New England may have been too scarce for such a tactic to succeed, and local mill owners were likely not yet well organized or even united in their attitude toward their workers.

Salome Lincoln's experience points to the duality at the heart of evangelical religion. Anthony F. C. Wallace, in his important study, *Rockdale,* sees such religion primarily as a manufacturer's tool to tame the refractory hand of labor. And so it sometimes was. But the evangelical churches had another side as well. With their emphasis on the primacy of the will and with their encouragement of self-taught preachers, they served as thoroughly democratic training grounds. Men of no formal education and women who had never spoken before in public gatherings learned oratorical skills and confidence. Such talents, developed within the bosom of the evangelical churches, would find direct expression in the labor conflict of the nineteenth century.

Primarily known to posterity because of her preaching skills, Salome Lincoln was one of a number of women in the early nineteenth century who assumed public roles through evangelical religion. Her work as a weaver and her role in the 1828 Taunton strike makes her a rare and important figure. Eventually she married, and left her mill work. She died an early death in 1841, the result of complications attendant on the birth of her second child. (GK)

CHAPTER III.

MISCELLANEOUS.

"Women are not for rule designed,
Nor yet for blind submission."

HAVING in the two former chapters noticed her conversion, and some of the reasons why she supposed God had called her to preach his everlasting gospel, the reader will pardon me, if I now call his attention back to the period where we closed the first chapter.

At the time of her conversion in 1822, she was at work in the weaving room in the factory at Hopewell, Taunton, for the Richmond Company. When she commenced her labors in the factory we are unable to say; but it was probably about the year 1821. And here I would remark, that combined with piety and talent, Salome was industrious. After she began to travel and preach; as she received but little from others, she was accustomed to work with her hands to clothe herself; and then go out on her missions of love, till it was expended ; being too sensitive to say anything in relation to her circumstances, and the church too *covetous* to inquire. *

*

Salome continued to work in the factory at Hopewell, until the first of May, 1829, when an unforeseen circumstance occurred which deprived her of work. The circumstance is briefly this :

For some cause, the corporation reduced the wages in the weaving department, where Salome was then at work. The girls indignant at this, bound themselves under an obligation, not to go back into the mill, until the former prices were restored ; and this not being granted, they formed themselves into a procession, and marched through the streets, to the green in front of the Court house. The procession started from Hopewell, about the middle of the forenoon. They were in uniform, — having on black silk dresses, with red shawls, and green calashes. They then went into a hall near the common, in

* This remark will not apply to every place — as there were honorable exceptions—and the Lord will reward every one according to their deeds.

order to listen to an address. Salome was selected as the orator of the day. She then took the stand, and in her own peculiar style, eloquently addressed them at considerable length, on the subject of their wrongs; after which they quietly returned to their homes.

For one inducement, and another, nearly all who had turned out at this time, returned into the factory again, and resumed their work. But not so with Salome! —She manfully refused to violate her word; but chose rather, to leave business — and break up all the social and religious ties she had formed; than to deviate from the paths of rectitude. After this she never worked in the factory again at Taunton; but sought employment elsewhere, and was successful.

After leaving the factory at Hopewell, she returned home, where she remained about two weeks, and then went to Easton, Mass. to work in the mill for Mr. Barzilla Dean. * She worked for him at two different periods. First from May 19th, 1829, to May 29, 1830;

* Mr. Dean's factory is located in Easton, about one mile from the Meeting-house, and ten miles north of Taunton Green.

when she left for a short season, but returned in the fall of the same year, and continued to work for him until February 4, 1831.

While living at Easton, she boarded with her employer. Says Mr. Dean, — "When in my family, she lived a devoted christian life: All her leisure moments, she spent in reading the Bible." Says another individual; a member of the same family: — "She always used to kneel before retiring. It looked strange to me then; but it made no difference if all the girls were present; and among her associates in the mill, there were some, who were very rude; but she would frequently check them."

*

Voice of Industry

Letter from the Editor (1846)

The *Voice of Industry,* founded in Fitchburg, Massachusetts, in May 1845, was published longer than any other labor paper of the 1840s. First edited by William F. Young, a Fitchburg mechanic, the paper moved to Lowell in October 1845, becoming the organ of the New England Workingmen's Association. Organized in 1844, the association advanced the causes of the ten-hour work day, trade unions, land reform, and cooperation. Young remained the paper's editor until April 1846, though he would return again to oversee its eventual demise in 1847. Sarah Bagley, a mill worker and the founder of the Female Labor Reform Association, edited the paper briefly, as did John Allen, an advocate of the utopian communitarianism of Charles Fourier. It was Allen who was editor at the time the following appeared, and he is presumed to be its author.

Allen's letter sketches the connections between the textile industry, immigration, and Rhode Island politics. The editor refers to the consequences of the Dorr War of 1842, an armed insurrection led by men intent on capturing state power in the name of democratic rights. Prior to 1841, the state of Rhode Island was governed by a series of

"Letter from the Editor," *Voice of Industry,* 18 September 1846; this article was reprinted in part in John Commons et al., *Documentary History of American Industrial Society,* volume 7 (Cleveland, 1916), pp. 142–143, in which article was incorrectly dated 1848. For background on the *Voice of Industry,* see Norman Ware, *The Industrial Worker* (Boston, 1924), pp. 212–213; and Helena Wright, "Sarah C. Bagley: A Biographical Note," *Labor History* 20 (Summer 1979), 398–413. The most comprehensive study of the "Dorr War" to date is Marvin E. Gettleman, *The Dorr Rebellion, A Study in American Radicalism, 1833–1849* (New York: Random House, 1973). For the role of immigrants in organized labor in late nineteenth-century New England, see David Montgomery, *Beyond Equality, Labor and the Radical Republicans, 1862–1872* (New York: Vintage, 1967); and Paul Buhle, "The Knights of Labor in Rhode Island," *Radical History Review,* volume 17 (Spring 1978), pp. 39–73.

conservative coalitions in which the old landed interests of Newport and Washington counties held disproportionate power. That power was based on the Royal Charter of 1663, which, as amended, restricted suffrage to those men who held real property valued at $134 or more.

Sometimes too easily compared to the Chartist movement in England in the 1840s, the Dorr War more closely resembled the European bourgeois revolutions of the 1830s. The Dorrites were led by liberal elements of the urban middle class (such as the lawyer Thomas Dorr, for whom the movement was named) as well as by artisans, and they attracted a substantial number of mill workers to their side. Mill towns such as Pawtucket and Woonsocket were centers of the movement. After two brief tragicomic skirmishes, the Dorrites were crushed. But their threat prompted the passage of a constitution that granted the vote to all American-born males, and, through reapportionment, increased the power of textile mill owners in the northern half of the state. Foreign-born males could not vote unless they owned real property worth at least $134. With the vast migration of Irish peasants into Rhode Island factories in the late 1840s, the state's textile mill owners emerged as the primary political beneficiaries of the new constitution.

The editor's belief that immigrant workers were more passive and dependent than Americans should be read with some skepticism. The statement says more about the critical rupture that occurred in the American radical tradition during this period than it does about the actual facts of Irish-American behavior. With a wholesale change in the ethnic composition of the textile mill work force in the years from 1845 to 1855, the lessons and traditions of the first generation of American factory workers could not be immediately transmitted. But by 1857, "wee bits" of Irish women were striking in Holyoke, Massachusetts, just as their counterparts in Pawtucket and Lowell had a generation earlier. By the 1860s, significant links would be established, through the mediation of the Fenian movement, between Irish and American workers, and the New England labor movement of the 1870s and 1880s would be led in part by Irish textile workers. (GK)

I have just closed a course of lectures in Blackstone, a town set off
from Mendon last winter, and containing three factory villages. I 1
had no idea of the extent of factory operations on the brave little
river from which this town derived its name. All the way between
Worcester and Providence it is tugging at the wheels of Corporations, and summons its thousands of operatives to serve and slave
under its despotism of machinery.

And I have seen no factory tyranny in Lowell, nor anywhere
else in New England, that would compare with that existing on
this river, especially in Rhode Island. The Algerine revolution and 2
the new constitution have destroyed what little freedom there once
was in this little State. By a provision in the constitution, no foreigner is allowed to vote, unless he owns a hundred and thirty-
four dollars' worth of dirt! The result of this rule is, to induce the
manufacturer to turn off the native citizens and employ foreigners
in their stead. As corporations have monopolized the waterfall,
and all the lands and houses surrounding them, there is but little
chance for a foreigner to become a voter. And if the American
citizen votes contrary to the will of his employer, he very quietly
tells him, "we want your tenement;" and he, with his dependent
family is driven into the steets to beg, unless he is fortunate enough
to get a situation and employment in some other place.

I was informed by the Postmaster of Woonsocket, that the character of the population in that village had entirely changed since the adoption of the new constitution. So many persons, he remarked, have moved in from other countries, that cannot write or read, that it makes a difference in the income of the Post Office of some hundreds of dollars per year. And in conversation, in stage coaches and in the streets, you seemingly hear more persons speak in a foreign accent, than in the "Yankee tongue."

Besides, our system of "protection" contributes to increase the number of foreign laborers. The tariff closes the avenues for the sale of foreign labor; for example, shut out French boots, and invites to our shores French boot makers. It shuts out English texture, lessens the demand for labor abroad, and brings to our country English bones and sinews to be wrought up into American texture. Thus wages are reduced in Europe, and by competition with foreign operatives in our own manufactories, are cut down to nearly the same level at home. The tariff enables a few manufacturers and monopolists to get rich by the premium paid on American productions, by the producers themselves; but it does not better, in the least, the condition of the American operative. It is a protection to capital and monopoly, but not to the laboring classes, whether native or foreign.

These causes combined have brought into Rhode Island a large foreign population . . . . The manufacturers of Rhode Island seem to prefer foreign laborers, not only because there is no prospect of their exercising the right of suffrage, but because being strangers and more dependent than native operatives, they are more submissive under corporation tyranny. And the factory despotism is therefore increasing here faster than in any other portion of New England.

1. Mill owners sought the creation of a separate town of Blackstone, an effort opposed by local farmers. Mill owners sought town divisions here and elsewhere in New England in order to increase their political power. For other examples, see Gary Kulik, "Patterns of Resistance to Industrial Captialism, Pawtucket Village and the Strike of 1824," in Milton Cantor, ed., *American Workingclass Culture* (Westport, Connecticut, 1979), pp. 209–239; and Jonathan Prude, "The Coming of Industrial Order: A Study of Town and Factory Life in Rural Massachusetts, 1813–1860," Ph.D. dissertation, Harvard University, 1976.

2. Refers to the aftermath of the Dorr War. The term "Algerine" was first used to ridicule a law passed in Rhode Island in 1842 prescribing penalites for the "subversion" of the state. It was later used to refer to the constitution drafted by the state's conservative landholders. The specific reference is to a contemporary tyrant, the Dey of Algiers.